T0205766

Communications in Computer and Information Science 2029

Rationale

The CCIS series is devoted to the publication of proceedings of computer science conferences. Its aim is to efficiently disseminate original research results in informatics in printed and electronic form. While the focus is on publication of peer-reviewed full papers presenting mature work, inclusion of reviewed short papers reporting on work in progress is welcome, too. Besides globally relevant meetings with internationally representative program committees guaranteeing a strict peer-reviewing and paper selection process, conferences run by societies or of high regional or national relevance are also considered for publication.

Topics

The topical scope of CCIS spans the entire spectrum of informatics ranging from foundational topics in the theory of computing to information and communications science and technology and a broad variety of interdisciplinary application fields.

Information for Volume Editors and Authors

Publication in CCIS is free of charge. No royalties are paid, however, we offer registered conference participants temporary free access to the online version of the conference proceedings on SpringerLink (http://link.springer.com) by means of an http referrer from the conference website and/or a number of complimentary printed copies, as specified in the official acceptance email of the event.

CCIS proceedings can be published in time for distribution at conferences or as post-proceedings, and delivered in the form of printed books and/or electronically as USBs and/or e-content licenses for accessing proceedings at SpringerLink. Furthermore, CCIS proceedings are included in the CCIS electronic book series hosted in the SpringerLink digital library at http://link.springer.com/bookseries/7899. Conferences publishing in CCIS are allowed to use Online Conference Service (OCS) for managing the whole proceedings lifecycle (from submission and reviewing to preparing for publication) free of charge.

Publication process

The language of publication is exclusively English. Authors publishing in CCIS have to sign the Springer CCIS copyright transfer form, however, they are free to use their material published in CCIS for substantially changed, more elaborate subsequent publications elsewhere. For the preparation of the camera-ready papers/files, authors have to strictly adhere to the Springer CCIS Authors' Instructions and are strongly encouraged to use the CCIS LaTeX style files or templates.

Abstracting/Indexing

CCIS is abstracted/indexed in DBLP, Google Scholar, EI-Compendex, Mathematical Reviews, SCImago, Scopus. CCIS volumes are also submitted for the inclusion in ISI Proceedings.

How to start

To start the evaluation of your proposal for inclusion in the CCIS series, please send an e-mail to ccis@springer.com.

Fuchun Sun · Jianmin Li
Editors

Cognitive Computation and Systems

Second International Conference, ICCCS 2023
Urumqi, China, October 14–15, 2023
Revised Selected Papers

Editors
Fuchun Sun
Tsinghua University
Beijing, China

Jianmin Li
Tsinghua University
Beijing, China

ISSN 1865-0929 ISSN 1865-0937 (electronic)
Communications in Computer and Information Science
ISBN 978-981-97-0884-0 ISBN 978-981-97-0885-7 (eBook)
https://doi.org/10.1007/978-981-97-0885-7

This Springer imprint is published by the registered company Springer Nature Singapore Pte Ltd.
The registered company address is: 152 Beach Road, #21-01/04 Gateway East, Singapore 189721, Singapore

Paper in this product is recyclable.

Preface

This volume contains the papers from the Second International Conference on Cognitive Computation and Systems (ICCCS 2023), which was held in Urumqi, Xinjiang, China, on October 14–15, 2023. ICCCS is an international conference on the related fields of cognitive computing and systems which was initiated by the Technical Committee on Cognitive Computing and Systems of the Chinese Association of Automation in 2022. ICCCS 2023 was hosted by the Chinese Association of Automation and organized by the Technical Committee on Cognitive Computing and Systems of the Chinese Association of Automation, and Xinjiang University.

Cognitive computing is an interdisciplinary field of research that focuses on developing intelligent systems that can process and analyze large amounts of complex data in a way that mimics human cognition. It is based on the idea that intelligent systems can be designed to learn, reason, and interact with humans in a more natural way, and that these abilities can be used to solve complex problems that would otherwise be difficult or impossible for humans to solve alone. By enabling agents to understand and process data in a more human-like way, cognitive computing can help organizations make better decisions, improve customer service, develop more effective treatments for diseases, and even predict and prevent natural disasters.

ICCCS aims to bring together experts from different areas of expertise to discuss the state of the art in cognitive computing and intelligent systems, and to present new research results and perspectives on future development. It is an opportunity to promote the research and development of cognitive computing and systems, and provides a face-to-face communication platform for scholars, engineers, teachers, and students engaged in cognitive computing and systems or related fields to promote the application of artificial intelligence technology in industry.

ICCCS 2023 received 68 submissions, all of which were written in English. After a thorough single-blind peer reviewing process, 26 papers were selected for presentation as full papers, resulting in an approximate acceptance rate of 38%. The accepted papers addressed challenging issues in various aspects of cognitive computation and systems, including not only theories and algorithms in cognition, machine learning, computer vision, decision making, etc., but also systems and applications in autonomous vehicles, computer games, intelligent robots, etc.

We would like to thank the members of the Advisory Committee for their guidance, and the members of the Program Committee and additional reviewers for reviewing the papers. We would also like to thank all the speakers and authors, as well as the participants for their great contributions that made ICCCS 2023 successful. Finally, we thank Springer for their trust and for publishing the proceedings of ICCCS 2023.

December 2023

Fuchun Sun
Jianmin Li

Organization

Conference Committee

Honorary Chairs

Bo Zhang	Tsinghua University, China
Nanning Zheng	Xi'an Jiaotong University, China
Deyi Li	CAAI, China

Advisory Committee Chairs

Qionghai Dai	Tsinghua University, China
Lin Chen	Institute of Biophysics, CAS, China
Wu Shou Er Si La Mu	Xinjiang University, China
Jifu Ren	University of Electronic Science and Technology of China, China
Yaochu Jin	University of Surrey, UK

General Chairs

Fuchun Sun	Tsinghua University, China
Zhenhong Jia	Xinjiang University, China
Jianwei Zhang	University of Hamburg, Germany

Organizing Committee Chairs

Liejun Wang	Xinjiang University, China
Huaping Liu	Tsinghua University, China
Zhongyi Chu	Beihang University, China

Program Committee Chairs

Dewen Hu	NUDT, China
Angelo Cangosi	University of Manchester, UK

Program Committee

Fakhri Karray	University of Waterloo, Canada
Haruki Ueno	National Institute of Informatics, Japan
Huaping Liu	Tsinghua University, China
Jianmin Li	Tsinghua University, China
Stefan Wermter	University of Hamburg, Germany
Witold Pedrycz	University of Alberta, Canada
Xiaolin Hu	Tsinghua University, China
Zengguang Hou	Institute of Automation, CAS, China
Zhongyi Chu	Beihang University, China

Publications Chairs

Quanbo Ge	Tongji University, China
Janmin Li	Tsinghua University, China

Finance Chair

Qianyi Sun	Tsinghua University, China

Registration Chair

Yixu Song	Tsinghua University, China

Local Arrangements Chair

Zhidong Ma	Xinjiang University, China

Contents

Perception and Learning

A Fuzzy Logic Based Top-Down Attention Modulation Framework for Selective Observation ... 3
Tao Jiang, Bingbing Kang, Xuming Wang, Jian Cao, and Jie Liang

A Tongue Image Classification Method in TCM Based on Multi Feature Fusion ... 15
Zhifeng Guo, Saisai Feng, Lin Wang, and Mingchuan Zhang

An Obstacle Detection Method for Visually Impaired People Based on Semantic Segmentation ... 28
Zhuo Chen, Xiaoming Liu, Dan Liu, Xiaoqing Tang, Qiang Huang, and Tatsuo Arai

MixStyle-Based Dual-Channel Feature Fusion for Person Re-Identification ... 34
Jian Fu, Xiaolong Li, and Zhu Yang

MS-MH-TCN: Multi-Stage and Multi-Head Temporal Convolutional Network for Action Segmentation ... 48
Zengxin Kang and Zhongyi Chu

Real-Time Visual Detection of Anomalies in Densely Placed Objects ... 61
Chunfang Liu, Jiali Fang, and Pan Yu

Robust License Plate Recognition Based on Pre-training Segmentation Model ... 74
Yanzhen Liao, Hanqing Yang, Ce Feng, Ruhai Jiang, Jingjing Wang, Feifan Huang, and Hongbo Gao

Predicting Cell Line-Specific Synergistic Drug Combinations Through Siamese Network with Attention Mechanism ... 87
Xin Bao, XiangYong Chen, JianLong Qiu, Donglin Wang, Xuewu Qian, and JianQiang Sun

Pressure Pain Recognition for Lower Limb Exoskeleton Robot with Physiological Signals ... 96
Yue Ma, Xinyu Wu, Xiangyang Wang, Jinke Li, Pengjie Qin, Meng Yin, Wujing Cao, and Zhengkun Yi

Research on UAV Target Location Algorithm of Linear Frequency Modulated Continuous Wave Laser Ranging Method 107
Yanqin Su and Jiaqi Liu

A Flexible Tactile Sensor for Robots Based on Electrical Impedance Tomography 123
Zhiqiang Duan, Lekang Liu, Jun Zhu, Ruilin Wu, Yan Wang, Xiaohu Yuan, and Longlong Liao

Joint Domain Alignment and Adversarial Learning for Domain Generalization 132
Shanshan Li, Qingjie Zhao, Lei Wang, Wangwang Liu, Changchun Zhang, and Yuanbing Zou

Value Creation Model of Social Organization System 147
Shuming Chen, Cuixin Hu, and Xiaohui Zou

Decision Making and Systems

Mixed Orientation ProMPs and Their Application in Attitude Trajectory Planning 159
Jian Fu, Zhu Yang, and Xiaolong Li

Multi-population Fruit Fly Optimization Algorithm with Genetic Operators for Multi-target Path Planning 174
Ke Cheng, Qingjie Zhao, Lei Wang, Wangwang Liu, Shichao Hu, and Kairen Fang

Task Assignment of Heterogeneous Robots Based on Large Model Prompt Learning 192
Mingfang Deng, Ying Wang, Lingyun Lu, Huailin Zhao, and Xiaohu Yuan

Quickly Adaptive Automated Vehicle's Highway Merging Policy Synthesized by Meta Reinforcement Learning with Latent Context Imagination 201
Songan Zhang, Lu Wen, Hanyang Zhuang, and H. Eric Tseng

Visual Inertial Navigation Optimization Method Based on Landmark Recognition 212
Bochao Hou, Xiaokun Ding, Yin Bu, Chang Liu, Yingxin Shou, and Bin Xu

A Novel Approach to Trajectory Situation Awareness Using Multi-modal Deep Learning Models 224
Dai Xiang, Cui Ying, and Lican Dai

A Framework for UAV Swarm Situation Awareness 233
Weixin Han, Yukun Wang, and Jintao Wang

The Weapon Target Assignment in Adversarial Environments 247
Rui Gao, Shaoqiu Zheng, Kai Wang, Longlong Zhang, Zhiyuan Zhu, and Xiaohu Yuan

A Distributed Vehicle-Infrastructure Cooperation System Based on OpenHarmony 258
Jingda Chen, Hanyang Zhuang, and Ming Yang

A Practical Byzantine Fault Tolerant Algorithm Based on Credit Value and Dynamic Grouping 272
Haonan Zhai and Xiangrong Tong

Data Exchange and Sharing Framework for Intermodal Transport Based on Blockchain and Edge Computing 292
Li Wang, Xue Zhang, Lianzheng Xu, Deqian Fu, and Jianlong Qiu

Internet Of Rights(IOR) in Role Based Blockchain 304
Yunling Shi, Jie Guan, Junfeng Xiao, Huai Zhang, and Yu Yuan

Tibetan Jiu Chess Intelligent Game Platform 322
Yandong Chen, Licheng Wu, Jie Yan, Yanyin Zhang, Xiali Li, and Xiaohua Yang

Author Index 335

Perception and Learning

A Fuzzy Logic Based Top-Down Attention Modulation Framework for Selective Observation

Tao Jiang(✉), Bingbing Kang, Xuming Wang, Jian Cao, and Jie Liang

Navy Aviation University, Yantai 264001, Shandong, China
hai_jt@qq.com

Abstract. The work presents a framework for top-down modulating the visual process under intention which is typically represented by words or short sentences. A fuzzy logic mapping process is developed to ground the word-form intention into a suitable weight vector to combine different visual feature channels during local saliency representation. Since the knowledge about the time-frequency characteristic of each feature extractor and their potential performance contribution to the higher-level qualitative intention is embedded, such a mapping solution is then feasible. Employing this top-down fuzzy logic mapping process, a novel task-oriented visual attention model for selective observation is then proposed. Implementing the model with a computational one, related experiments on traffic scene images has been done. Their result illustrates that the goal of selective surveillance on certain vehicle speed can be validly achieved. It implies that this framework is a competent model for visual selective attention, which expands the way to implement a computational mechanism for top-down modulating the bottom-up process especially in the case of task-related attentional action in machine vision systems.

Keywords: Selective Observation · Top-down modulation · Local Saliency Representation · Fuzzy Logic · Task-related Visual Attention · Computational Model

1 Introduction

Active vision research has a quite not short research span since it was proposed [1] to improve the flexibility, adaptivity and most importantly the computational efficiency for a general machine vision system. Therefore, how to develop a competent visual attention model plays a critical role in the related research field. Among these models, the one of multiresolution local saliency representation on parallel feature space [2] has been thoroughly discussed and then widespread adopted by many researchers especially in computer vision field. In a heuristic way by applying quite much psychological founding on human visual attention behavior like stimulus onset capture and visual search, and hints from their neural correlation, this model adopts a combination process to compute a single saliency map from multiple conspicuous maps of different parallel feature

F. Sun and J. Li (Eds.): ICCCS 2023, CCIS 2029, pp. 3–14, 2024.
https://doi.org/10.1007/978-981-97-0885-7_1

channels. However, it is actually of the bottom-up pre-attention model driven by input stimulus, not suitable for the top-down attentional activity if without further modification. For task-related visual activities like intentionally observing certain scene parts or searching some objects, there are some modified models have been developed [3, 4] based on it. The common critical issue is how to adjust the weight for combining the result from different visual processes, e.g. mainly weight the result from different feature channels, according the task on hand.

There are two approaches in defining or choosing the modulating manner. The first approach naturally adopts heuristics to set the weight vector or select related feature channels based on prior task knowledge, e.g. its probability distribution [5]. Recently another different approach has been promoted to learn the proper weight vector from human performance like the fixation location on a set of scene images each containing similar object(s) belonging to one class [3], or more generally to learn a suitable modulating strategy directly from human gaze annotated video dataset in daily activities [6] or to train the modulating process, e.g. how to assign weights to desired feature and adjust weights to direct attention for desired feature by other high-level and often task-oriented process like object recognition [4]. All the learning processes are implemented off-line in a supervised manner. Differing to them, some research, mainly on modelling object-based attention [7], promotes an additional top-down process to control the attentional action, while the bottom-up attention process including the saliency maps building one for each feature remains unaffectedly, providing its output to the upper process [7, 8]. Hence, it indeed implements no modulation on the bottom-up attention module.

Although their proposed modulation mechanisms are competent in many cases, it still may be sound if a theoretical solution for top-down modulation of bottom-up processing can be given by imbedding more knowledge about its visual process for visual agents. It's the motivation for this work. Though fuzzy logic is one of the best formal method to represent qualitative concept, there are few works to apply it in developing the top-down attention modulation model. In this work, the language instruction is used to represent the task-level intention, and a fuzzy logic framework is developed to ground the intention represented by qualitative concepts into the feature selection process, typically the adjusting of their combination weights.

2 The Framework

2.1 General Procedure for Selective Observation

Let us define selective observation as a process that a cognitive system tends to effectively percept certain spatial or spatiotemporal forms, e.g. a special pattern, an interesting object or an abnormal event, in a scene which is ordinally dynamic. The general procedure proposed by us to meet this need is illustrated by Fig. 1. It grounds the intentional action into related visual processes like the one of local saliency representation by three sequential steps.

Step 1. Adopt Qualitative Concept to Represent the Intention. That means the intention system select proper concepts, typically represented with qualitative words or short sentences, to define the task requirements. For example, "pay attention to fast moving cars" or just "fast-moving" in brief.

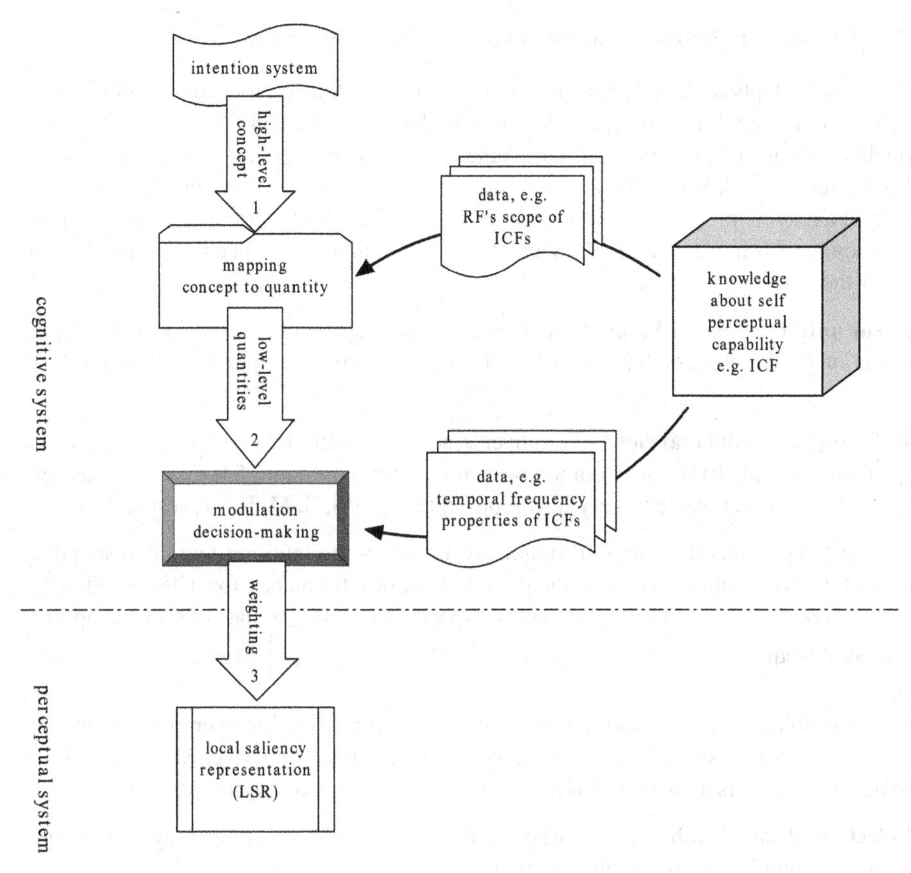

Fig. 1. The general process for selective observation with three main steps.

Step 2. Ground the Concept into Quantities Related to Bottom-up Process. For example, by applying the knowledge about the time response characteristics of each independent component filter (ICF, each is used as a feature extractor for one feature channel during low-level visual processing), it can easily decide to which extend the optimal temporal frequency for one ICF may deserve its response to certain visual stimulus that varies in its received field (RF) with certain qualitative speed just mentioned by the intention system.

Step 3. Make a Decision on Modulating the Bottom-up Process. The cognitive system then applies the mapping relation acquired from step 2 to make decision on suitable modulation, while using knowledge about the structure of the computational model for perception. For example, a parallel feature space structure leads to a choice of directly setting a proper weight vector for combining the saliency maps from varies of feature channels.

2.2 Ground Qualitative Concepts into Low-Level Quantities

This work adopts a fuzzy logic solution to ground qualitative concepts into quantitate values. When applying the fuzzy set theory, the value domain for a class of related qualitative concepts is treated as a universe set. Then, concepts are refined into several lexical terms, each responding to one fuzzy set related to the concept class. With the help of a class of properly selected or designed membership functions, each lexical term can be mapped into the value domain. We take the qualitative speed as an example to show the mapping procedure.

Assumption on Visual Perception. The perceptual system has a local saliency representation (LSR) process that uses 125 ICFs, each owning a temporal RF with 9 frames size.

Defining the Universal Set. The universe set now is the related value scope of the qualitative speed. To simplify our analysis, it is decomposed into 5 lexical terms as very slow, slow, middle, fast and very fast, represented by VS, S, M, F, VF, respectively.

Let f represents the temporal frequency, based on the intuition that the responding value for one qualitative speed is the optimal temporal frequency for ICFs, noted as f_t^*, the related value domain for the concept of speed then is the set that consists the optimal temporal frequency for all ICFs, noted as $\left\{f_{t,i}^* | i = 1, 2, \ldots, k\right\}$, where k is the number of ICFs.

To simplify the representation, it always take a normalization operation on any universal set to fix its domain into the interval of [0.1], noted an universal set as X. Each lexical term on X is then treated as a fuzzy set, written respectively as $\tilde{S}_i$, $i = 1, 2, \ldots, 5$.

Selection of the Membership Function. For each fuzzy set $\tilde{S}_i$, there may exist several choices to build its membership function s_i.

Gaussian Distribution Function. One choice is to adopt the Gaussian distribution function as below to build the membership function.

$$g_i(f; \mu_i, \sigma_i) = exp\left\{-\frac{(f - \mu_i)^2}{2\sigma_i^2}\right\} \tag{1}$$

where $f \in X$.

Then the membership function s_i is given by below formulas.

$$\begin{cases} a = 5 \\ \mu_i = \frac{i-1}{a}, \sigma_i = \frac{1}{2a} \\ s_i(f) = \frac{g_i(f;\mu_i,\sigma_i)}{\int_0^1 g_i(f;\mu_i,\sigma_i)df} \end{cases} \tag{2}$$

Delta Function. When the number of ICFs is quite small, another feasible choice is to use a delta function form as below.

$$s_i(f) = \delta(f - i) \tag{3}$$

The general δ function is given by below function.

$$\delta(f) = \begin{cases} 1, f = 0 \\ 0, f \neq 0 \end{cases} \tag{4}$$

where $f \in \{f_i | f_i \in [0, l \oslash 2], i = 1, 2, \ldots, k\}$, l is the frequency band width of the RF for one ICF, and the symbol $\oslash$ means a dividing operation on l with 2 then followed by a right round operation on the quotient.

Knowledge Applied During Grounding. The knowledge applied mainly in this step includes 1) the relationship of speed to the optimal temporal frequency for each ICFs, and 2) the effective optimal temporal frequency band as 0–5*fps* (frames per second) since the temporal RF for all the ICFs is 9 frames assumed before.

2.3 Make Decision on Effective Modulation

Effective modulation can only be achieved by applying the knowledge about the actual structure of the perceptual system, which may limit the possible ways to modulate its on-line processing or using its processing result flexibly according the ongoing task. In our framework illustrated in Fig. 1, a LSR process is embedded for bottom-up visual information processing. So, as a simplest choice, to modulate means to decide a proper weight value, $\omega_i, i = 1, 2, \ldots, k$, to combine result from each feature channel.

Weighting for Qualitative Speed. One weighting manner is computed by below function.

$$\omega_i = A^*_{f_{t,i}} s_i(f^*_{t,i}) \tag{5}$$

where $A^*_{f_{t,i}}$ is the optimal amplitude for the i-th ICF, and $f^*_{t,i}$ is its optimal respond frequency.

If the confidence about the optimal frequency respond is omitted, the weight can be directly computed from the value of the membership for each ICF as below.

$$\omega_i = s_i(f^*_{t,i}) \tag{6}$$

Knowledge Applied During Decision-Making. The knowledge applied mainly in the third step includes 1) the optimal respond time frequency for each ICFs, 2) the optimal amplitude for each ICF, i.e. the confidential level for one temporal frequency, and 3) implicitly the parallelization of the feature space and hierarchical structure of the local saliency representation process.

3 Case Study on Traffic Scene

3.1 Data Set

The traffic video data set [9] is adopted to illustrate how selective observation can be achieved in such a routine automatic visual surveillance task. For comparison the TV program video data set [10] is also adopted especially during training the IFCs. All the video segments have the same sample rate as 25 fps and in gray scale.

3.2 Algorithms

In implementing the upper framework, there are three steps of processing shall be done before in order to acquire the needed ICFs, to compute the frequency response of the learned ICFs, and to obtain saliency maps from each feature channel, respectively.

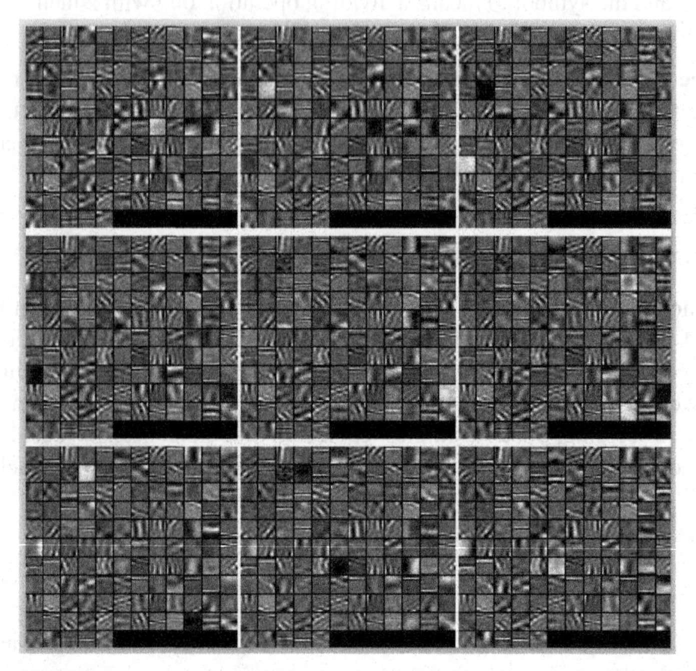

Fig. 2. The 125 ICFs learned with FastICA algorithm with $11 \times 11 \times 9$ sized RF. The spatial RF, i.e. the smallest image patches, for all 125 ICFs in one moment are arranged one by one into one of the nine middle sized patches. From the left to right, top to bottom, each frame of the ICFs, i.e. the time course of the spatial RF for each ICF, is illustrated.

Obtain the ICFs. A statistical learning approach is adopted to propose our attention model, hence feature extractors are learned from real scene video data. It assumes that a more task-specific feature space is more competent in real applications.

The ICFs are learning from an independent component analysis (ICA) method [11], more specifically in this work trained from a FastICA algorithm [12]. Different to other ICFs learned by ICA, our trained ICFs are 3-dimensional spatiotemporal feature extractors that could be directly applied on dynamic scene. Two of the three dimensions are used to extract spatial feature and the last for temporal or dynamic feature. Figure 2 illustrates the ICF that trained on the two data sets.

Frequency Response Analysis of the ICFs. The fast Fourier transform (FFT) algorithm is used to compute the amplitude and phase responses, then the optimal amplitude and temporal frequency for each ICF can be acquired.

Suppose we obtain the amplitude spectrum through FFT for one ICF, noted as $P[f_x, f_y, f_t]$, where x, y represent the two spatial dimension and t the temporal one, for the i-th ICF, its optimal temporal frequency, $f^*_{t,i}$, is given by the below formula.

$$f^*_{t,i} = argmax\left\{\sum\nolimits_{f_x, f_y} P_i[f_x, f_y, f_t] | f_t \in \{0, \Omega_t, 2\Omega_t, \ldots, k(l)\Omega_t\}\right\} \quad (7)$$

where $\Omega_t = \frac{1}{T_s l}$(Hz) is the elementary temporal frequency, T_s is the sample rate in time axis, l the total number of the frames for each ICF, and $k(l)$ a nearest round operation on the dividing result of l by 2.

Moreover, the corresponding optimal amplitude for the ICF is computed as below.

$$A^*_{f_t,i} = max\left\{\sum\nolimits_{f_x, f_y} P_i[f_x, f_y, f_t] | f_t \in \{0, \Omega_t, 2\Omega_t, \ldots, k(l)\Omega_t\}\right\} \quad (8)$$

Table. 1 presents our analysis result on the 125 ICFs shown in Fig. 2. In our experiments, the sample rate is 25Hz ($T_s = \frac{1}{25}s$), $l = 9$, so $\Omega_t \approx 2.87$ Hz.

Table 1. Some statistical result about the optimal temporal frequency of the 125 ICFs. The unit of f_t^* is circles/360ms since a sample rate of the data set is 25 frames per second.

f_t^*	number	average $A^*_{f_t}$	variance	minimal	maximum
0	48	0.7282	0.0046	0.4561	0.8328
1	43	0.4379	0.0067	0.2833	0.6422
2	21	0.3501	0.0012	0.2877	0.4244
3	8	0.3399	0.0038	0.2882	0.4519
4	5	0.3323	0.0016	0.3007	0.3924

Implement Local Saliency Representation. In this work, the algorithm based on our approach to model locally spatiotemporal saliency representation (LSTSR) [13] is adopted. Shared the basic structure of the model from [2], our computational model replaces its feature space with the trained ICFs as feature extractors and expands both spatial and temporal/dynamic saliency representation into a single coherent scheme, since the RFs of the ICFs are formed during a statistic learning process by feeding with video samples not with static images. Therefore, it is more task-oriented and inherently more suitable for dynamic visual input than the heuristic feature space embedded ones as [2].

3.3 Experiments on Qualitative Speed Surveillance

These experiments are designed to verify to which extent the top-down modulated local saliency representation process be competent for car speed surveillance. The simple modulation manner is adopted by only adjusting the combination weight of each feature channel before a united saliency map is composed. The saliency map is built in each frame; hence a dynamic saliency representation emerges.

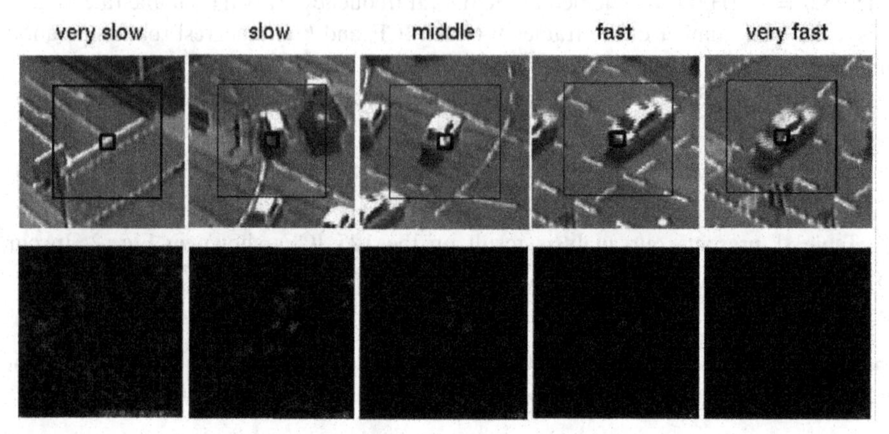

Fig. 3. Saliency maps acquired under different speed instruction by adopting the delta function as the membership during concept grounding. The scene showed in the upper row are segments from the date "dt" in [9]. The fixation points for the focus of attention according to related qualitative speed are illustrated by small rectangles. For the convivence of seeing the result clearly, a larger rectangle is added around each fixation point and the images are enlarged. The below row shows one shot of their computed saliency map for each scene respectively.

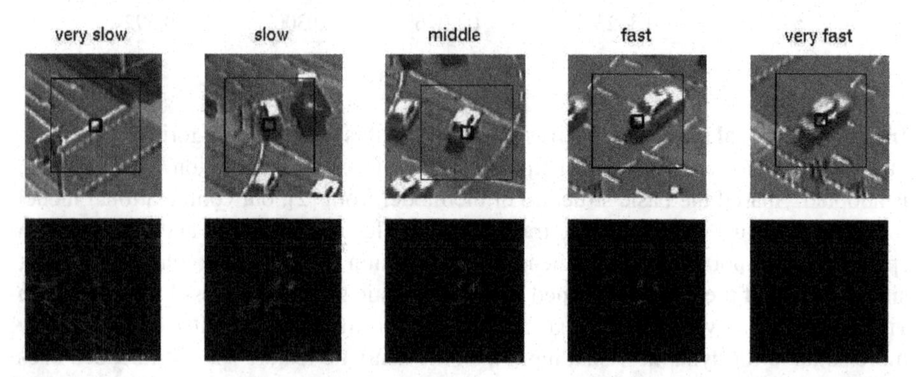

Fig. 4. Saliency maps acquired under different speed instruction by adopting the Gaussian function as the membership during concept grounding. The scene showed in the upper row are also segments from the date "dt" in [9] as same as in Fig. 3.

Based on the general framework that described in Sect. 2, unrelated feature channels can be masked or depressed according the value of their optimal frequency that maps their correlation to certain speed with membership function given in Eq. (1) or (3). The correlation result is represented as a novel weight vector in order to form the LSR result.

Figure 3 shows the result of the saliency map on different traffic scene by adopting the δ function as the membership function. As a comparison, Fig. 4 gives the result of the Gaussian one. Both of them illustrate that the modulation is feasible for qualitative speed surveillance. Interesting, the result of saliency representation reveals that the fast the speed the intention demands, the rarer or smaller the area is salient will be, which is true in many cases. It also verifies that a single LSR process may be competent to direct the visual agent onto the location according to intention if a proper grounding process has been added into its cognitive system.

4 Discussion

4.1 Justification for the Feasibility

Ontological Assumption. It means that the top-down intentional modulation on a shared low-level visual process is necessary for an active visual agent.

Upper experiments justify the feasibility of our framework for selective observation, although it just provides an initial test with limited data set on one typical task. It reveals that the low-level saliency representation can be effective modulated with a ground process from high-level concept to low-level quantities by using a fuzz logic mapping solution. The first underlying ontological assumption is that for a visual agent, no matter a robot vision system or just a computer vision software, its active processing model ought to embed at least two stages of progresses, one for bottom-up visual processing driven by the input data, another for top-down intention processing oriented to the ongoing task. The two stages shall share a common module in sake of a low implementation cost and in favor of a unified scheme to simplify its structure. The LSR process is such an excellent candidate. More specifically, our LSTSR model is more competent than the local spatial one as in [2] and can directly be applied to dynamic scenes. Different to our approach, additional motion saliency representation module always needs to be inserted into the original locally spatial saliency representation (LSSR) model or a distinct model for motion detection is involved like [14].

The Role of Saliency Representation. The saliency representation process is intrinsically an implicit and valid segmentation process no matter data-driven or task-oriented.

In order to percept the surrounding in an active manner, the less the information the agent shall to pay attention on. That is in general a segmentation process to keep meaningful information or even enhance related information in the input. Saliency representation can provide the agent such a segmentation result, albeit typically it computes the saliency map in grayscale, not a binary one. The upper experiments on qualitative speed surveillance reveal that due to a feasible modulation procedure the resulting saliency map can direct the agent to focus on task-related visual form and location.

Concept Grounding by a Fuzzy Logic Procedure. At least limited in the scope of qualitative concepts, fuzzy logic is qualified to map high-level intention to low-level quantities.

Our framework illustrates the feasibility of fuzzy logic as a useful tool to ground qualitative concept into low-level quantities that may further be used to modulate the low-level process like LSR. Our result even shows that the different choices of membership function less obviously affect the saliency representation result in our task. One reason may be that we have not compared different tasks that need different representation complexity during grounding the intention.

Embodiment of Knowledge for Decision-Making on Modulation. Knowledge about the structure of visual processing and the performance characteristics of low-level processes is necessary in order to benefit the decision-making process to modulate effectively and efficiently the common visual process.

Our framework and experiments involve the applying of these two sorts of knowledge about its ongoing processing for an active visual agent. It is one fundamental task for the cognitive system whenever it tends to control or direct the perceptual system to achieve more flexible and intention-sensitive perception. The cognitive agent needs to understand about itself before making intentional action. Hence, effective embodiment of the knowledge about its visual perception destinates the performance for intention-oriented active behavior, not limited in the scope of visual processing.

4.2 Beyond Feature Modulation

Ways for Modulation. The possible ways to modulate the low-level processing and the best choice for an effective operation shall be decided both on the structure of the visual processing model and on the performance relevance of each detailed processing contributing to the high-level task. Take the parallel feature space model adopted by our work [13] and others [2] as an example, weights for combining feature channels are adjustable, so are other parameters related to low-level processes such as the domain of the scale for a pyramid multi-resolution representation, the width of the window for local contrast inhibiting operators [3], etc.

From Feature Weighting to Feature Selection. Feature weighting, though easy but quite effective in simple cases, is a direct way to modulate the low-level processing especially by biasing the saliency representation result for a visual computational model that implements a parallelized feature processing structure.

To achieve the required efficiency especially for real-time applications, feature selection is one proper method. Based on the computed value of the weight vector, it can mask the unrelated feature channel, only enable the most or the quite related one(s) by setting a threshold. One way to obtain the suitable threshold value is to do experiments with the help of some training or supervised learning algorithms.

Cross the Gap between Qualitative Concepts and Quantities. Our results justify the validity of using a fuzzy logic solution to ground qualitative concept into quantities corresponding to low-level processing. But it is still far from applying to more complex tasks, e.g. finding certain kind of objects, paying attention to certain kind of their behavior. The main limitation may not come from the fuzzy logic mapping procedure which in principle owns the potential to adapt to more complex tasks if giving more dedicated design, but from the model structure for visual information processing. That means for more complex task-related intention grounding, the first choice is to update or redesign the common processing, then the mapping procedure. One solution is to embody capabilities like object perception and event inference according the task on hand, then design the grounding procedure according knowledge about the updated structure and related processing details. For such a general-purpose cognitive agent, its visual process hierarchy shall consist of a cascade of pattern processing modules, typically from the bottom to the top, for processing low-level features, basic spatial and dynamic forms, objects, actions, events, etc. To achieve attentional behavior, models on different levels of attention, e.g. object-based attention, task-oriented attention, even natural language represented intention guided attention, shall be embedded depending on the complexity of the demanded tasks.

5 Conclusion

In this paper a framework for selective observation of dynamic scene by effective modulating the LSTSR process is proposed. The fuzzy logic solution is developed to ground qualitative concept into low-level quantities relating to the low-level processing. Experiments on traffic scene reveals the intention for flexible qualitative speed surveillance can be feasibly solved with this framework. It seems that the fuzzy logic solution for top-down attention modelling can pave a theoretic foundation for more complex task, like language directed visual navigation behavior [15]. Generally, our framework illustrates how to connect the intention system coherently with the visual saliency representation system, so a cognitive system for task-oriented real-time visual applications can then be obtainable. On the other hand, it justifies that LSR shall be a competent middle-level process for purposive or active vision systems. How to modulate its processing including other low-level or bottom-up visual processes more efficiently, e.g. considering the real-time implementation issues such as computational cost and computational complexity, shall be one meaningful work.

One of our future work on this proposed framework is to carry more experiments on more complex visual tasks, and making comparison on varieties of dataset. In order to map effectively the more complex intention or even interpret the language represented intention into the process of visual modulation, a Bayesian inference process is planned to be added into the current model, then our framework can be expanded.

References

1. Aloimonos Y.: Purposive and qualitative active vision, In: 1st Proceeding of International Conference on Pattern Recognition, pp. 346–360 (1991)
2. Itti, L., Koch, C., Niebur, E.: A model of saliency-based visual-attention for rapid scene analysis. IEEE Trans. Pattern Anal. Mach. Intell. **20**, 254–259 (1998)
3. Borji, A., Ahmadabadi, M.N., Araabi, B.N.: Cost-sensitive learning of top-down modulation for attentional control. Mach. Vis. Appl. **22**, 61–76 (2011)

4. Benicasa, A.X., Quiles, M.G., Zhao, L., Romero, R.A.F.: Top-down biasing and modulation for object-based visual attention. In: Lee, M., Hirose, A., Hou, Z.-G., Kil, R.M. (eds.) ICONIP 2013. LNCS, vol. 8228, pp. 325–332. Springer, Heidelberg (2013). https://doi.org/10.1007/978-3-642-42051-1_41
5. Chen, H.Z., Tian, G.H., Liu, G.L.: A selective attention guided initiative semantic cognition algorithm for service robot. Int. J. Autom. Comput. **15**(5), 559–569 (2018)
6. Ma, K.T., Li, L., Dai, P., Lim, J.H., Shen, C., Zhao, Q.: Mutl-layer model for top-down modulation of visual attention in natural egocentric vision. In: IEEE ICIP 2017, pp. 3470–3474 (2017)
7. Sun, Y.R.: Hierarchical object-based visual attention for computer vision. Artif. Intell. **146**(1), 77–123 (2003)
8. Ozeki, M., Kashiwagi, Y., Inoue, M., Oka, N.: Top-down visual attention control based on a particle filter for human-interactive robots. In: Proceeding of 4th International Conference on Human System Interactions, HSI 2011, pp. 188–194 (2011)
9. Traffic Dataset from Universität Karlsruhe. http://i21www.ira.uka.de/image_sequences
10. TV Shows Dataset from Hateren, J. H., Department of Neurobiophysics, Groningen University, Netherland. ftp://hlab.phys.rug.nl/pub
11. Hyvärinen, A., Patrik, O., Hoyer, P.O.: Topographic independent component analysis as a model of V1 organization and receptive fields. Neurocomputing **38**(40), 1307–1315 (2001)
12. Haykin, S.: Neural Networks: A Comprehensive Foundation, 2nd edn. Prentice-Hall Inc., Upper Saddle River (1999)
13. Jiang, T., Jiang, X.: Locally spatiotemporal saliency representation: the role of independent component analysis. In: Wang, J., Liao, X., Yi, Z. (eds.) ISNN 2005. LNCS, vol. 3496, pp. 997–1003. Springer, Heidelberg (2005). https://doi.org/10.1007/11427391_160
14. Belardinelli, A., Pirri, F., Carbone, A.: Motion saliency maps from spatiotemporal filtering. In: Paletta, L., Tsotsos, J.K. (eds.) WAPCV 2008. LNCS (LNAI), vol. 5395, pp. 112–123. Springer, Heidelberg (2009). https://doi.org/10.1007/978-3-642-00582-4_9
15. Lindi, F., Baraldi, L., Cornia, M., et al.: Multimodal attention networks for low-level vision-and-language navigation. Comput. Vis. Image Underst. **210**, 103255 (2021)

A Tongue Image Classification Method in TCM Based on Multi Feature Fusion

Zhifeng Guo, Saisai Feng, Lin Wang, and Mingchuan Zhang(✉)

School of Information Engineering, Henan University of Science and Technology, Luoyang 471003, China
{210321050414,210321050372}@stu.haust.edu.cn,
{Linwang,zhang_mch}@haust.edu.cn

Abstract. Traditional Chinese medicine distinguishes tongue features such as tongue color, fur color, tongue shape and crack mainly through the visual observation and empirical analysis of traditional Chinese medicine doctors. Therefore, the judgment standard will be affected by the subjective factors of doctors and surrounding environment. These factors restrict the application and development of tongue diagnosis. Therefore, objectifying tongue diagnosis information and standardizing diagnosis is an important direction of tongue diagnosis automation research. This paper presents a classification method of TCM tongue image based on multi feature fusion. By constructing a multi feature fusion model, two sub networks are used to classify the different features of the tongue image, so as to realize the task of multi feature classification of the tongue image. The model classifies the tongue image into tongue color classification, fur color classification, tongue shape classification and crack classification, and outputs the color parameters of tongue color and fur color while outputting the classification results. The model adds the method of transfer learning, which can reduce the demand for the amount of tongue image data, At the same time, the accuracy of the model is improved.

Keywords: Tongue diagnosis · Deep learning · Transfer learning

1 Introduction

Since its development, traditional Chinese medicine has made outstanding contributions to the treatment, prevention and health preservation of the general public. It is not only the treasure of Chinese traditional culture, but also a precious spiritual wealth. Traditional Chinese medicine uses the four diagnostic methods of "seeing, hearing, asking and examining", and makes diagnosis

---

This work was supported in part by the National Natural Science Foundation of China (NSFC) under Grants No. 62002102 and 62072121, and in part by the Scientific and Technological Innovation Team of Colleges and Universities in Henan Province under Grants No. 20IRTSTHN018, and in part by the Key Technologies R & D Program of Henan Province under Grants No. 212102210083, 222102310565, 232102210028.

F. Sun and J. Li (Eds.): ICCCS 2023, CCIS 2029, pp. 15–27, 2024.
https://doi.org/10.1007/978-981-97-0885-7_2

according to the characteristics of the patient's condition. Tongue diagnosis is a clinical diagnostic method to understand the physiological function and pathological changes of the body by observing the changes of the tongue and analyzing the differences of tongue perception.

At present, tongue diagnosis is gradually combined with the field of image vision in artificial intelligence [1–3], and people have made new discoveries and progress in the clinical diagnosis of tongue image and the judgment of disease. For example, in the period of the most serious outbreak of COVID-19, telemedicine is particularly important. Doctors can judge the patient's condition by observing the changes of the patient's tongue through electronic equipment, and can also provide reference and treatment basis for the prevention and treatment of covid-19 [4]. In other aspects, it has also developed a lot of technical research on diagnosis and treatment according to tongue image, especially the research on the objectification of tongue image has also achieved many remarkable results. Researchers classify according to the characteristics of tongue image, which is usually divided into five categories: tongue color classification, tongue shape classification, fur color classification, tongue pattern classification and tongue vein classification. Because tongue vein refers to the vein under the tongue, it is difficult to collect tongue image and identify and extract features. At present, researchers mainly focus on the first four categories.

The tongue color is usually divided into light white, light red, red, crimson and cyan. The tongue color of healthy people is mostly light red. When the body has a disease, the blood concentration will often change, and these changes will be reflected in the tongue color. Wang et al. [5] proposed a tongue color space for mathematical description of diagnostic feature extraction, established a database for storing tongue image, divided the tongue color region, and studied it from three aspects: tongue color gamut, tongue surface center and color distribution of typical regions. Hou et al. [6] proposed a tongue color classification method based on deep learning, modified the parameters of neural network to make it suitable for the training of its model, and further deepened the objective research on tongue color classification. In terms of coating color research, Fu et al. [7] used depth neural network to quantitatively study the characteristics of tongue coating, combined with basic image processing technology, and verified the effectiveness of this method through a standard and balanced tongue image data set. Li et al. [8] also improved feature extraction and classification methods based on convolutional neural network. The extracted tongue coating features replaced manual features, and used multi instance support vector machine to solve the problem of uncertain tongue coating position. In the field of tongue shape classification, Huang et al. [9] proposed a classification method for automatic recognition and analysis of tongue shape based on geometric features. The method corrects tongue deflection by applying three geometric criteria, and then classifies tongue shape according to seven geometric features defined by various measurements of tongue length, area and angle. Tongue pattern classification is generally accompanied by other tongue features. Li et al. [10] proposed a detector based on wide line statistical shape features to identify whether there are cracks in the tongue. First, a wide line is detected in the tongue image, and then the shape features are extracted. Chen et al. [11] used the correlation of gray level co-occurrence matrix to screen the extracted tongue cracks based on the tongue crack gray level and texture extraction algorithm in the tongue image, so as to realize the automatic extraction of tongue cracks.

Although scholars at home and abroad have done a lot of research on the classification of tongue image, and achieved some results. However, there are still some problems to be solved in the existing methods and systems in the modernization of tongue diagnosis. First of all, most studies on the extraction and classification of tongue image features are often single classification. In reality, traditional Chinese medicine students use tongue diagnosis to judge the patient's condition only by integrating a variety of tongue image features. Therefore, the tongue image features of single classification have high limitations and are difficult to play a role in real life. In addition, the tongue image of traditional Chinese medicine, like other medical images, has the problems of scarce training sample data and large amount of tongue image feature information. If there is not enough training samples to support, the accuracy of classification will inevitably be reduced, which will affect the subsequent development and use.

To solve the above problems, this paper proposes a tongue image classification network based on multi feature fusion, and constructs an overall model, which includes two sub network models to realize the classification task of four tongue image features. In the whole training process, the classification task of tongue color and fur color uses ResNet- 34 network model based on transfer learning, and the classification task of tongue shape and crack uses VggNet-16 network model based on transfer learning. Finally, the method of transfer learning is unified to improve the training speed of the model, improve the performance of the network model and reduce the time spent in the process of training the model.

2 System Model

2.1 Tongue Image Classification Model Based on Multi Feature Fusion

In image classification, shape features are usually divided into contour features and regional features. Contour features can also be considered as edge features, which are generally used to classify images with obvious features, while regional features are used to classify one of the shape regions. VggNet [12] has a simple structure and uses only one loss function, so the model has strong expansibility and good generalization ability. It is very suitable for extracting contour features and regional features. Tongue color and fur color belong to color features. In general application scenarios, color features are divided into global features. ResNet [13] is a cross layer connected network, which can make the whole model overlay a deeper network, and the network will not degrade with the increase of depth. Therefore, using ResNet to extract and classify tongue color and fur color features can achieve better results.

The system model extracts, fuses and classifies the single feature of the tongue according to the classification task, and then combines the two separate models to realize multi feature fusion classification. The structure of this model is shown in Fig. 1.

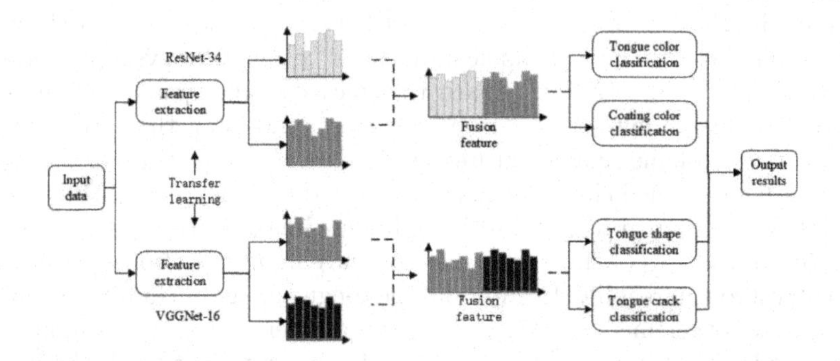

Fig. 1. Tongue image classification model based on multi feature fusion.

The multi feature fusion network model structure combines ResNet-34 and VggNet-16 networks. In the classification task, multi feature fusion is used to describe the tongue image, realize information complementarity, reduce redundant information, and then improve the effect of tongue image classification.

Color features have several distinct characteristics: feature extraction is relatively easy, the scale of features does not change with convolution operation in the extraction process, and color features do not change with image rotation. They can be expressed by low-order matrix features.

The first moment of tongue color feature is expressed as:

$$\delta_i = \frac{1}{N}\sum_{j-1}^{N} P_{ij} \tag{1}$$

where i represents the number of color channels, the tongue image of this experiment is initially RGB three channels, N represents the number of pixels in the image, δ_i represents the average number of color channels, and P_{ij} represents the color component in the image.

The second moment of color features is expressed as:

$$\lambda_i = \sqrt{\frac{1}{N}\sum_{j-1}^{N} (P_{ij} - \delta_i)^2} \tag{2}$$

where λ_i is the calculated standard deviation.

The third moment of color features is expressed as:

$$\beta_i = \sqrt[3]{\frac{1}{N}\sum_{j-1}^{N} (P_{ij} - \delta_i)^3} \tag{3}$$

where β_i is the calculated deviation.

VggNet-16 is used to classify tongue shape and crack. In order to improve the efficiency and accuracy of classification and eliminate the interference of other features such as color, this experiment chooses to gray the RGB parameters of tongue image before feature extraction. Here, the common method is the average method to average the three brightness of color tongue image, so as to obtain a gray value.

$$f(i,j) = \frac{R(i,j) + G(i,j) + B(i,j)}{3} \tag{4}$$

The shape feature of tongue image is obtained by solving the gradient histogram, which is expressed as follows:

$$\theta(i,j) = \tan^{-1}\left[\frac{I(i,j+1) - I(i,j-1)}{I(i+1,j) - I(i-1,J)}\right] \tag{5}$$

The texture feature can be extracted according to the gray value of image spatial distribution, and it usually has no rotation deformation and strong anti noise ability. Therefore, the texture feature of tongue image can be obtained by solving the contrast, correlation, second-order moment and entropy of gray level co-occurrence matrix. The contrast expression is:

$$f_1 = \sum_{n=0}^{L-1} n^2 \sum_{i=0}^{L-1} \sum_{j=0}^{L-1} \hat{p}_d(i,j) \tag{6}$$

f_1 is the contrast,(i,j) is the gray level of the pixel, d is the spatial position interval between two pixel points, $p_d(L-1)$ is the number of occurrences of pixels of $(L-1)$.f_2 is the correlation degree, and the expression is:

$$f_2 = \frac{\sum_{i=0}^{L-1} \sum_{j=0}^{L-1} ij\hat{p}_d(i,j) - \mu_1\mu_2}{\sigma_1^2\sigma_2^2} \tag{7}$$

The expression of second-order matrix f_3 is:

$$f_3 = \sum_{i=0}^{L-1} \sum_{j=0}^{L-1} p_d^2(i,j) \tag{8}$$

f_4 is the entropy of image gray level co-occurrence matrix, and the expression is:

$$f_4 = -\sum_{i=0}^{L-1} \sum_{j=0}^{L-1} \hat{p}_d(i,j) \log \hat{p}_d(i,j) \tag{9}$$

The parallel fusion method is adopted in this experiment.Let A and B represent the two feature spaces of tongue color and fur color in tongue image respectively, or the two feature spaces of tongue shape and crack, and α and β represent the feature vectors of the two feature spaces respectively, then the parallel fusion can be expressed as:

$$\eta = \alpha + n\beta \tag{10}$$

where η is the composite eigenvector and n is the parameter. Finally, feature classification is carried out. In the network, the full connection layer is usually placed together with Softmax. According to the data output from the previous layer, the formula is used to convert it into probability distribution, and the maximum value is taken as the classification result of the current output.

$$P_{soft\max(m_j)} = \frac{e^{m_i}}{\sum_{i=1}^{n} e^{m_i}} \tag{11}$$

Algorithm 1. Tongue image classification algorithm based on multi-feature fusion

Input: Tongue image
Output: Tongue color, tongue coating color, tongue shape, tongue crack, classification, Color parameters

1: Input tongue image data training set D
2: Initialize training parameters and current accuracy
3: **while** $x \in d$ **do**
4: Maximum pooling
5: **for** $x \leq d$ **do**
6: Calculate $\delta_i, \lambda_i, \beta_i$
7: **for** $\alpha \in A$, $\beta \in B$ **do**
8: $\eta = \alpha + n\beta$
9: $\mathrm{P}_{soft\max(m_j)} = \frac{e^{m_i}}{\sum_{i=1}^{n} e^{m_i}}$
10: **end for**
11: **for** $x \leq d$ **do**
12: $f(i,j) = \frac{R(i,j)+G(i,j)+B(i,j)}{3}$
13: Calculate the gradient histogram of tongue image $E(i,j)$
14: **end for**
15: **end for**
16: **for** $\alpha \in A$, $\beta \in B$ **do**
17: $\eta = \alpha + n\beta$
18: $\mathrm{P}_{soft\max(m_j)} = \frac{e^{m_i}}{\sum_{i=1}^{n} e^{m_i}}$
19: **end for**
20: **if** $\mathrm{P}_{soft\max(m_j)} \neq 0$ **then**
21: $H(p,q) = -\sum_x (p(x) \ln q(x))$
22: **end if**
23: **end while**

Where $\mathrm{P}_{soft\max(m_j)} = \frac{e^{m_i}}{\sum_{i=1}^{n} e^{m_i}}$ is the probability of the corresponding type, n is the number of output nodes, and m_i is the output value of the $i-th$ node.

After obtaining the predicted output value, the cross entropy loss function is used to calculate the difference between the predicted output and the actual output. The greater the loss value, the greater the difference between the two probability distributions.Where $p(x)$ is the real tag value and $q(x)$ is the predicted value.

$$H(p,q) = -\sum_x (p(x) \ln q(x)) \tag{12}$$

The specific process of the whole algorithm is shown in Algorithm 1.

3 Algorithm Analysis

3.1 Time Complexity Analysis

The time complexity of the algorithm in this chapter can be analyzed by the circular structure of tongue image pixels running in the feature extraction process. Assuming that it represents the total number of pixels, each node of the algorithm needs to run once in the feature extraction, then the two networks fuse the features respectively, and the network node features run again. Therefore, the total time algorithm complexity is $O(n^2) + O(n^2) = O(2n^2)$, The coefficient is removed, so the time complexity of the algorithm in this chapter is $O(n^2)$.

3.2 Spatial Complexity Analysis

The time complexity of the algorithm in this chapter can be analyzed by the circular structure of tongue image pixels running in the feature extraction process. Assuming that it represents the total number of pixels, each node of the algorithm needs to run once in the feature extraction, then the two networks fuse the features respectively, and the network node features run again.

4 Experimental Results and Analysis

4.1 Data Set

In this experiment, the data set we use is the tongue image obtained from traditional Chinese medicine medical institutions, which is reclassified according to various features and tested experimentally. Starting from the two characteristics of tongue color and moss color, the tongue color is divided into five categories [14]: light white, light red, red, crimson and cyan. The moss color is divided into three categories: white moss, yellow moss and gray black moss. Based on the two characteristics of tongue shape and crack, the tongue shape is divided into five categories: strip tongue, hammer tongue, triangular tongue (long triangular tongue, short triangular tongue), circular tongue and square tongue. The cracks are divided into two categories: with cracks and without cracks. Figure 2 shows four categories of tongue image data.

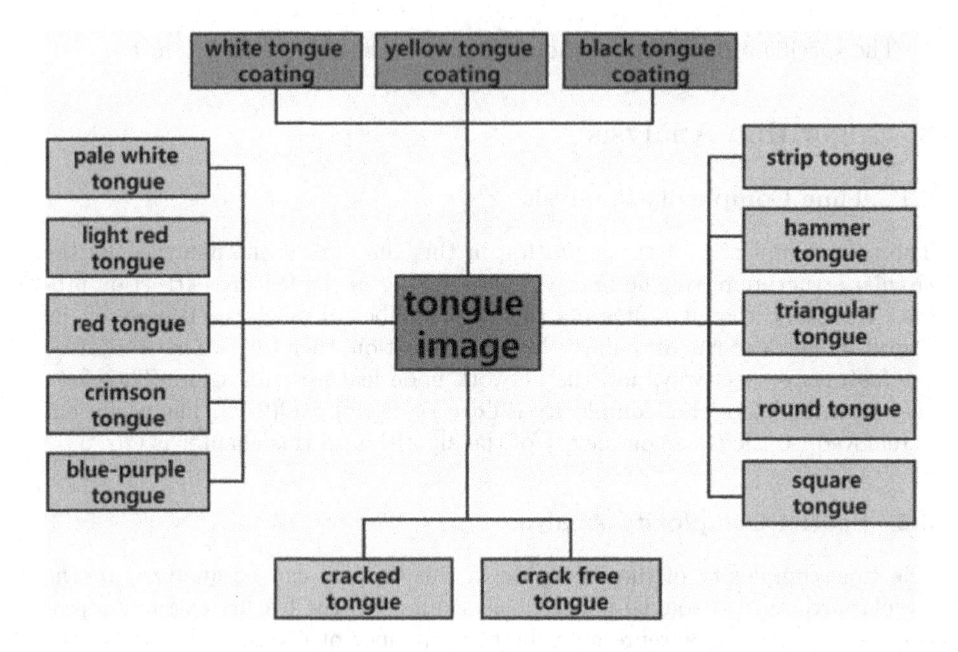

Fig. 2. Four categories of tongue image.

4.2 Evaluation Method

Because it is a parallel classification of multiple features of tongue image, the commonly used evaluation method is to show the classification results of each kind, and the accuracy rate in each evaluation is taken as the average value. Finally, the average calculation results of various features are obtained. According to the experimental results, whether the parameters adjusted by the network model are appropriate can also reflect the rationality of the algorithm design. The calculation formula is as follows.

$$Accurate\_ave = \frac{1}{n}\sum_{k=0}^{n} Accurate_k \tag{13}$$

4.3 Feature Extraction Graph

The ResNet-34 model trained by transfer learning can extract the feature map of tongue color and fur color, and the VggNet-16 network model based on transfer learning can extract the feature map of tongue shape and crack in training. Figure 3 shows the shallow characteristic diagrams of tongue color, fur color, tongue shape and crack respectively. The characteristics of tongue shape and crack do not need to use color and other characteristics, so there is a grayscale pretreatment before training, so as to reduce the influence of other factors in the training process, and improve the speed and accuracy of training to a certain extent.

Through the feature extraction diagram, it can be seen that in the classification of tongue color and coating color, the tongue itself has been separated from the surrounding environment, and the brightness and color RGB values of the tongue area are significantly different from those around, which shows that after the fine-tuning of migration learning, the ResNet-34 network structure is more accurate for the color recognition and extraction in the tongue image, and the characteristics of coating color are also more obvious, which can see the separation of tongue coating and tongue, The extraction and recognition of moss color also achieved good results.

In the feature extraction of crack and tongue shape, due to the gray processing method, the interference of the surrounding environment and the color of the tongue itself is greatly reduced in the training process of the model, and the concentration and accuracy of training are improved from the side. Since VggNet-16 network model contains only one loss function and has stronger expansibility, it is very appropriate to adjust its structure to classify tongue shape and crack. Then, migration learning is added to adjust the parameters of the whole connection layer, which reduces the amount of parameters and improves the training speed without affecting the overall accuracy. On the whole, good results are achieved.

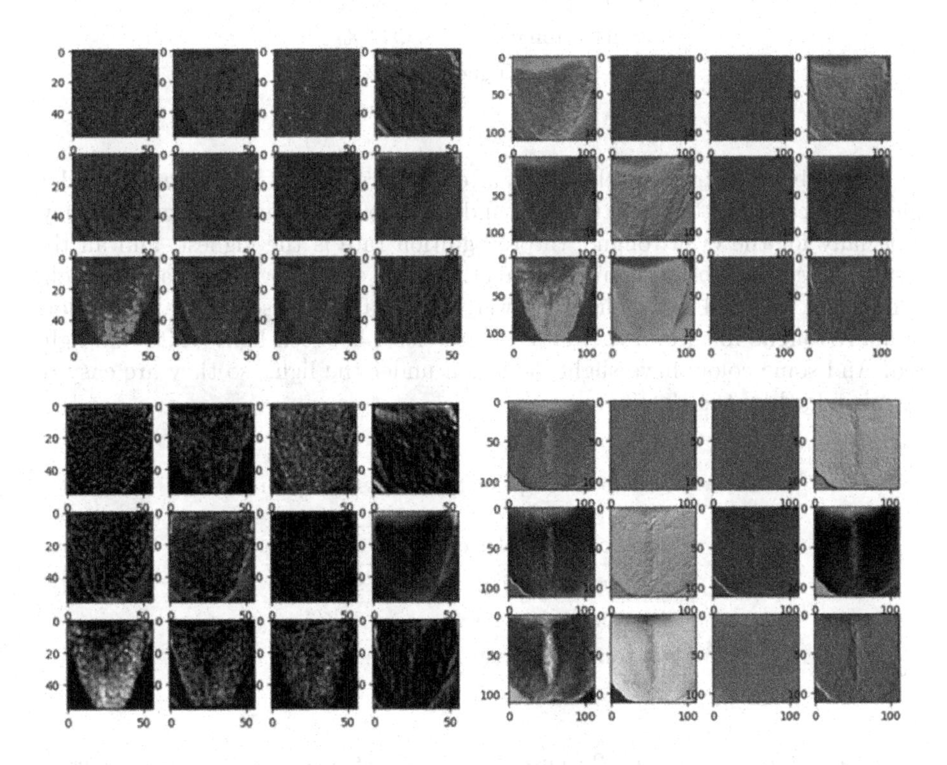

Fig. 3. Four categories of tongue image.

4.4 Experimental Result

In order to make some rare tongue images reach the same training scale, the collected tongue image data are enhanced. In this experiment, ResNet-34 and VggNet-16 network models are used. In order to improve the training efficiency, both networks add the method of transfer learning to classify various features of tongue images. After many times of training, the average accuracy after evaluation is taken.

The following experiments are all in the same data set. According to the different types, the test set is put into the newly constructed network model after integrating two kinds of transfer learning fine-tuning for training. The accuracy of each sample test is listed in the table.

Table 1. Table of tongue color classification results

Tongue color category	Accuracy
pale white tongue	90.8%
light red tongue	87%
red tongue	93.7%
crimson tongue	94.5%
blue-purple tongue	97%

It can be seen from Table 1 that after the training of the network model, the classification of test tongue color and fur color has achieved good accuracy, especially for the cyan tongue, the recognition rate is the highest, and all the tested images can be recognized correctly, while the recognition rate for light white, light red and red is slightly lower. The analysis reason is that the colors of many tongue images in the data set are close, between light white and light red, And some colors have slight deviation under the light, so they are easy to be misjudged by the classifier.

Table 2. Table of tongue coating classification results

Tongue coating color category	Accuracy
white tongue coating	91.2%
yellow tongue coating	87.5%
black tongue coating	96.4%

It can be seen from Table 2 that in the classification of fur color, the accuracy of gray black fur is the highest, while the accuracy of white fur and yellow fur is slightly worse. This is because some tongue like white fur covers a large area of the tongue body, and the tongue fur is thicker, so some white fur will be misjudged as yellow fur, or some yellow fur is thinner, and there is more body fluid on the tongue, which makes the brightness of fur color on the tongue image high, and the classifier will be mistaken as yellow fur.

Table 3. Table of tongue shape classification results

Tongue shape category	Accuracy
strip tongue	82.5%
hammer tongue	82.5%
triangular tongue	94.7%
round tongue	87.5%
round tongue	90%

It can be seen from Table 3 that in the classification of tongue shape, the accuracy of strip tongue and hammer tongue is low. This is because the two tongue shapes are relatively similar, and the model is prone to misjudgment. The accuracy of triangular tongue is the highest, up to 94.7%, indicating that the characteristics of triangle are obvious and the recognition rate is high.

Table 4. Table of tongue crack classification results

Tongue crack category	Accuracy
cracked tongue	98.2%
crack free tongue	97.5%

It can be seen from Table 4 that the classification accuracy of cracked tongue and non cracked tongue is very high, 98.2% and 97.5% respectively, indicating that the grayed tongue image can better detect the characteristics of cracks.

On the whole, the model structures of the two networks are integrated, and their respective network advantages are brought into play. They have achieved good classification results, and can better complete the task of tongue image classification.

5 Conclusion

Tongue image classification is of great significance for the objectification of tongue diagnosis. This paper analyzes the four main features of tongue image, combines computer image processing technology with TCM tongue diagnosis knowledge, and completes the tongue image classification function based on multi feature fusion through the parallel combination of network model on the basis of realizing the single feature classification of tongue image. Before classifying the tongue image, the traditional method needs to carry out the steps of tongue image acquisition, target detection and segmentation, color correction by comparing the color card, manual feature extraction and so on. The method proposed in this paper does not need to describe and extract the target tongue image features manually, but puts the collected tongue image into the network model for training after preprocessing. Four categories of tongue images and corresponding parameters can be obtained by inputting tongue images in the prediction stage. In the process of collecting tongue images, professional equipment is used to reduce the color difference caused by the external lighting environment, which affects the classification results. On the whole, the integration of collection, segmentation and classification functions is realized. This paper also adds the method of transfer learning to the network model, migrates the model parameters trained by ImageNet massive data set to the task of this study, and solves the problems that are not conducive to tongue classification, such as poor classification accuracy and easy model fitting, which are caused by less tongue image data and insufficient training samples, so as to greatly improve the training speed of the network.

References

1. Zuchun, W.: Discussion on tongue image acquisition method and Application in tongue diagnosis Objectification research [D]. Guangzhou University of Chinese Medicine (2011). (in Chinese)
2. Shenhua, J., Jiang, L.: Research progress on objectification of tongue image and nature of tongue coating. Shanghai J. Tradit. Chinese Med. **50**(07), 94–97 (2016). (in Chinese)
3. Cai, Y.-H., Hu, S.-B., Guan, J., et al.: Development analysis and application of Objectified tongue diagnosis technology in Chinese medicine. World Sci. Technol. Modernization Chinese Med. **23**(07), 2447–2453 (2021). (in Chinese)
4. Pang, W., Zhang, D., Zhang, J.: Tongue features of patients with coronavirus disease, a retrospective cross-sectional study. Integr. Med. Res. **9**(3), 100493 (2019)
5. Xingzheng, W., Bob, Z., Zhimin, Y., et al.: Statistical analysis of tongue images for feature extraction and diagnostics. IEEE Trans. Image Process. **22**(12), 5336–5347 (2013)
6. Hou, J., Su H., Yan, B., et al.: Classification of Tongue Color Based on CNN. In: 2017 IEEE 2nd International Conference on Big Data Analysis, pp. 725–729 (2017)
7. Fu, S., Zheng, H., Yang, Z., et al.: Computerized tongue coating nature diagnosis using convolutional neural network. In: 2017 IEEE 2nd International Conference on Big Data Analysis, pp. 730–734 (2017)
8. Xiaoqiang, L., Yonghui, T., Yue, S.: Tongue coating classification based on multipleinstance learning and deep features. In: Gedeon, T., Wong, K., Lee, M. (eds.) ICONIP 2019. CCIS, vol. 1142, pp. 504–511. Springer, Cham (2019). https://doi.org/10.1007/978-3-030-36808-1_55
9. Huang, B., Wu, J., Zhang, D., Li, N.: Tongue shape classification by geometric features. Inf. Sci. **180**, 312–324 (2010)

10. Xiaoqiang, L., Dan, W., Qing, C.: WLDF: effective statistical shape feature for cracked tongue recognition. J. Electr. Eng. Technol. **12**(1), 420–427 (2017)
11. Chen, F., Xia, C., et al.: Extraction of tongue crack based on gray level and texture. DEStech Trans. Comput. Sci. Eng. 1–11 (2018)
12. Simonyan, K., Zisserman, A.: Very deep convolutional networks for large-scale image recognition. arXiv preprint arXiv:1409.1556 (2014)
13. He, K., Zhang, X., Ren, S., Sun, J.: Deep residual learning for image recognition, pp. 770–778 (2015)
14. Gao, H., Wang, Z., Li, Y., Qian, Z.: Overview of the quality standard research of traditional chinese medicine. Front. Med. **5**(2), 195–202 (2011)

An Obstacle Detection Method for Visually Impaired People Based on Semantic Segmentation

Zhuo Chen[1], Xiaoming Liu[1](✉), Dan Liu[1], Xiaoqing Tang[1], Qiang Huang[1], and Tatsuo Arai[1,2]

[1] School of Mechatronical Engineering, Beijing Institute of Technology, Beijing 100081, China
liuxiaoming555@bit.edu.cn

[2] The University of Electro-Communications, Tokyo 182-8585, Japan

Abstract. Using low-cost visual sensors to assist indoor and outdoor navigation is an important method to solve the problem of visually impaired people living and going out. To this end, we proposed an obstacle-detection method for visually impaired people based on semantic segmentation. We use the semantic segmentation method to determine which targets in the camera view field need to be noticed and use the related information to establish a real-time local map. At the same time, we propose a method to fuse semantic information with local point clouds, achieving obstacle detection based on probability fusion. Finally, the distance between the interested target and the camera will be returned and sent to the user. The proposed method can achieve visual navigation with more than ten frames per second (fps), lower than 0.3 m detection accuracy, and smaller than 4 MB generated model, which is also compatible with multiple cameras and control terminals.

Keywords: Obstacle Detection · Semantic Segmentation · Assistance for Visually Impaired People

1 Introduction

Visual impairment can seriously affect the life quality of patients, but this kind of disease is widespread worldwide [1]. Visually impaired individuals often find it inconvenient to go outdoors, mainly due to the ever-changing population, vehicles, and obstacles. Fortunately, many software programs and wearable devices have already been developed to assist the daily lives of visually impaired people, especially in reading and writing [2], using electronic devices, and other indoor activities [3]. However, compared to typical daily life scenarios, assisting visually impaired people in walking and going out poses more challenges. Because of weak visual ability, people with visual impairment face additional cognitive pressure when using auditory or tactile to obtain external information [4]. Using external methods to gather and analyze the content that visually impaired individuals need to perceive while going outdoors and expressing the aggregated information to them as much as possible through a single sound or mechanical stimulus can help alleviate the enormous psychological pressure faced by blind people when traveling.

F. Sun and J. Li (Eds.): ICCCS 2023, CCIS 2029, pp. 28–33, 2024.
https://doi.org/10.1007/978-981-97-0885-7_3

Using wearable sensors to assist visually impaired people in determining whether there are obstacles on their way is a reliable alternative [5]. Ultrasonic sensors [6], electromagnetic wave radar [7], LiDAR [8], low-cost visual sensors [9], and infrared sensors [10] are several commonly used sensor systems for obstacle detection and avoidance. However, radar and LiDAR are expensive and bulky, making them unsuitable for modification of wearable devices. Ultrasonic sensors and infrared sensors have limited detection distances and are severely affected by environmental and human noise. Low-cost visual sensors are an ideal tool for assisting visual obstacle avoidance due to their low cost and portability.

Two key issues need to be addressed when using visual sensors to avoid obstacles. The first step is to determine which objects are obstacles or moving traffic targets, helping visually impaired individuals understand the environmental conditions. Secondly, it is necessary to clarify the distance between each obstacle or traffic target and the sensor in order to avoid the obstacle avoidance function being activated too frequently or not working, which may cause unnecessary trouble for users.

Therefore, we propose an obstacle detection method for visually impaired people based on semantic segmentation. We use the semantic segmentation method based on Densely Connected Convolutional Networks [11] to determine which targets in the field of camera view need to be noticed. Expressly, we referred to the simultaneous localization and mapping methods in robotics to establish a real-time local map. Finally, the measurement of the distance between obstacle targets of interest and the sensor can be realized in real time. The proposed method can achieve real-time visual navigation and is compatible with multiple cameras and control terminals. The algorithm was tested using the CamVid dataset [12] and can ensure reliable running speed.

2 Methods

2.1 Semantic Segmentation for Obtaining Obstacle Information

There are already many image recognition methods based on deep neural networks. However, considering that outdoor traffic conditions are complex, which means identifying targets is not sufficient in navigation for visually impaired people. Therefore, the segmentation results for the target need to be precise to the pixel level, which is the semantic segmentation task. It is worth noting that in practical applications, it is necessary to increase the processing speed as much as possible to ensure real-time performance while ensuring roughly correct classification results in order to face objects approaching high speeds. We established the segmented network based on Densely Connected Convolutional Networks [11], which does not require pre-training and has a lower number of network layers (see Fig. 1.).

The input image will be converted into 16 feature layers and iterated n (take n in the order of 4,5,7,10) times, named dense block (DB). Then, all obtained features are performed transporting down (TD) with batch standardization, convolution, and pooling. These operations are regarded as a dense block of downward transporting (DB-TD) and repeat four times. When going up, features are performed with transporting up (TU), a transposed convolution with a stride of 2 first, and then the DB processes taking n in symmetric order to DB-TD (take n in the order of 10,7,5,4). Typically, each layer is connected to other layers through a feedforward form. Like the structure of the ResNet [13], these connections can encourage the reuse of features, which means there is no need to relearn every parameter and thus reduce the number of parameters. Besides, the feature maps extracted from all preceding layers will also be merged in the new layer, which can enhance the feature propagations and improve the accuracy.

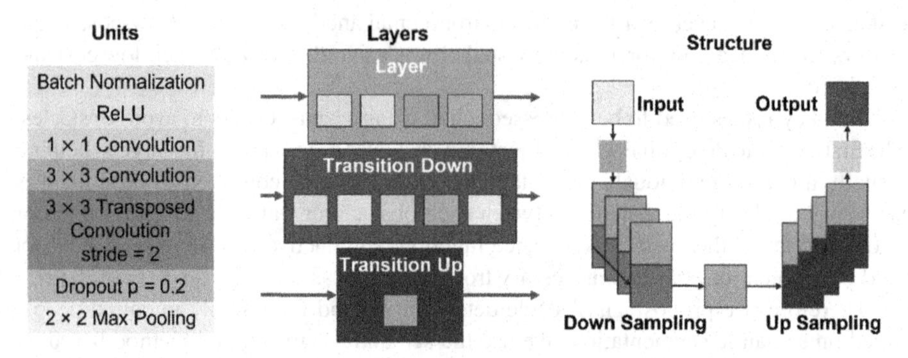

Fig. 1. Semantic segmentation network structure

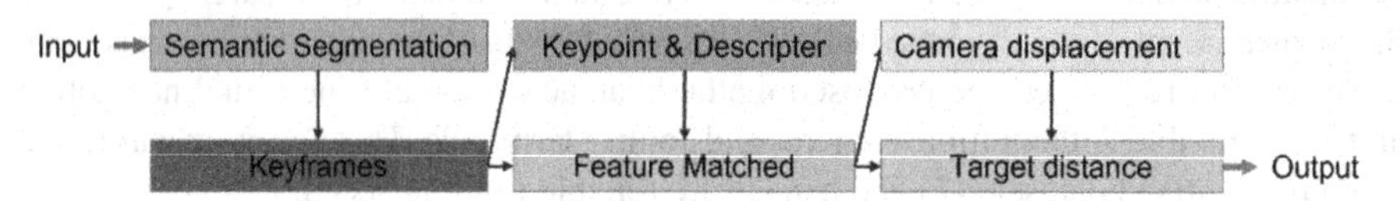

Fig. 2. Generation flow of the local map with semantic information

2.2 Real-Time Local Map with Semantic Information Generation

Compared to a stable indoor environment, the outdoor environment is dynamic and almost infinitely large. Therefore, it is not feasible to establish a global map like SLAM. Some devices such as RGB-Depth cameras, can be used to establish local maps in the form of depth images directly. However, these cameras cannot be simply used for navigation because of the light instability and the inconvenient calibration work. It is necessary to propose a universal method for various cameras to establish the current local map.

Here, a local map based on the keypoint detection is adopted (see Fig. 2.). Firstly, obstacle targets are detected based on the work cycle of semantic segmentation. The corresponding image is treated as a keyframe when obstacle targets are detected. Extract the ORB feature points [14] in the keyframe and classify the features against the interest targets and reference targets in the semantic segmentation results. Among them, the feature points on reference targets are regarded as the reference for the visual odometer to calculate the displacement of the camera. Then, according to the camera displacement, the distance between the interest target and the camera can be calculated by affine transformation. Considering that object movement can cause deformation of the global map and that the global map will consume many computing resources for maintenance over time, only a local semantic map is generated, and the centroid of the feature point distribution on the object of interest is calculated, ultimately obtaining the position of the object of interest.

For indoor scenes, establishing a global map helps with long-term applications and improves the efficiency of repetition. However, the computer controller has not yet been able to align the results of the single semantic segmentation with the identified terrain. This issue involves mapping two-dimensional semantic information to three-dimensional space and associating the generated semantic labels with point clouds. For outdoor scenes, although some devices, such as binocular cameras and RGB-D cameras, can quickly create point clouds. However, for monocular cameras that are more suitable for wearable device applications, point clouds are established through affine transformations. Since there is no global map shown here, semantic information cannot be directly mapped. When establishing a local map, semantic information is combined with positional information through probability fusion.

Probability fusion is the mapping of corresponding semantic segmentation results by pixels in the same way when calculating the pose of obstacles using source images. This can obtain the probability of a certain position in space being classified as an object. For indoor scenes that establish a global map, the probabilities of spatial 3D points are optimized together with loop detection. However, due to the lack of loop detection in local maps, multiple images were selected to be extracted near keyframes, and the probability of belonging points was determined through multiple semantic segmentation to determine the position of obstacles.

3 Results

We use an embedded computing platform (Nvidia Jetson Xavier, NVIDIA) as the control terminal. Firstly, the effectiveness of semantic segmentation was tested on the CamVid dataset. The results can be seen in Table 1. Due to the generated model being very small and not requiring pre-training, the time required for the segmentation of each image on our device is less than 5 ms, fully meeting the real-time requirements. The evaluation of the mean intersection to union ratio (Mean IoU) and global accuracy of all results can also meet the needs of the application.

Figure 3. Shows the results of segmentation and distance detection. After adding a module for local maps, the basic functions can be implemented. Since, in practical use, voice broadcasting or tactile feedback can provide sufficient time for backend computing processing, this scheme is completely feasible in low-speed and relatively simple situations.

We have also discussed the performance of the obstacle detection. Table 2. Shows the comparison of the tested result between the mono views and views with depth under the TUM RGB-D SLAM database [14]. Due to the minor computational complexity in mapping and semantic information fusion and the more accurate depth information calculation compared to monocular information obtained through affine transformation and calibration, RGB-D cameras showed better performance.

Table 1. Semantic segmentation results on the database.

Name	Parameters (M)	Mean IoU	Global accuracy	Time (ms)
62_Layer_DenseNet	3.2	62.1%	85.2%	<5

Table 2. Performance of obstacle detection.

Name	Detection Performances	
	Mean Error (m)	Execute speeds (fps)
Fr1_mono	0.4	22.4
Fr2_mono	0.21	21.1
Fr3_mono	0.05	14.5
Fr1_rgbd	0.24	23.1
Fr2_rgbd	0.10	19.8
Fr3_rgbd	0.01	16.3

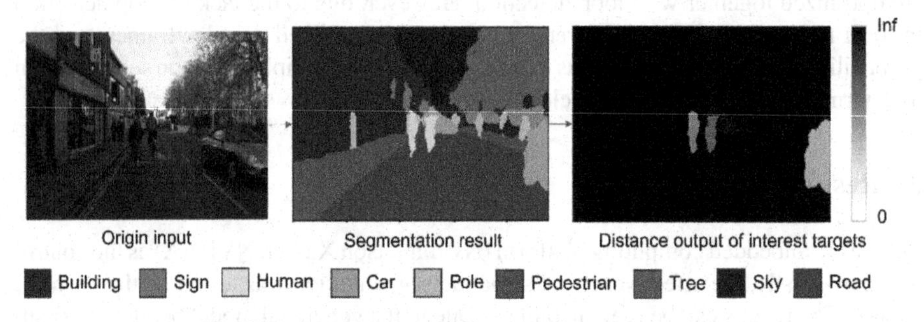

Fig. 3. Results of segmentation and distance detection

4 Conclusion

We proposed an obstacle detection method for visually impaired people based on semantic segmentation. We use the semantic segmentation method to determine which targets in the camera view field need to be noticed and use the related information to establish a real-time local map. Finally, the distance between the interested target and the camera will be returned and sent to the user. The proposed method can achieve real-time visual navigation and is compatible with multiple cameras and control terminals. The algorithm was tested using the CamVid dataset and can ensure reliable running speed. Through testing by the TUM RGB-D SLAM database, our obstacle detection function can achieve an accuracy of less than 0.3 m and a response speed of over ten fps using complete RGB D camera information, confirming the feasibility of this scheme.

References

1. Bourne, R.R.A., Steinmetz, J.D., Flaxman, S., et al.: Trends in prevalence of blindness and distance and near vision impairment over 30 years: an analysis for the global burden of disease study. Lancet Glob. Health **9**(2), e130–e143 (2021)
2. Juneja, S., Joshi, P.: Design and development of a low cost and reliable writing aid for visually impaired based on Morse code communication. Technol. Disabil. **32**(2), 59–67 (2020)
3. Isaksson, J., Jansson, T., Nilsson, J.: Desire of use: a hierarchical decomposition of activities and its application on mobility of by blind and low-vision individuals. IEEE Trans. Neural Syst. Rehabil. Eng. **28**(5), 1146–1156 (2020)
4. Xiong, Z., Huang, X.: Comparison of the static and dynamic vibrotactile interactive perception of walking navigation assistants for limited vision people. iEEE Access **10**, 42261–42267 (2022)
5. Joseph, A.M., Kian, A., Begg, R.: State-of-the-art review on wearable obstacle detection systems developed for assistive technologies and footwear. Sensors **2023**(23), 2802 (2023)
6. Adarsh, S., Kaleemuddin, S.M., Bose, D., Ramachandran, K.I.: Performance comparison of infrared and ultrasonic sensors for obstacles of different materials in vehicle/ robot navigation applications. IOP Conf. Ser. Mater. Sci. Eng. **149**(1), 012141 (2016)
7. Marti, E.D., de Miguel, M.A., Garcia, F., Perez, J.: A Review of sensor technologies for perception in automated driving. IEEE Intell. Transp. Syst. Mag. **11**(4), 94–108 (2019)
8. Fang, Z., Zhao, S., Wen, S., Zhang, Y.: A real-time 3d perception and reconstruction system based on a 2d laser scanner. J. Sensors **2018**, 2937694 (2018)
9. Yu, H., Zhu, J., Wang, Y., Jia, W., Sun, M., Tang, Y.: Obstacle classification and 3D measurement in unstructured environments based on ToF cameras. Sensors **2014**(14), 10753–10782 (2014)
10. Discant, A., Rogozan, A., Rusu, C., Bensrhair, A.: Sensors for obstacle detection—a survey. In: Proceedings of the 2007 30th International Spring Seminar on Electronics Technology (ISSE), Cluj-Napoca, Romania (2007)
11. Jégou, S., Drozdzal, M., Vazquez, D,, Romero, A., Bengio, Y.: The one hundred layers tiramisu: fully convolutional densenets for semantic segmentation. In: 2017 IEEE Conference on Computer Vision and Pattern Recognition Workshops (CVPRW), pp. 1175–1183. Honolulu, HI, USA (2017)
12. Jain, S.D., Xiong, B., Grauman, K.: FusionSeg: learning to combine motion and appearance for fully automatic segmentation of generic objects in videos. In: 2017 IEEE Conference on Computer Vision and Pattern Recognition (CVPR), pp. 2117–2126. IEEE, Honolulu, USA (2017)
13. He, K., Zhang, X., Ren, S., & Sun, J. Deep Residual Learning for Image Recognition. 2016 IEEE Conference on Computer Vision and Pattern Recognition (CVPR). IEEE, Las Vegas, USA, pp. 770–778 (2016)
14. Sturm, J., Engelhard, N., Endres, F., Burgard, W., Cremers, D.: A benchmark for the evaluation of RGB-D SLAM systems. In: Proceedings of the International Conference on Intelligent Robot Systems (IROS) (2012)

MixStyle-Based Dual-Channel Feature Fusion for Person Re-Identification

Jian Fu, Xiaolong Li(✉), and Zhu Yang

School of Automation, Wuhan University of Technology, Wuhan 430070, China
fujian@whut.edu.cn, 1628887613@qq.com

Abstract. The problem of Person Re-Identification is still a big challenge, as the complex network structure and unsatisfactory generalization performance of widely used deep neural networks make them unsuitable for application to real-world problems. In this paper, we propose a global feature-based person re-identification network with strong generalization. The extracted features part contains two channels of feature fusion: the feature extraction module and the feature generalization module. The feature generalization module is a new MixStyle module added to the feature extraction module, which can effectively mix the style information of images under different domains or even the same domain to form multiple potential domain features, thus improving the generalization performance of the model. In addition, this paper also makes some improvements to the loss function by adding a new constraint on the positive sample pair distance, which makes it possible to maximizes the reduction of intra-class distance in addition to pushing the distance between different classes during the training process. Experimental results on two datasets, Market1501 and DukeMTMC, demonstrate that the method proposed in this paper exhibits strong generalization performance for the person re-identification problem and outperforms current global feature-based person re-identification methods.

Keywords: Person re-Identification · MixStyle · Feature fusion · Loss function improvement

1 Introduction

The method of retrieving a specific pedestrian from a surveillance system relying on the recognition of face biometrics has major limitations, and person re-identification [1], as a derivative of face recognition, greatly complements the face recognition technology. With the application of deep learning to the field of person re-identification, the recognition accuracy of person re-identification models has been increasing year by year. However, the scenes taken by different cameras, the lighting conditions, weather factors, and even the occlusion, distance from the camera, and posture of the pedestrian from the same camera all

This work was supported by the National Natural Science Foundation of China under Grant 61773299.

F. Sun and J. Li (Eds.): ICCCS 2023, CCIS 2029, pp. 34–47, 2024.
https://doi.org/10.1007/978-981-97-0885-7_4

lead to very different styles in the final photos, which makes it difficult to correspond the same person from the pictures taken by different cameras, and this is also a difficult factor that limits the further improvement of the accuracy of the person re-identification model. Therefore a person re-identification model with strong generalization performance and high recognition accuracy is of practical research significance.

In the field of person re-recognition, how to design a feature extraction method with good performance is a key factor to decide whether the model performance is good or bad. The current methods for extracting features for pedestrian re-recognition are broadly classified into global feature-based and local feature-based methods [2]. Global features refer to the network extracting only one feature for each image, and this feature contains all the information in the image, but with the interference of the complex background style factors of the image, the feature extracted only based on the global extraction is not enough to accomplish a high accuracy rate. Local features are extracted by splitting the image into multiple parts, which can effectively focus on the important information of the human body while filtering out the interfering elements, but the extraction method based on local features suffers from high computational complexity as well as the difficulty of how to align the corresponding poses [3] of the human body in the two images. In order to solve the problem of difficult local feature alignment, Spindle Net [4] of CVPR17 proposes an approach based on the combination of global and local features, where the Spindle Net network first extracts 14 human key points through the network of skeleton key point extraction, and later extracts 7 human structure ROIs using these key points, and these 7 ROI regions and the original image enter the same CNN network to extract features. The original image is passed through the complete CNN to get a global feature, three large regions are passed through the FEN-C2 and FEN-C3 subnetworks to get three local features, and four limb regions are passed through the FEN-C3 subnetwork to get four local features. After that these eight features are concatenated at different scales to finally obtain a person re-identification feature that fuses global features and local features at multiple scales. Different from Spindle Net, a Global-Local-Alignment Descriptor(GLAD) is proposed in the literature [5]. GLAD uses the extracted key points of the human body to divide the picture into three parts: head, upper body and lower body. After that, the whole picture and the three local pictures are fed together into a parameter sharing CNN network, and the final extracted features fuse the global and local features. Although these methods solve the problem of local feature alignment of two images, the complex network structure and high computational cost also become its shortcomings. Based on this, this paper proposes a two-channel feature fusion method based on global features, which reduces the model complexity and improves the generalization performance of the network at the same time.

In the process of training, the prerequisite for updating the model towards a better direction is the selection of a suitable loss function, which plays a decisive role in guiding the ability of the model to learn semantic features in the picture.

In the field of person re-identification, loss functions can be classified into representation loss [6] and metric loss [7] according to the scenario in which they are used. A representative method of representation loss is the cross-entropy loss [8] calculated based on categorical information. Unlike the representation methods, metric loss is a process of continuously optimizing the mapping function based on the distance between features, and the network trained by it has a smaller distance between image features for the same category than for different categories. The commonly used metric loss functions for performing metric learning are validation loss [9], and triplet loss [10,11]. The validation loss reduces the intra-class loss and increases the inter-class loss by inputting sample pairs. Triplet loss, on the other hand, trains the model with a set of one anchor (a), one positive sample (p), and one negative sample (n), constantly adjusting the distance between pairs of samples. Based on the triplet loss, the literature [12] proposed a triplet loss based on hard sample mining, where the farthest positive sample image and the closest negative sample image are selected to train the network and enhance its generalization ability. We found that the triplet loss is realized by narrowing the distance difference between positive and negative samples to the anchor point to push far away from the distance between classes, but it can only make the relative distance difference between positive and negative samples to the anchor point larger than a threshold, but can not guarantee that the distance between positive samples to the anchor point is very small, which makes the effect is not obvious enough in the process of clustering. Therefore, this paper improves the original triplet loss by adding a new constraint on the distance between the positive samples to the anchor point, which makes the trained model push far away from the features of different classes of samples while at the same time bring closer the distance between the features of the same class of samples. In addition, compared with the model using a single loss, the literature [13] achieves better results by jointly training the network with cross-entropy loss and validation loss together.

In summary, the main contributions of this paper are as follows:

(1) We design a feature generalization module based on MixStyle as a channel for feature extraction, which can achieve the effect of expanding the potential domain of the original features, thus improving the generalization performance of the network.
(2) We make further optimization on the basis of triplet loss to set an optimized constraint for the positive samples on the distance, which makes the clustering effect clearer.
(3) We propose a two-channel feature fusion based on global features for person re-recognition model, and the experimental results show that our proposed method is superior to the existing global feature-based algorithms.

2 MixStyle-Based Style Blending Study

The background of an image has been considered as a distraction in the field of image recognition, in fact, by obtaining information about the style of the image it is possible to help us in our image recognition work. The stylistic information of the image largely determines the domain in which the image is divided, but when the image is fed into a deep network, this information has completely disappeared at the output. Recent style transfer studies have shown that the underlying CNN [14] retains all the information about the style of the images through instance-level feature statistics, based on which the literature [15] proposed the method of MixStyle, which mixes the style information of two images to form a new style of image, at which the image only has a change in style information, and its semantic information remains the same, which means that we can use this method to form images of various styles [16] for training models. The experimental results also show that MixStyle has strong generalization performance in category classification, instance retrieval, and reinforcement learning. In this paper, inspired by MixStyle, the dataset of person re-identification is derived from images of different environments under different cameras with completely different style information, which can be regarded as images under different domains, so it is reasonable to introduce this method into the training model of person re-identification.

Most of the existing person re-identification models are based on GoogleNet [17], ResNet [18] and DensNet [19] as the backbone networks to learn a better feature extraction network from the training set images, so that the features have a better classification effect in the training set feature space, and then the trained feature extraction network is used to extract features from the test set images. Due to the large variation in styles between images, feature extraction networks that perform well in the training set may not achieve the desired results in the test set. Trying to improve the generalization of the model by exhausting all pictures in all domains during training is almost impossible, in order to learn as much information about as many domains as possible from a limited number of training images, we introduce the MixStyle module. Because style information exists only in the underlying neural network, therefore, we insert the MixStyle module behind the first and second residual blocks of the ResNet-50 network to mix the image styles of different domains to form a new style of images, providing the trained network model with more sets of images of potential domains. The purpose of this paper is to achieve the goal of improving the generalization ability of the model by adding a new feature generalization module, which makes it possible to recognize more features of different potential domains during training and achieve better feature fitting during testing. We also found that using only the feature generalization module for feature extraction tends to lose valuable style information of the source domain, so we kept the initial feature extraction module and combined the features of both parts to train our model. The experimental results show that the method extracts diverse style information, which is beneficial to improve the generalization ability of the model and outperforms the traditional feature extraction methods.

3 Proposed Person Re-Identification Method

In this section, we describe in detail the proposed person re-identification method based on MixStyle dual-channel feature fusion. First, we describe each component of the model as a whole and mainly explain the specific method of feature fusion. Then the principles of the implementation of the feature generalization module are derived and elaborated. Finally, based on the shortcomings of the previous triplet loss in training, this paper proposes an improved triplet loss and introduces the overall loss function used by the model accordingly.

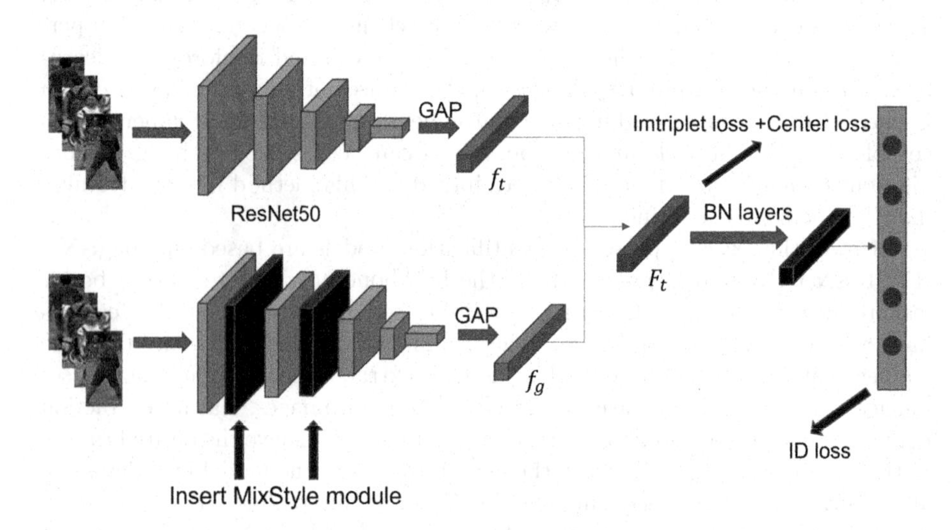

Fig. 1. The proposed overall network structure based on the two-channel feature fusion model.

3.1 Overall Network Structure

As in Fig. 1, it is the general network structure diagram of the proposed method in this paper. Based on the close connection between the information of the domain where the image features are located and the image style information, we design the feature genericization module to mix the style information of different images, hoping to obtain images with diverse style information without changing the semantic information of the images, which is used to train a model with stronger generalization performance. However, we found that if only one feature generalization module was used to extract the features of the images, it would lead to the problem of losing valuable original style images, which is undoubtedly harmful for a model under training, so we kept the original feature extraction module to obtain the features of the original images.

The backbone network used in this paper is ResNet-50, which consists of one convolutional layer and four residual blocks. In this paper, the stride of the last convolutional layer of the residual block is reduced to one [20], so that more feature map information can be obtained by increasing the computational cost by a small amount and without increasing the number of training parameters. The feature vectors from the feature extraction module and the feature generalization module are fused according to Eq. (1) to obtain the final feature vectors for calculating the triplet loss and the center loss.

$$F_t = \eta f_g + (1 - \eta) f_t \tag{1}$$

where f_t is the feature vector from the feature extraction module, f_g is the feature vector from the feature generalization module, and η is the hyperparameter used to adjust the degree of fused generalized features.

Since the objects targeted by triplet loss and ID loss are different in the embedding space, if triplet loss and ID loss are allowed to optimize the same feature variables without any processing, the inconsistency of their respective evaluation criteria will lead to the loss sum not being minimized because the optimization direction of both is not the same. Therefore, a batch normalization layer [20] is passed before the classification result is output from the fully connected layer, and the normalized features are used to calculate the classification loss, and the triplet loss is computed using the features before normalization, by which the ID loss is unbound and they each converge faster towards optimization. In addition to this, normalization makes the feature variables homogeneous in all dimensions, and the processed features are distributed in a Gaussian fashion over the surface of the hypersphere, and this distribution also allows the ID loss to converge faster.

3.2 Feature Generalization Module

The feature extraction network has been roughly introduced in the previous subsection, and this section will elaborate the implementation details and basic principles of the feature generalization module. As shown in Fig. 1, the feature generalization module is based on the feature extraction module, and the MixStyle module is inserted after the first and second residual blocks of the ResNet-50 network. It is found that the best position to insert the MixStyle module differs according to the task, and the current insertion position is the most suitable position for the person re-identification task.

Instead of explicitly generating images with a new style, the MixStyle module extracts the underlying style information of the two sets of input images based on statistical feature data, mixes them and applies them to the images that have already been normalized to create a more diverse style. More specifically, for the person re-identification task the domain labels of the training images are unknown, and we use a random shuffle of the input batch of samples x to produce the auxiliary batch $\tilde{x}$. MixStyle will calculate the variance and mean of the corresponding positional images of the original batch and the generated auxiliary batch, respectively, and calculate the style information of the actual input images of the feature generalization module by Eq. (2).

$$\begin{aligned} \gamma_{\text{mix}} &= \lambda\sigma(x) + (1-\lambda)\sigma(\tilde{x}) \\ \beta_{\text{mix}} &= \lambda\mu(x) + (1-\lambda)\mu(\tilde{x}) \end{aligned} \tag{2}$$

Here γ_{mix} is the variance of the blended style information and β_{mix} is the mean of the blended style information. Then, the calculated blended style information is applied to the images belonging to the original batch samples that have undergone the destylization process according to Eq. (3), and finally the images of the potential domain are obtained.

$$\text{MixStyle}(x) = \gamma_{mix}\frac{x-\mu(x)}{\sigma(x)} + \beta_{mix} \tag{3}$$

3.3 Improved Loss Function

As we mentioned earlier, the strategy of joint training based on metric loss and representation loss has better performance in the current network structure, therefore, in this paper, we use the strategy of joint training of the model with ID loss, central loss and improved triplet loss. The total loss function is as in Eq. (4).

$$L = L_{imtri} + L_{id} + \alpha L_{cen} \tag{4}$$

For the original triplet loss function, given a set of samples $E_i = \{E_i^a, E_i^p, E_i^n\}$, mapping samples to feature space after model $f(E_i) = \{f(E_i^a), f(E_i^p), f(E_i^n)\}$, the inter-sample distance is calculated by the distance metric function, which can be expressed as:

$$L_{tri} = \left(D\left[f\left(E_i^a\right), f\left(E_i^p\right)\right] - D\left[f\left(E_i^a\right), f\left(E_i^n\right)\right] + \theta\right)_+ \tag{5}$$

Here, $(.)_+ = \max(0, .)$, $D[.]$ is defined as the L2-norm distance, From way 1 of Fig. 2, it can be found that the original triplet loss can only constrain the distance difference between positive and negative pairs within θ, and does not optimize the distance inside the positive pair. To solve this problem, the improved triplet loss in this paper is as Eq. (6).

$$\begin{aligned} L_{imtri} = &\left(D\left[f\left(E_i^a\right), f\left(E_i^p\right)\right] - D\left[f\left(E_i^a\right), f\left(E_i^n\right)\right] + \theta\right)_+ \\ &+\delta\left(D\left[f\left(E_i^a\right), f\left(E_i^p\right)\right] - \varepsilon\right)_+ \end{aligned} \tag{6}$$

Here, δ is the weight parameter to balance the distance between positive pairs, which is set to 0.3 in our experiments. The improved triplet loss not only limits the distance difference between positive and negative pairs, but also constrains the distance within positive pairs to ε, with ε much smaller than θ. As can be seen from way 2 of Fig. 2, it is pushing the distance between negative pairs farther while still shortening the distance between positive pairs.

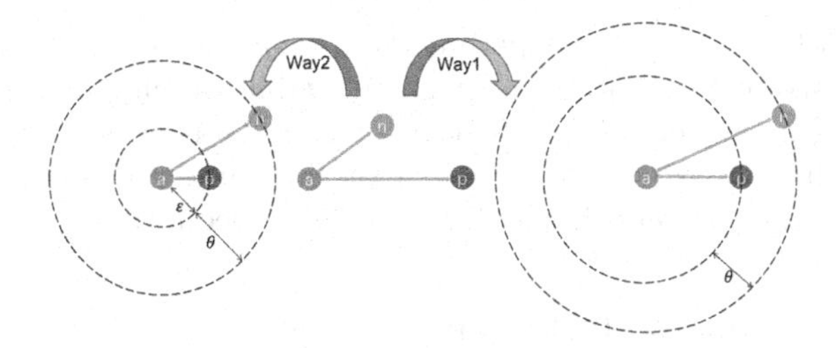

Fig. 2. The goal of triplet loss during training. way 1: The original triplet loss constrains the positive and negative pairwise distance differences to θ. way 2: Our improvement adds a new constraint on the distance of the positive pair as well, while ε is much smaller than θ.

4 Experiments

4.1 Datasets and Evaluation Indicators

We will conduct experiments to evaluate the performance of our model on two datasets, Market1501 [21] and DukeMTMC-reID [22]. The Market1501 dataset is sourced from 1501 pedestrians under six cameras on the campus of Tsinghua University, with a total of 12936 images and 751 pedestrian categories in the training set. The test set images are divided into 3368 retrieved images and 15913 candidate images, and the test set contains a total of 750 pedestrian categories. The DukeMTMC-reID dataset is collected from eight outdoor cameras at Duke University, with a total of 1812 pedestrian categories. Among them, the training set images have a total of 16522 images of 702 pedestrians, and the test set consists of 2228 retrieved images and 17661 candidate images, containing a total of 702 pedestrian categories. Unlike Market1501, it is a pedestrian cropping frame with annotation performed by hand, and there are also 408 interfering pedestrians, thus making it more difficult to identify than the Market1501 dataset.

To better measure the effectiveness of the method, the evaluation metrics used in this paper are mainly Rank-1 accuracy and mean Average Precision (mAP).

4.2 Implementation Details and Parameter Settings

Before training starts, ResNet-50 is initialized using the pre-training parameters on ImageNet [23]. One of our training batches consists of 16 pedestrians with 4 images each, for a total of 64 images. In the image preprocessing part, we perform random erasures of different sizes for each image with a probability of 0.5 to achieve data enhancement. For the hyperparameters in the experiments: In training, we decide with some probability whether to activate the MixStyle module or not to have more randomness, and in testing, we don't use the MixStyle, the activation probability of the MixStyle module is set to 0.5, the positive and negative sample pair distance difference θ is set to 0.3, and the positive pair distance ε is set to 0.01. Our experiments are conducted on RTX3090, using Adam update strategy to optimize our model and a warmup strategy to set the learning rate. Our model is trained for a total of 120 epochs, and the learning rate increases from 3.5×10^{-5} to 3.5×10^{-4} at the initial 10 epochs, and decays to one-tenth at the 40th epoch and 70th epochs respectively.

4.3 Results Compared to Baselines

To verify the effectiveness of our proposed method, we compare our approach with state-of-the-art methods in Table 1. These methods include IDE [24], SVD-Net [25], TriNet [26], AWTL [27] and Baseline [28], which achieve better results, based on global features, and PCB-RPP [29], BFE [30], which are based on local features, and Pyramid [31] based on stripes. By observing the performance on two datasets, Market1501 and DukeMTMC-reID, our proposed algorithm obtains the best results among all global feature-based methods, obtaining 87.8% mAP, 94.9% Rank-1 accuracy and 78.1% mAP, 87.7% Rank-1 accuracy, respectively. Compared to the best performing strong baseline model, it also obtained 1.9% mAP and 1.7% mAP improvement in the two datasets, respectively. The reason why it achieves slightly better results than ours compared to Pyramid is that 21 different local features are collected, which are much less effective than our method when it extracts only global feature. Therefore, it can be said that our proposed method effectively improves the retrieval accuracy, which shows that our model can greatly improve the performance of generalization to unknown domain images.

Table 1. Experimental evaluations on Market1501 dataset and DukeMTMC dataset.

Method	Market1501		DukeMTMC	
	mAP	r = 1	mAP	r = 1
IDE	59.9	79.5	-	-
SVDNet	62.1	82.3	56.8	76.7
TriNet	69.1	84.9	-	-
AWTL	75.7	89.5	63.4	79.8
Baseline	85.9	94.5	76.4	86.4
Pyramid	88.2	95.7	79.0	89.9
PCB-RPP	81.6	93.8	69.2	83.3
BFE	85.0	94.5	75.8	88.7
Ours	87.8	94.9	78.1	87.7

4.4 Ablation Experiment

In this section, in order to further investigate how the feature generalization module and each part of our model is beneficial to the performance, the feature generalization module is inserted into different locations and ablation experiments are performed on the Market1501 and DukeMTMC-reID datasets for the split network structure, respectively. For the experiments the network structure is described as follows: 1) w/o L_{imtri}: our model removes the improved triplet loss from the loss function. 2) w/o fgm: our model removes the feature generalization module. 3) w/o fem: our model removes the feature extraction module. 4) add1: insert the feature generalization module after the first residual block. add12: Insert feature generalization modules after both the first and second residual blocks and so on.

Table 2 shows the results of the ablation experiment on the two benchmark datasets separately, and our method improves 1.3% and 1.1% in mAP accuracy metrics compared to the result of w/o L_{imtri}, indicating that our improved triplet loss has a more significant improvement for constraining the distance between positive sample pair, which is conducive to improving the clustering effect of features in the feature space. The result of w/o fgm illustrate that our strategy of fusing generalized features in the process of feature extraction is successful, and the feature generalization module is effective for improving the generalization performance of the model. Compared to our method, w/o fem has a slight performance degradation, which we speculate is due to the fact that training with the generalized features alone leads to the loss of the original feature information, which is detrimental to the model fitting process thus affecting the retrieval accuracy of the model.

Table 2. Performance study of partial network structure elimination.

Method	Market1501		DukeMTMC	
	mAP	r = 1	mAP	r = 1
w/o fgm	86.9	94.6	77.2	86.8
w/o fem	87.1	94.8	77.5	86.9
w/o L_{imtri}	86.5	94.0	77.0	86.5
add1	87.3	94.8	77.8	87.0
add2	87.4	94.7	77.7	87.2
add4	85.8	93.4	76.6	85.4
add14	86.3	93.8	76.8	85.9
add12(Ours)	87.8	94.9	78.1	87.7

In the ablation experiments where the feature generalization module is inserted in the location, we found that inserting it in more than one bottom layer gives better results. In the task of this paper, the model performance of add12 is better than that of add1, and it is encouraging to note that both of them have higher recognition accuracies than without the feature generalization module. However, we also found that applying the feature generalization module after the fourth residual block leads to a performance degradation of the model. This is not unfounded, as the fourth residual block is closer to the prediction layer, which holds more semantic information about the image than style information, and thus the MixStyle module applied here does not achieve the goal of improving the generalization performance. In summary, the reasonable application of the feature generalization module to multiple underlying network structures can effectively improve the performance of the model.

4.5 Influence Analysis of Parameter η

The parameter η is a hyperparameter in Eq. (1) that measures the degree of fusion of the output features of the feature generalization module. To explore the effect of parameter η on model performance, in this subsection, we set η between 0.1 and 0.9 to conduct comparison experiments on the Market1501 and DukeMTMC-reID datasets, respectively. According to Fig. 3, we find that the

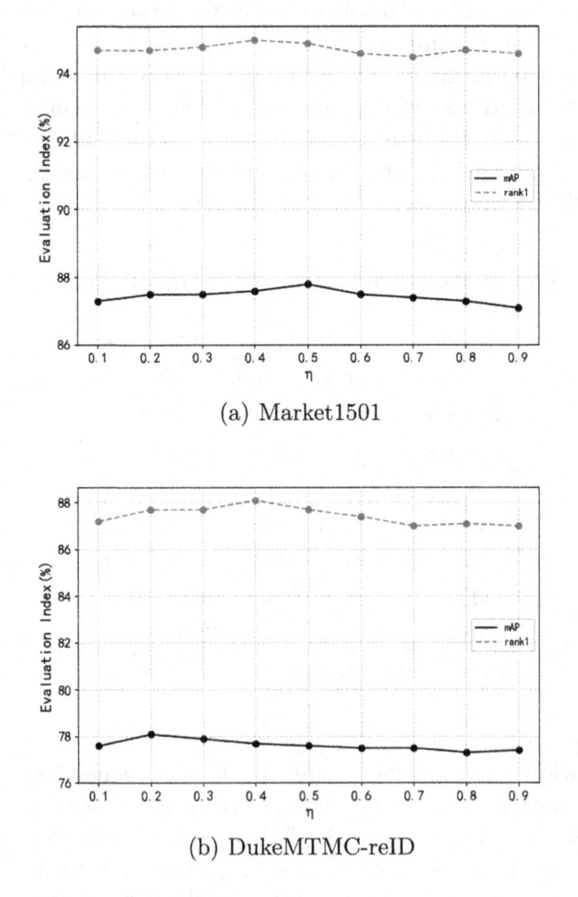

(a) Market1501

(b) DukeMTMC-reID

Fig. 3. Performance impact of different η values on two datasets.

effect of parameter η on Rank-1 accuracy is more obvious on the DukeMTMC-reID dataset than on the Market1501 dataset because when the model performance reaches a certain level, those images that are relatively easy to recognize have already been retrieved and there is not enough room for further improvement. For the mAP accuracy, the highest mAP is obtained when $\eta = 0.5$ on the Market1501 dataset, while the best is achieved when $\eta = 0.2$ on the DukeMTMC-reID dataset, according to our analysis, because the DukeMTMC-reID dataset itself has larger perturbations, which may form invalid style features when the degree of fusion increases and this may affect the training of the model. Therefore, the experimental result show that choosing the best training parameter η for different datasets is helpful for the performance improvement of the model.

5 Conclusions

In this paper, we propose a person re-identification method based on global feature fusion, which can effectively extend the domain classes of the training set by using the fusion of the generalized global features with the initial global features, and thus improve the generalization ability of the model. In addition, for the problem that the original triplet loss only optimizes the distance difference between positive and negative sample pairs, we design an improved triplet loss that adds a constraint on the distance between positive sample pairs as well, making it possible to simultaneously close the distance between positive sample pairs during the training process, which is beneficial to the clustering distribution of different person features in the feature space. It is shown through experiments that our method outperforms existing global feature-based methods.

References

1. Ye, M., Shen, J., Lin, G., et al.: Deep learning for person re-identification: a survey and outlook. IEEE Trans. Pattern Anal. Mach. Intell. **44**(6), 2872–2893 (2021)
2. Luo, H., Jiang, W., Fan, X., et al.: A survey on deep learning based person re-identification. Acta Automatica Sinica **45**(11), 2032–2049 (2019)
3. Zhang, X., Luo, H., Fan, X., et al.: Alignedreid: surpassing human-level performance in person re-identification. arXiv preprint arXiv:1711.08184 (2017)
4. Zhao, H., Tian, M., Sun, S., et al.: Spindle net: Person re-identification with human body region guided feature decomposition and fusion. In: Proceedings of the IEEE Conference on Computer Vision and Pattern Recognition, pp. 1077–1085 (2017)
5. Wei, L., Zhang, S., Yao, H., et al.: GLAD: global-local-alignment descriptor for scalable person re-identification. IEEE Trans. Multimedia **21**(4), 986–999 (2018)
6. Zheng, L., Yang, Y., Hauptmann, A.G.: Person re-identification: past, present and future. arXiv preprint arXiv:1610.02984 (2016)
7. Shi, H., Yang, Y., Zhu, X., et al.: Embedding deep metric for person re-identification: a study against large variations. In: Leibe, B., Matas, J., Sebe, N., Welling, M. (eds.) ECCV 2016 Part I. LNCS, vol. 9905, pp. 732–748. Springer, Cham (2016). https://doi.org/10.1007/978-3-319-46448-0_44
8. Ho, Y., Wookey, S.: The real-world-weight cross-entropy loss function: modeling the costs of mislabeling. IEEE Access **8**, 4806–4813 (2019)

9. Yi, D., Lei, Z., Liao, S., et al.: Deep metric learning for person re-identification. In: 2014 22nd International Conference on Pattern Recognition, pp. 34–39. IEEE (2014)
10. Dong, X., Shen, J.: Triplet loss in siamese network for object tracking. In: Proceedings of the European Conference on Computer Vision (ECCV), pp. 459–474 (2018)
11. Cheng, D., Gong, Y., Zhou, S., et al.: Person re-identification by multi-channel parts-based CNN with improved triplet loss function. In: Proceedings of the IEEE Conference on Computer Vision and Pattern Recognition, pp. 1335–1344 (2016)
12. Hermans, A., Beyer, L., Leibe, B.: In defense of the triplet loss for person re-identification[J]. arXiv preprint arXiv:1703.07737 (2017)
13. Zheng, Z., Zheng, L., Yang, Y.: A discriminatively learned CNN embedding for person reidentification. ACM Trans. Multimedia Comput. Commun. Appl. (TOMM) **14**(1), 1–20 (2017)
14. Huang, X., Belongie, S.: Arbitrary style transfer in real-time with adaptive instance normalization. In: Proceedings of the IEEE International Conference on Computer Vision, pp. 1501–1510 (2017)
15. Zhou, K., Yang, Y., Qiao, Y., et al.: Domain generalization with mixstyle. arXiv preprint arXiv:2104.02008 (2021)
16. Zhou, K., Liu, Z., Qiao, Y., et al.: Domain generalization in vision: a survey. arXiv preprint arXiv:2103.02503 (2021)
17. Szegedy, C., Liu, W., Jia, Y., et al.: Going deeper with convolutions. In: Proceedings of the IEEE Conference on Computer Vision and Pattern Recognition, pp. 1–9 (2015)
18. He, K., Zhang, X., Ren, S., et al.: Deep residual learning for image recognition. In: Proceedings of the IEEE Conference on Computer Vision and Pattern Recognition, pp. 770–778 (2016)
19. Huang, G., Liu, Z., Van Der Maaten, L., et al.: Densely connected convolutional networks. In: Proceedings of the IEEE Conference on Computer Vision and Pattern Recognition, pp. 4700–4708 (2017)
20. Luo, H., Gu, Y., Liao, X., et al.: Bag of tricks and a strong baseline for deep person re-identification. In: Proceedings of the IEEE/CVF Conference on Computer Vision and Pattern Recognition Workshops (2019)
21. Zheng, L., Shen, L., Tian, L., et al.: Scalable person re-identification: a benchmark. In: Proceedings of the IEEE International Conference on Computer Vision, pp. 1116–1124 (2015)
22. Ristani, E., Solera, F., Zou, R., et al.: Performance measures and a data set for multi-target, multi-camera tracking. In: Hua, G., Jégou, H. (eds.) ECCV 2016 Part II. LNCS, vol. 9914, pp. 17–35. Springer, Cham (2016). https://doi.org/10.1007/978-3-319-48881-3_2
23. Deng, J., Dong, W., Socher, R., et al.: Imagenet: a large-scale hierarchical image database. In: 2009 IEEE Conference on Computer Vision and Pattern Recognition, pp. 248–255. IEEE (2009)
24. Zhang, T., Yi, Z.M., Li, X.: Improved algorithm for person re-identification based on global features. Laser Optoelectron. Progress **57**(24), 241503 (2020)
25. Sun, Y., Zheng, L., Deng, W., et al.: Svdnet for pedestrian retrieval. In: Proceedings of the IEEE International Conference on Computer Vision, pp. 3800–3808 (2017)
26. Schroff, F., Kalenichenko, D., Philbin, J.: Facenet: a unified embedding for face recognition and clustering. In: Proceedings of the IEEE Conference on Computer Vision and Pattern Recognition, pp. 815–823 (2015)

27. Ristani, E., Tomasi, C.: Features for multi-target multi-camera tracking and re-identification. In: Proceedings of the IEEE Conference on Computer Vision and Pattern Recognition, pp. 6036–6046 (2018)
28. Xu, L.Z., Peng, L.: Person reidentification based on multiscale convolutional feature fusion. Laser Optoelectron. Progress **56**(14), 141504 (2019)
29. Sun, Y., Zheng, L., Yang, Y., et al.: Beyond part models: Person retrieval with refined part pooling (and a strong convolutional baseline). In: Proceedings of the European Conference on Computer Vision (ECCV), pp. 480–496 (2018)
30. Dai, Z., Chen, M., Zhu, S., et al.: Batch feature erasing for person re-identification and beyond. arXiv preprint arXiv:1811.07130 (2018). 1(2), 3
31. Zheng F, Deng C, Sun X, et al. Pyramidal person re-identification via multi-loss dynamic training[C]//Proceedings of the IEEE/CVF conference on computer vision and pattern recognition. 2019: 8514–8522

MS-MH-TCN: Multi-Stage and Multi-Head Temporal Convolutional Network for Action Segmentation

Zengxin Kang and Zhongyi Chu(✉)

School of Instrumentation and Optoelectronic Engineering, Beihang University, Beijing 100083, China
{kangzengxin,chuzy}@buaa.edu.cn

Abstract. The segmentation of manual action in long video has important application value in the fields of demonstration programming and human-computer interaction. The state-of-the-art manual action segmentation method uses multi-stage temporal convolution. Although it can capture the long temporal dependence between actions, there are still excessive segmentation errors in the predicted results. In this paper, we propose a multistage and multi-head temporal convolutional network (MS-MH-TCN) for improving the performance on the action segmentation task. A multi-head calculation is performed at each stage, and the results of the multi-head are pooled to average and fed into the next stage. This approach improves the model's prediction and generalization of input information, because different heads can focus on different aspects of the input and can adapt to learn how to combine them to generate final action predictions. We also propose a new segment smoothing loss function to punish over-segmentation errors. An extensive evaluation showed the effectiveness of the proposed model in capturing long-term dependencies and identifying action segments. Our model achieved the most advanced results on the 50salad dataset.

Keywords: Action segmentation · Temporal convolutional network · Multi-head attention · Action segmentation

1 Introduction

Action segmentation from video has important research value in computer vision field. Previously, most of the studies focused on the classification of trimmed videos with a single activity [1,6,7,24,28], but now scholars have paid attention to long videos with many action segments. This is because in many applications, such as surveillance and robotics, temporally segmentation of activities in long untrimmed videos is crucial.

The original temporal action segmentation method attempts to clip the video by combining with sliding window [10,22] and [19]. Through the temporal window of different scales, the fragment features are extracted, and the action

F. Sun and J. Li (Eds.): ICCCS 2023, CCIS 2029, pp. 48–60, 2024.
https://doi.org/10.1007/978-981-97-0885-7_5

is detected and classified. However, such approaches are not suitable for long videos. In addition, [11,15] and [21] used Markov model for frame-wise action segmentation. But these methods are still unsatisfactory because it only considers the probability of transition between the current state and the previous state, and does not take into account longer historical information. This results in incorrect action segmentation in the model.

Recently, many researchers have applied models based on temporal convolutional networks (TCN) to action segmentation tasks [2,14,16]. But these methods rely on the temporal pooling layer to increase the receptive domain, resulting in the loss of much of the fine-grained information needed for recognition. In order to overcome the limitations of the above methods, [3] proposed a MS-TCN. The model consists of several stages, each of which outputs a prediction and is refined in the next stage. It runs at the full temporal resolution of the video and gets better results. However, in MS-TCN, the upper receptive field is very large and the lower receptive field is very small, so it is still difficult to avoid the loss of fine-grained information in the prediction process.

In order to better capture different features of information, we introduce a multi-head attention mechanism [27] based on the multi-stage temporal convolutional network. At each stage, we apply a multi-head structure, which enables the model to have richer feature extraction capabilities. Because the multi-head attention mechanism can focus on different locations at the same time, so it can better capture the features information of different locations in the input sequence. Each head can focus on a different aspect, improving the model's ability to understand and express data. The output of multiple heads is combined by the average pooling layer and passed to the next stage for refinement. Compared to previous methods, the proposed model achieves better results. Figure 1 shows an overview of the proposed MS-MH-TCN model. Furthermore, we propose a segment smoothing loss during training which penalizes over-segmentation errors in the predictions.

The evaluation demonstrated the effective improvement of our model in capturing long-term dependencies between action classes and producing high-quality predictions. Our approach achieved state-of-the-art results on a challenging action segmentation benchmark: 50salad [26].

2 Related Work

Action segmentation has important application value in video surveillance, robot assembly and other fields, and has been widely concerned by scholars. In the early methods, the sliding window method and non-maximum suppression were adopted [10,22]. However, the model needs to make predictions in a variety of window scales, resulting in high computational costs. Other methods model actions based on changes in the state of objects and materials [5] or interactions between hands and objects [4] to identify different classes of actions. Although these methods were successful at the time, their performance was limited because they could not capture the context of the video.

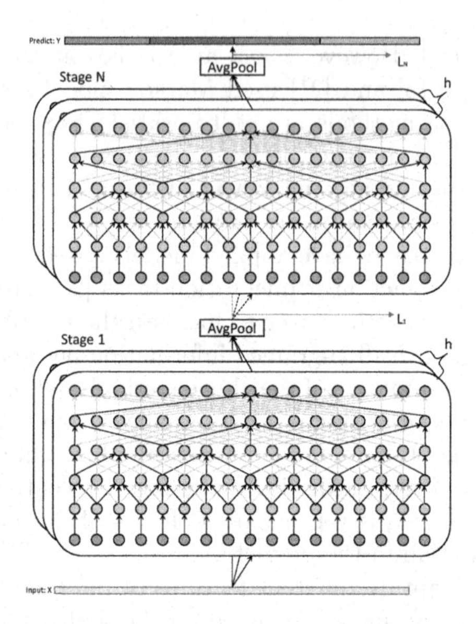

Fig. 1. Overview of the multi-stage and multi-head temporal convolutional network (MS-MH-TCN).

To alleviate this problem, the hidden Markov model (HMM) is widely used. In [11], feature vectors of every frame of video are extracted and modeled by HMM to determine the most likely action sequence. In [12], hmm is combined with the Gaussian mixture model (GMM) as a frame-level classifier. Later, Richard et al. [21] and Kuehne et al. [13] improved the classifier using a GRU instead of the GMM used in [12,22]. Gammulle [8] presents a semi-supervised generative adversarial network architecture for continuous fine-grained action segmentation. Temporal context information is captured via a Gated Context Extractor (GCE) module, which directs the queued context information through the generator model to enhance action segmentation. Although these methods take into account the context information of the video, they still have the obvious problem of excessive segmentation because they cannot capture the dependency of the long video.

In order to better capture the features of long video sequences, Lea [14] use a hierarchy of temporal convolutional networks (TCN) to perform fine-grained action segmentation and detection. Their Encoder-Decoder TCN uses pooling and up-sampling to effectively capture long-range temporal patterns, while the expanded TCN uses dilated convolutions. Ding and Xu [2] add lateral connections to the encoder-decoder TCN [14] and propose a weakly-supervised sequence model to align action sequences. Abu Farha and Gall [3] introduce a multi-stage TCN (MS-TCN) architecture for the action segmentation task. Each stage features a set of dilated temporal convolutions to generate an initial prediction which is refined by the next one. Li [17] builds on top of [3] a dual dilated layer that combines both large and small receptive fields decoupling the first stage from the refining stages to address the different requirements of these stages. These articles show that temporal convolution can effectively extract long-range dependencies of actions in videos.

3 Temporal Action Segmentation

3.1 Single-Stage TCN

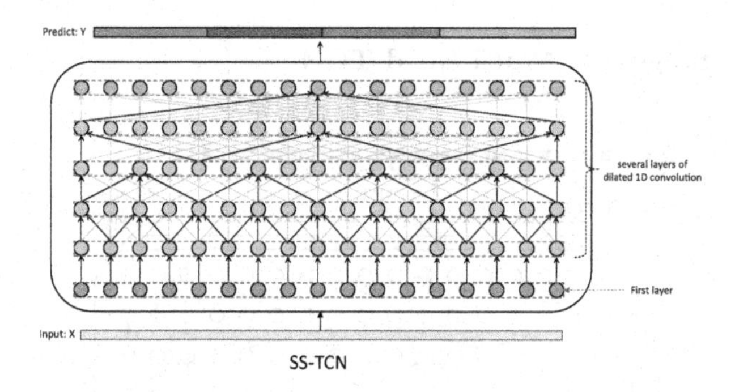

Fig. 2. Overview of the single-stage temporal convolutional network (SS-TCN).

The single-stage model is called the single-stage temporal convolutional network (SS-TCN) [3]. As shown in Fig. 2, the first layer of SS-TCN is the 1×1 convolution layer, which adjusts the dimensions of the input features to match the number of feature mappings in the network. Then, this layer is followed by several layers of dilated one-dimensional convolution. The dilation factor that is doubled at each layer, i.e. 1, 2, 4, ..., 512. All these layers have the same number of convolutional filters with kernel size 3. Each layer applies a dilated convolution with ReLU activation to the output of the previous layer. Then residual connections are further used to facilitate gradients flow. The set of operations at each layer can be formally described as follows:

$$\hat{H}_l = ReLU\left(W_d * \hat{H}_{l-1} + b_d\right) \tag{1}$$

$$H_l = \hat{H}_{l-1} + W_l * \hat{H}_l + b_l \tag{2}$$

where H_l is the output of layer l, $*$ denotes the convolution operator, $W_d \in \mathbb{R}^{3\times D\times D}$ are the weights of the dilated convolution filters with kernel size 3 and D is the number of convolutional filters, $W_l \in \mathbb{R}^{1\times D\times D}$ are the weights of a 1×1 convolution, and $b_d, b_l \in \mathbb{R}^D$ are bias vectors. The receptive field at each layer is determined using by

$$ReceptiveField(l) = 2^{l+1} - 1 \tag{3}$$

where $l \in [1, L]$ is the layer number. To get the probabilities for the output class, a 1×1 convolution is applied over the output of the last dilated convolution layer followed by a softmax activation, i.e.

$$Y_t = Softmax(W * h_{L,t} + b) \tag{4}$$

where Y_t contains the class probabilities at time t, $h_{L,t}$ is the output of the last dilated convolution layer at time t, $W \in \mathbb{R}^{C \times D}$ and $b \in \mathbb{R}^C$ are the weights and bias for the 1×1 convolution layer. C is the number of classes.

3.2 Single-Stage and Multi-Head TCN

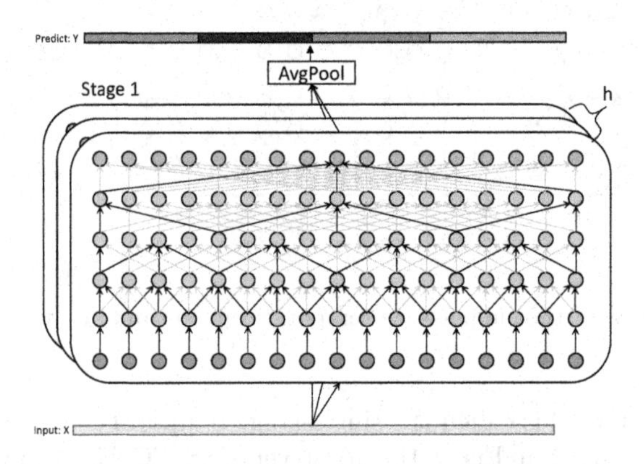

Fig. 3. Overview of the single-stage and multi-head temporal convolutional network (SS-MH-TCN).

In order to better capture different features of the input, we introduce multi-head attention mechanism, we introduce a single-stage and multi-Head temporal convolutional network (SS-MH-TCN) for the temporal action segmentation task. In this SS-MH-TCN model as shown in Fig. 3, the input of the multi-head is the frame-wise features of the video as follows

$$Y^0 = x_{1:T} \tag{5}$$

$$\begin{cases} MultiHead(Y^0) = Concat(head_1, \ldots, head_h), \\ \qquad where\ head_i = \mathcal{S}_i(Y^0) \end{cases} \tag{6}$$

$$Y^1 = AvgPool(MultiHead(Y^0)) \tag{7}$$

where $\mathcal{S}$ is the SS-TCN discussed in Sect. 3.1 and h is the head number. Each of the headers has input Y^0 and their outputs are concatenated as $MultiHead$. Finally, $MultiHead$ is combined by the average pooling layer into Y^1 as output of the stage. Multi-head perform several independent temporal convolution operations on the input Y^0, allowing the model to focus on different parts of the input simultaneously to capture more relevant information. This approach improves the model's representation and generalization of input information because different heads can focus exclusively on different aspects of the input and can adapt to learn how to combine them to produce the final representation.

3.3 Multi-Stage and Multi-Head TCN

The multi-stage and multi-head model is shown in Fig. 1.

$$Y^s = \mathcal{MH}(Y^{s-1}) \tag{8}$$

where Y^s is the output at stage s and $\mathcal{MH}$ is the SS-MH-TCN discussed in Sect. 3.2.

3.4 Loss Function

To further solve the problem of over-segmentation, we propose a new smoothing loss function called segment smoothing loss. For this loss, we use a segmented mean squared error over the frame-wise log-probabilities

$$\mathcal{L}_{S-MSE} = \frac{1}{TC}\sum_{t,c}(\triangle_{t,c}^2 * \delta_t) \tag{9}$$

$$\triangle_{t,c} = |log\ \hat{y}_{t,c} - log\ \hat{y}_{t-1,c}| \tag{10}$$

$$\delta_t = 1 - sign(|y_t - y_{t-1}|) \tag{11}$$

where T is the video length, C is the number of classes, and $\hat{y}_{t,c}$ is the probability of class c at time t. y_t is the numeric label of ground truth at time t. δ_t is the segment indicator parameter as shown in the Fig. 4. Note that the gradients are only computed with respect to $\hat{y}_{t,c}$, whereas $\hat{y}_{t-1,c}$ is not considered as a function of the model's parameters. This loss is similar to the truncated mean squared error loss ($\mathcal{L}_{T-MSE}$) [3], as shown in Eq. 12 and Eq. 13. However, we found that the $\mathcal{L}_{S-MSE}$ reduces the over-segmentation errors more. This is because in $\mathcal{L}_{T-MSE}$) you need to manually set the hyperparameter τ to set the segment to be smoothed, while in our $\mathcal{L}_{S-MSE}$ you generate the segment to be smoothed directly according to the label. This avoids the uncertainty associated with manually setting parameters.

$$\mathcal{L}_{T-MSE} = \frac{1}{TC}\sum_{t,c}\hat{\triangle}_{t,c}^2 \tag{12}$$

$$\hat{\triangle}_{t,c}^2 = \begin{cases} \triangle_{t,c}^2 & : \triangle_{t,c}^2 \leq \tau \\ \tau & : otherwise \end{cases} \tag{13}$$

We will compare the $\mathcal{L}_{T-MSE}$ loss and the proposed loss $\mathcal{L}_{S-MSE}$ in the experiments.

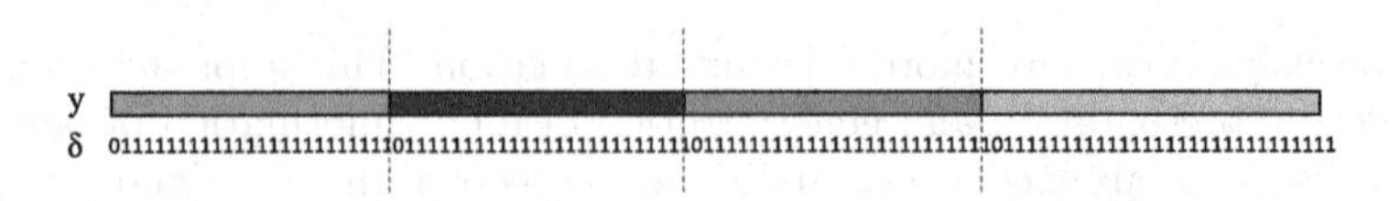

Fig. 4. The relationship between the segment indicator parameter δ and the ground truth label y.

To train the model, we use a combination of a classification loss and a segment smoothing loss. For the classification loss, we use a cross entropy loss

$$\mathcal{L}_{cls} = \frac{1}{T}\sum_{t} -log(y_{t,c}) \tag{14}$$

where $y_{t,c}$ is the predicted probability for the ground truth label c at time t. To further improve the quality of the predictions, we use an additional segment smoothing loss to reduce such over-segmentation errors.

The final loss function for a single stage is a combination of the above mentioned losses

$$\mathcal{L}_s = \mathcal{L}_{cls} + \beta\mathcal{L}_{S-MSE}, \tag{15}$$

where β is a model hyper-parameter to determine the contribution of the different losses. Finally to train the complete model, we minimize the sum of the losses over all stages

$$\mathcal{L} = \sum_{s} \mathcal{L}_s. \tag{16}$$

4 Experiments

4.1 Implementation Details

All stages are the same for MS-MH-TCN. Each stage in MS-MH-TCN contains ten dilated convolution layers. Dropout is used after each layer with probability 0.5. We set the number of filters to 64 in all layers of the model and the filter size is 3. In all experiments, we use Adam optimizer with a learning rate of 0.0005. We evaluate the proposed models on the challenging dataset - 50Salads, which contains 50 videos with 17 action classes. The videos were recorded from the top view. On average, each video contains 20 action instances and is 6.4 min long. As the name of the dataset indicates, the videos depict salad preparation activities. These activities were performed by 25 actors where each actor prepared two different salads. For evaluation, we use five-fold crossvalidation and report the average as in [26]. Note that all the reported results are obtained using the I3D features.

For evaluation, we report the framewise accuracy (Acc), segmental edit distance (Edit) and the segmental F1 score at overlapping thresholds 10%, 25% and 50%, denoted by F1@{10, 25, 50}, which is a measure of the quality of the prediction as proposed by [14].

4.2 Effect of the Number of Heads on MS-MH-TCN

To verify the effectiveness of the multi-head structure, we evaluated the effects of MS-MH-TCN architecture at different stages and different heads with $\mathcal{L}_{T-MSE}$ [3]. Table 1 shows the comparison results, where h = 1 represents a head structure, which is equivalent to SS-TCN or MS-TCN. As shown in the single stage, all of these models achieve a comparable frame-wise accuracy. Nevertheless, the quality of the predictions is very different. Looking at the segmental edit distance and F1 scores of these models, We can see that the over-segmentation generated by the SS-MH-TCN model gradually decreases as the number of heads increases. That is, using a multi-head structure reduces these errors and increases Edit and F1 scores. This effect is clearly visible when we use two, three, four and five stages. However, by adding the fifth heads, we can see that the performance starts to degrade. This might be an over-fitting problem as a result of increasing the number of parameters. As shown in Table 1, in each stage, for different head structures, the results of the optimal index are expressed in bold. The best results of all experiments are indicated with an asterisk. Taking all the experimental results into consideration, in the rest of the experiments we use a MS-MH-TCN with four stages and four heads. The effect of a multi-head architecture can also be seen in the qualitative results shown in Fig. 5. In a four-stage structure, four heads work best.

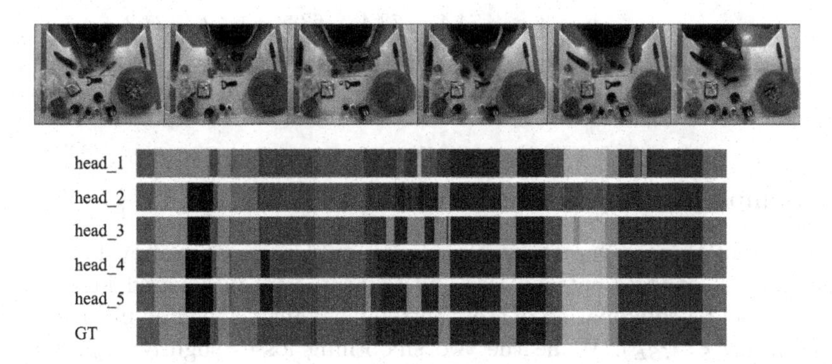

Fig. 5. Qualitative result from the 50Salads dataset for comparing different number of heads with 4 stages.

Table 1. Effect of the number of heads on the 50Salads dataset

MS-MH-TCN		F1@ {10, 25, 50}			Edit	Acc
Single Stage	h = 1 [3]	27.0	25.3	21.5	20.5	78.2
	h = 2	26.2	24.5	20.9	20.8	78.4
	h = 3	28.0	26.6	23.0	20.6	**79.8**
	h = 4	29.23	27.8	23.7	23.0	79.7
	h = 5	**31.1**	**29.0**	**25.3**	**24.6**	78.6
Two Stages	h = 1 [3]	55.5	52.9	47.3	47.9	79.8
	h = 2	60.3	57.3	50.5	52.1	79.2
	h = 3	60.3	58.0	51.4	51.7	80.3
	h = 4	**62.2**	**59.8**	**52.4**	**53.3**	**81.0**
	h = 5	59.2	56.6	49.3	50.4	79.2
Three Stages	h = 1 [3]	71.5	68.6	61.6	64.0	78.6
	h = 2	66.9	65.3	56.1	60.3	78.9
	h = 3	74.1	71.6	62.9	64.6	81.0
	h = 4	70.4	67.2	59.0	62.8	78.1
	h = 5	**75.1**	**73.3**	**65.5**	**68.2**	**81.7***
Four Stages	h = 1 [3]	76.3	74.0	64.5	67.9	80.7
	h = 2	75.2	73.1	65.3	67.4	80.3
	h = 3	75.8	74.2	66.8	68.9	80.6
	h = 4	**78.6***	**77.0***	**68.7***	**71.5**	**81.5**
	h = 5	76.9	74.6	66.6	69.8	81.4
Five Stages	h = 1 [3]	76.4	73.4	63.6	69.2	79.5
	h = 2	73.9	71.6	61.1	66.9	77.3
	h = 3	74.4	72.4	63.5	67.2	77.2
	h = 4	78.5	**76.4**	67.0	**71.7***	79.7
	h = 5	**78.6***	**76.4**	**67.7**	70.9	**80.9**

4.3 Comparing Different Loss Functions

We compared three different loss functions, only a cross entropy loss (L_{cls}), a combination of cross entropy loss and truncated mean squared loss ($\mathcal{L}_{cls} + \mathcal{L}_{T-MSE}$), and a combination of cross entropy loss and segmented mean squared loss ($\mathcal{L}_{cls} + \mathcal{L}_{S-MSE}$). While the two smoothing losses slightly improves the frame-wise accuracy compared to the cross entropy loss alone, we found that they produces much less over-segmentation errors. Table 2 shows a comparison of these losses. As shown in Table 2, the proposed loss achieves better F1 and edit scores with an absolute improvement. This indicates that our loss produces less over-segmentation errors compared to cross entropy and truncated mean squared loss, since it forces consecutive frames to have similar class probabilities, which results in a smoother output. Figure 6 shows qualitative comparison of these losses.

Both $\mathcal{L}_{S-MSE}$ and $\mathcal{L}_{T-MSE}$ measure the difference in the probability distribution between the preceding and following frames. However, the results show that the proposed $\mathcal{L}_{S-MSE}$ loss produces better results than the $\mathcal{L}_{T-MSE}$ loss as shown in Table 2. The reason behind this is the fact that the $\mathcal{L}_{T-MSE}$ loss due to the truncation factor τ does not penalize cases where the difference between the target probability and the predicted probability is very large. Whereas the proposed loss penalizes all (large or small) differences as well.

Table 2. The comparison of different loss functions. The optimal parameter $\lambda = 0.05$ in reference [3] is selected.

Four Stages Four Heads		F1@ {10, 25, 50}			Edit	Acc
$\mathcal{L}_{cls}$		76.1	73.7	65.2	68.6	81.1
$\mathcal{L}_{cls} + \lambda\mathcal{L}_{T-MSE}$	$\tau = 1$	73.9	71.2	64.2	65.8	80.6
	$\tau = 2$	**78.6**	**77.0**	**68.7**	**71.5**	**81.5**
	$\tau = 3$	76.2	74.0	65.8	68.8	79.6
	$\tau = 4$	74.2	71.4	64.1	67.9	80.4
$\mathcal{L}_{cls} + \beta\mathcal{L}_{S-MSE}$	$\beta = 1$	76.3	74.1	65.7	68.1	78.8
	$\beta = 2$	**80.8**	**78.3**	70.1	72.9	81.4
	$\beta = 2.5$	80.5	78.2	**70.3**	**74.3**	**82.1**
	$\beta = 3$	80.1	77.9	67.8	72.9	81.3
	$\beta = 4$	79.8	77.7	68.6	73.5	81.1
	$\beta = 5$	77.4	74.3	65.5	69.7	79.1

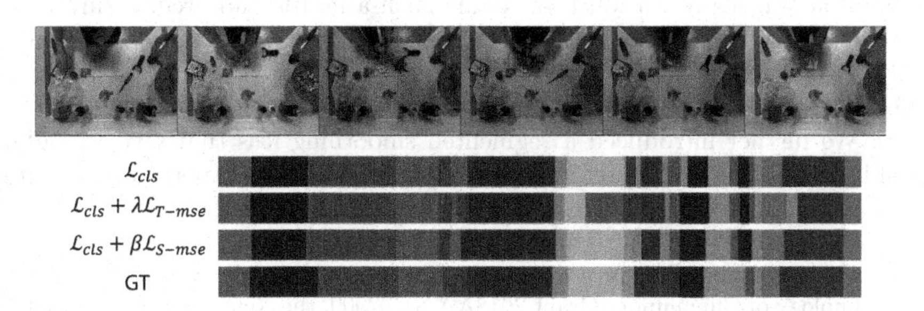

Fig. 6. Qualitative result from the 50Salads dataset for comparing different loss functions.

4.4 Comparison with the State-of-the-Art

In this section, we compare the proposed models to the state-of-the-art methods on 50Salads datasets. The results are presented in Table 3. As shown in the table, our models outperform the state-of-the-art methods and with respect to three evaluation metrics: F1 score, segmental edit distance, and frame-wise accuracy (Acc) with a large margin.

Table 3. The parsing result with different temporal methods.

50Salads	F1@ {10, 25, 50}			Edit	Acc
Spatial CNN [15]	32.3	27.1	18.9	24.8	54.9
IDT+LM [20]	44.4	38.9	27.8	45.8	48.7
Bi-LSTM [25]	62.6	58.3	47.0	55.6	55.7
DilatedTCN [14]	52.2	47.6	37.4	43.1	59.3
ST-CNN [15]	55.9	49.6	37.1	45.9	59.4
TUnet [23]	59.3	55.6	44.8	50.6	60.6
ED-TCN [14]	68.0	63.9	52.6	52.6	64.7
TResNet [9]	69.2	65.0	54.4	60.5	66.0
TRN [16]	70.2	65.4	56.3	63.7	66.9
TDRN+UNet [16]	69.6	65.0	53.6	62.2	66.1
TDRN [16]	72.9	68.5	57.2	66.0	68.1
LCDC+ED-TCN [18]	73.8	–	–	66.9	72.1
MS-TCN [3]	76.3	74.0	64.5	67.9	80.7
MS-MH-TCN	**80.5**	**78.2**	**70.3**	**74.3**	**82.1**

5 Conclusion

We propose a multi-stage and multi-head temporal convolution model for the temporal action segmentation task. Using such a multi-head architecture helps in providing more context to predict the class label at each frame. The experimental evaluation demonstrated the capability of our architecture in capturing temporal dependencies between action classes and reducing oversegmentation errors. We further introduced a segmented smoothing loss that gives an additional improvement of the predictions quality. Our models outperform the state-of-the-art methods on 50Salads datasets with a large margin.

Acknowledgements. This project was supported in part by the Ministry of Science and Technology of China under Grant 2018AAA0102900, the New Generation of Artificial Intelligence Technology Innovation 2030 Major Project, and in part by the National Natural Science Foundation of China under Grant U1913206 and Grant 51975021.

References

1. Carreira, J., Zisserman, A.: Quo vadis, action recognition? A new model and the kinetics dataset. In: Proceedings of the IEEE Conference on Computer Vision and Pattern Recognition, pp. 6299–6308 (2017)

2. Ding, L., Xu, C.: Weakly-supervised action segmentation with iterative soft boundary assignment. In: Proceedings of the IEEE Conference on Computer Vision and Pattern Recognition, pp. 6508–6516 (2018)
3. Farha, Y.A., Gall, J.: MS-TCN: multi-stage temporal convolutional network for action segmentation. In: Proceedings of the IEEE/CVF Conference on Computer Vision and Pattern Recognition, pp. 3575–3584 (2019)
4. Fathi, A., Farhadi, A., Rehg, J.M.: Understanding egocentric activities. In: 2011 International Conference on Computer Vision, pp. 407–414. IEEE (2011)
5. Fathi, A., Rehg, J.M.: Modeling actions through state changes. In: Proceedings of the IEEE Conference on Computer Vision and Pattern Recognition, pp. 2579–2586 (2013)
6. Feichtenhofer, C., Fan, H., Malik, J., He, K.: Slowfast networks for video recognition. In: Proceedings of the IEEE/CVF International Conference on Computer Vision, pp. 6202–6211 (2019)
7. Feichtenhofer, C., Pinz, A., Wildes, R.P.: Spatiotemporal residual networks for video action recognition. corr abs/1611.02155. arXiv preprint arXiv:1611.02155 (2016)
8. Gammulle, H., Denman, S., Sridharan, S., Fookes, C.: Fine-grained action segmentation using the semi-supervised action GAN. Pattern Recogn. **98**, 107039 (2020)
9. He, K., Zhang, X., Ren, S., Sun, J.: Deep residual learning for image recognition. In: Proceedings of the IEEE Conference on Computer Vision and Pattern Recognition, pp. 770–778 (2016)
10. Karaman, S., Seidenari, L., Del Bimbo, A.: Fast saliency based pooling of fisher encoded dense trajectories. In: ECCV THUMOS Workshop, vol. 1, p. 5 (2014)
11. Kuehne, H., Gall, J., Serre, T.: An end-to-end generative framework for video segmentation and recognition. In: 2016 IEEE Winter Conference on Applications of Computer Vision (WACV), pp. 1–8. IEEE (2016)
12. Kuehne, H., Richard, A., Gall, J.: Weakly supervised learning of actions from transcripts. Comput. Vis. Image Underst. **163**, 78–89 (2017)
13. Kuehne, H., Richard, A., Gall, J.: A hybrid RNN-HMM approach for weakly supervised temporal action segmentation. IEEE Trans. Pattern Anal. Mach. Intell. **42**(4), 765–779 (2018)
14. Lea, C., Flynn, M.D., Vidal, R., Reiter, A., Hager, G.D.: Temporal convolutional networks for action segmentation and detection. In: Proceedings of the IEEE Conference on Computer Vision and Pattern Recognition, pp. 156–165 (2017)
15. Lea, C., Reiter, A., Vidal, R., Hager, G.D.: Segmental spatiotemporal CNNs for fine-grained action segmentation. In: Leibe, B., Matas, J., Sebe, N., Welling, M. (eds.) ECCV 2016. LNCS, vol. 9907, pp. 36–52. Springer, Cham (2016). https://doi.org/10.1007/978-3-319-46487-9_3
16. Lei, P., Todorovic, S.: Temporal deformable residual networks for action segmentation in videos. In: Proceedings of the IEEE Conference on Computer Vision and Pattern Recognition, pp. 6742–6751 (2018)
17. Li, S.J., AbuFarha, Y., Liu, Y., Cheng, M.M., Gall, J.: MS-TCN++: multi-stage temporal convolutional network for action segmentation. IEEE Trans. Pattern Anal. Mach. Intell. **45**(6), 6647–6658 (2023)
18. Mac, K.N.C., Joshi, D., Yeh, R.A., Xiong, J., Feris, R.S., Do, M.N.: Learning motion in feature space: locally-consistent deformable convolution networks for fine-grained action detection. In: Proceedings of the IEEE/CVF International Conference on Computer Vision, pp. 6282–6291 (2019)
19. Oneata, D., Verbeek, J., Schmid, C.: The lear submission at thumos 2014 (2014)

20. Richard, A., Gall, J.: Temporal action detection using a statistical language model. In: Proceedings of the IEEE Conference on Computer Vision and Pattern Recognition, pp. 3131–3140 (2016)
21. Richard, A., Kuehne, H., Gall, J.: Weakly supervised action learning with RNN based fine-to-coarse modeling. In: Proceedings of the IEEE Conference on Computer Vision and Pattern Recognition, pp. 754–763 (2017)
22. Rohrbach, M., Amin, S., Andriluka, M., Schiele, B.: A database for fine grained activity detection of cooking activities. In: 2012 IEEE Conference on Computer Vision and Pattern Recognition, pp. 1194–1201. IEEE (2012)
23. Ronneberger, O., Fischer, P., Brox, T.: U-Net: convolutional networks for biomedical image segmentation. In: Navab, N., Hornegger, J., Wells, W., Frangi, A. (eds.) MICCAI 2015. LNCS, vol. 9351, pp. 234–241. Springer, Cham (2015). https://doi.org/10.1007/978-3-319-24574-4_28
24. Simonyan, K., Zisserman, A.: Two-stream convolutional networks for action recognition in videos. In: Advances in Neural Information Processing Systems, vol. 27 (2014)
25. Singh, B., Marks, T.K., Jones, M., Tuzel, O., Shao, M.: A multi-stream bi-directional recurrent neural network for fine-grained action detection. In: Proceedings of the IEEE Conference on Computer Vision and Pattern Recognition, pp. 1961–1970 (2016)
26. Stein, S., McKenna, S.J.: Combining embedded accelerometers with computer vision for recognizing food preparation activities. In: Proceedings of the 2013 ACM International Joint Conference on Pervasive and Ubiquitous Computing, pp. 729–738 (2013)
27. Vaswani, A., et al.: Attention is all you need. In: Advances in Neural Information Processing Systems, vol. 30 (2017)
28. Yang, G.Z., Pei, L., Xin, K.Z., Yi, C.Z.: Manual assembly action segmentation method based on spatiotemporal features. In: 2021 China Automation Congress, pp. 6696–6701 (2021)

Real-Time Visual Detection of Anomalies in Densely Placed Objects

Chunfang Liu(✉), Jiali Fang, and Pan Yu

Faculty of Information and Technology, Beijing University of Technology, Beijing, China
{cfliu1985,panyu}@bjut.edu.cn, FangJL@emails.bjut.edu.cn

Abstract. In the industrial production line, the operation of packing small packages that are densely arranged in vertical rows and move quickly is still mainly done manually. In this situation, it is a typical and difficult problem for robot to put small packages with various quantity and arrangement requirements into large packages in real time. In order to solve this problem, this paper proposes a real-time detection method based on the improved YOLOv5 deep neural network for identifying the precise quantity and abnormality of dense vertical bagged small packages on the assembly line. This method can quickly detect the position, category, quantity, and rotation angle of small packages. Combining depth information with bounding box rotation angle information, it can timely detect abnormal placement of small packages on the assembly line. Then the misjudgment of the number of small packages caused by blurred images is corrected by amending the width of the object with low reliability. This method effectively identifies abnormal situations in the industrial production line and provides an effective solution, which is crucial for robots to autonomously complete packaging tasks in real-time according to various quantities and arrangement requirements.

Keywords: Industrial production · Abnormal situations · Real-time detection

1 Introduction

In recent years, with the rapid development of artificial intelligence and robot technology, robots have been used in the real-time operation of moving objects on the assembly line, such as food sorting, picking, packaging and other processes. However, in most applications, the robot can only operate a single object at a time through object detection and trajectory planning algorithms, or operate on the object that can be clearly separated on the assembly line. For the operation of dense vertical and fast-moving small packages, it is still mainly done manually. It is a typical problem for robots to put the small packages with the various quantity and arrangement requirements into the large package in real time.

One of the main difficulties of this problem is the real-time detection of the quantity and abnormality of the closely arranged and fast-moving small packages

F. Sun and J. Li (Eds.): ICCCS 2023, CCIS 2029, pp. 61–73, 2024.
https://doi.org/10.1007/978-981-97-0885-7_6

on the assembly line: (1) The small packages are tightly arranged vertically, irregularly deformed and moving fast, resulting in blurred images captured by the camera. The traditional detection algorithm for a single object is difficult to distinguish the small packages in this case, which easily leads to misjudgment the number of objects and eventually causes the failure of the robotic grasping; (2) Occasionally, there are individual small packages that are placed in an abnormal mess on top of the moving small packages that in tight vertical arrangement. It is necessary to detect their position and posture in time, so that the robot can quickly remove them to prevent the normal independent packaging process from being affected.

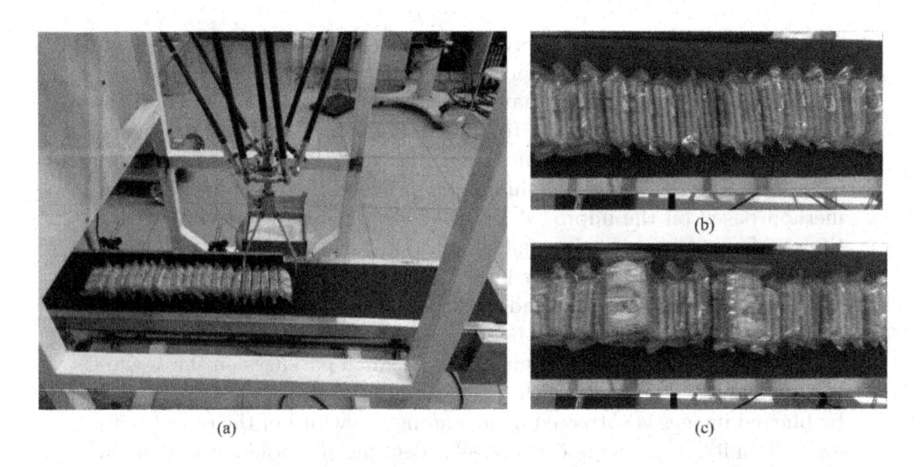

Fig. 1. (a) The packaging system developed in this paper. (b) The object in motion captured by the camera. (c) Abnormal placement of objects captured by the camera.

In view of the above problems, we build an automatic packaging system shown in Fig. 1 and propose a visual real-time detection method for the precise quantity and abnormality of small packages densely arranged vertically on the assembly line. This method uses the improved YOLOv5 deep neural network to detect the dense vertical small packages, and obtains the position, category, quantity and attitude angle of the objects. Then, the recognition result and the visual depth information are utilized to detect the position and posture of small packages that are abnormally placed on the assembly line for dealing with abnormally placed objects in time. In order to further accurately determine the number of objects, the reliability of the number detection is judged according to the width information of each detected bounding box. If the reliability is low, the number of misjudged objects is corrected for ensuring accurate recognition.

The main contributions of this paper are as follows:

(1) We apply the improved YOLOv5 neural network to detect the dense vertical small packages, and combine the results with the visual depth information to detect the position and posture of the abnormally placed objects on the

assembly line in real time, which is conducive to timely move abnormally placed food.

(2) According to the width information of each detected bounding box, the reliability of the number detection is judged. If the reliability is low, the number of misjudged objects is corrected for ensuring accurate recognition.

2 Related Work

Object detection has been extensively studied in the past 20 years, mainly divided into one-stage object detection and two-stage object detection, two-stage detection [4] usually feeds high-scoring region proposals obtained from the first-stage CNN to the second-stage CNN for final prediction. The output of the first stage [7,11–14,20] can be obtained by only one CNN operation, which is fast and accurate, and YOLO is one of them. The current advanced real-time object detection is mainly based on YOLO [12–14] and FCOS [16,17], they are [1,3,5,18,19,21]. These algorithms can quickly and accurately detect the object and frame the object with a bounding box, but the bounding boxes are all horizontal, the angle of the object cannot be obtained. For the abnormally placed small packages in the industrial production line, the robot cannot grasp them with a proper posture.

Some deep learning architecture have also been proposed to predict graspable locations in recent years, such as GraspNet, PointNet, etc. Mahler et al. [8] propose a Grasp Quality Convolutional Neural Network (GQ-CNN) model that can quickly predict the probability of grasp success from depth images, where grasps are designated as graspers plane position, angle and depth relative to the RGB-D sensor. And they verified on the robot, achieved a relatively high success rate. An approach is use to obtain graspable regions in images, which without representing object pose and shape [6,9,10,15]. [22] learned 3D point cloud representations of objects, and leveraged 3D point clouds for grasping based on critical models learned from simulated grasping data. [2] proposed a deep learning architecture for predicting graspable positions in robot operations, input RGB-D images of single or multiple objects, and output multiple grasps of single or multiple objects after designed neural network. The model was tested on the Cornell Dataset and achieved good results. These algorithms can obtain the best graspable locations of the objects, but cannot classify the objects.

3 Proposed Method

We propose an effective detection method for the real-time judgment of the quantity and abnormality of small packages closely arranged vertically on the conveyor belt. Figure 2 is a flowchart of this method.

The next two subsections describe the methods used in this paper, including the method that detects abnormally placed small packages and the method identifies quantity correction in small packages.

3.1 The Detection of Abnormally Placed Small Packages

In this section, an improved YOLOv5 deep neural network is applied to detect abnormal small packages densely arranged vertically in the industrial production line. Figure 3 shows the network structure of the YOLOv5. First, the camera captures picture of small packages on the assembly line, it can be seen from the picture that there are abnormally placed objects. Then the picture is sent to the improved YOLOv5 network, which can detect abnormally placed small packages in time.

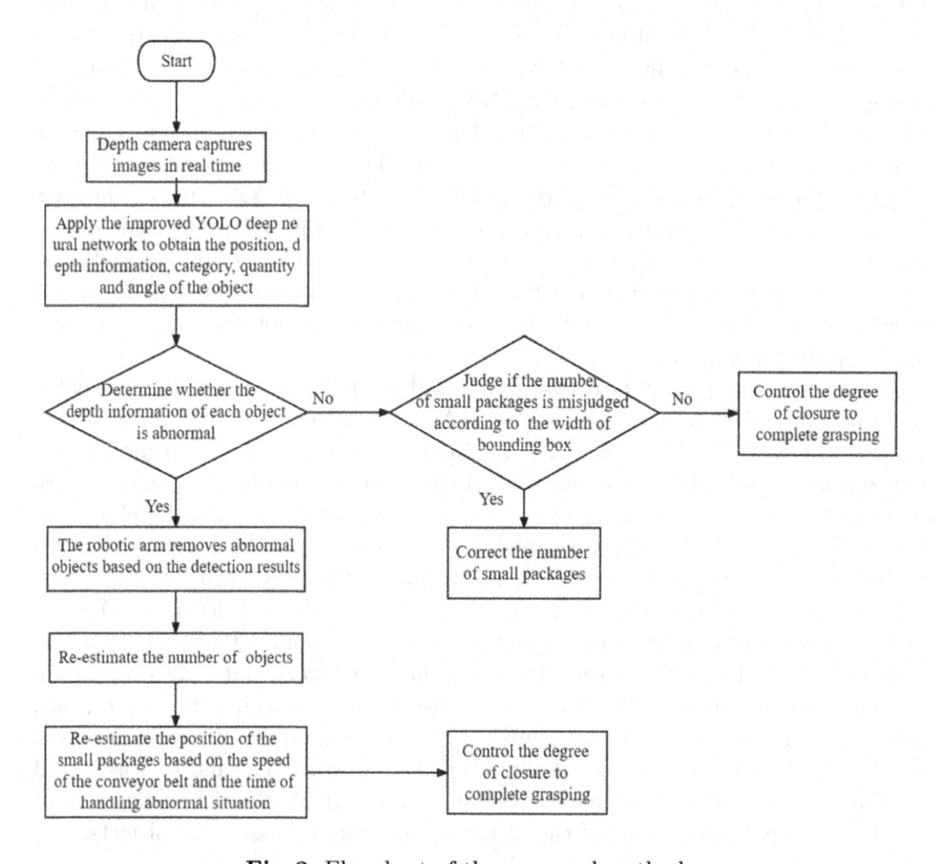

Fig. 2. Flowchart of the proposed method

YOLOv5 uses a deep convolutional neural network (CNN) to extract features from the input image, which are then used to predict the bounding boxes and class probabilities of objects in the image. The network architecture of YOLOv5 is based on a modified version of the CSPNet backbone, which is designed to improve the efficiency and accuracy of feature extraction. The model also uses

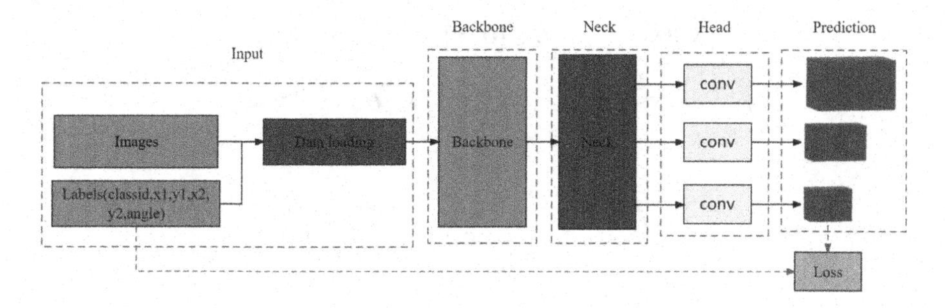

Fig. 3. Network structure of the proposed YOLOv5

a novel anchor-free detection method, which eliminates the need for predefined anchor boxes and improves the flexibility of object detection.

The improved YOLOv5 deep neural network adds a θ dimension to the data loading, which turns θ into a classification, and adds an angle classification loss to the loss function, which can detect the position, category, quantity and rotation angle of small packages. The loss function for the θ is:

$$l = -(\tilde{\theta} * log(\sigma(\theta) + (1 - \tilde{\theta}) * log(1 - \sigma(\theta)) \quad (1)$$

where $\tilde{\theta}$ is the angle value in the label, θ represents the angle value detected by the model we trained, σ is a commonly used activation function.

To balance the loss for the position, category, quantity, and rotation angle, we adopted a Relative Weighting approach, which uses distinct weights to each loss component based on their relative importance. This allowed us to emphasize specific aspects of the task over others. For instance, we allocated a higher weight to the position loss component to prioritize precise object localization.

Since the depth camera and the conveyor belt are fixed, the depth information of the moving small packages are approximative. In abnormal circumstances, there will be one or more objects above the densely arranged small packages (lying flat, messy on the densely arranged small packages), the depth information of the abnormal objects is lower than the normal small packages. Combining the depth information of objects with the YOLOv5 recognition results to identify abnormally placed small packages, and the position and rotation angle information of the abnormal objects will be transmitted to the robot in real time. It enables the robot to quickly grab abnormally placed snacks with a suitable posture and remove them.

As shown in Fig. 4, the angle in the object detection result starts from the horizontal axis, ends with the width to the short side, and turns clockwise. When grabbing abnormally placed food, θ can be the angle of rotation around the Z axis, which can be converted into a quaternion, so that the robot arm can

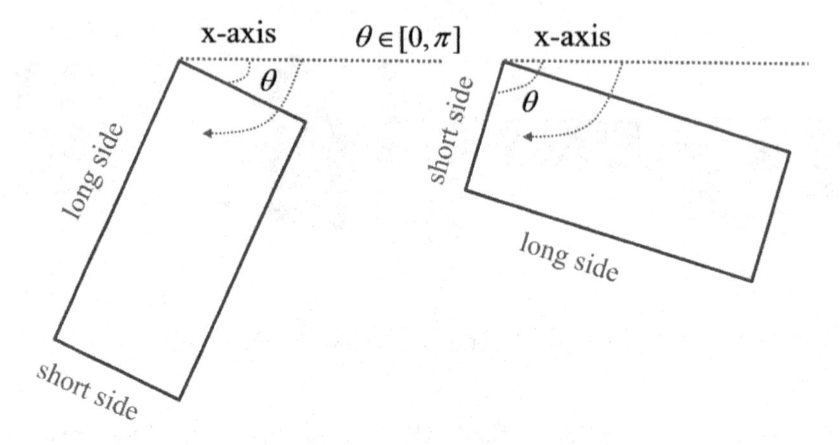

Fig. 4. Five-parameter method with 180° angular range

handle abnormal situations with a suitable posture. Among them, the conversion relationship from Euler angle to quaternion is:

$$q_0 = \cos\frac{\alpha}{2}\cos\frac{\beta}{2}\cos\frac{\theta}{2} + \sin\frac{\alpha}{2}\sin\frac{\beta}{2}\sin\frac{\theta}{2} \tag{2}$$

$$q_1 = \sin\frac{\alpha}{2}\cos\frac{\beta}{2}\cos\frac{\theta}{2} - \cos\frac{\alpha}{2}\sin\frac{\beta}{2}\sin\frac{\theta}{2} \tag{3}$$

$$q_2 = \cos\frac{\alpha}{2}\sin\frac{\beta}{2}\cos\frac{\theta}{2} + \sin\frac{\alpha}{2}\cos\frac{\beta}{2}\sin\frac{\theta}{2} \tag{4}$$

$$q_3 = \cos\frac{\alpha}{2}\cos\frac{\beta}{2}\sin\frac{\theta}{2} - \sin\frac{\alpha}{2}\sin\frac{\beta}{2}\cos\frac{\theta}{2} \tag{5}$$

where α, β, θ represent the angles rotated around the X, Y, and Z axes respectively. The angle in the object detection results is the angle of rotation around the Z axis, so the angles of the α and β axes are set to 0, which will be substituted into the above formula, we can get the corresponding quaternion:

$$q_0 = \cos\frac{\theta}{2} \tag{6}$$

$$q_1 = 0 \tag{7}$$

$$q_2 = 0 \tag{8}$$

$$q_3 = \sin\frac{\theta}{2} \tag{9}$$

According to the quaternion, the robot arm can handle abnormally placed small packages in an appropriate posture. This method has a small amount of calculation and good real-time performance. Since the objects placed abnormally are in any direction, the improved YOLO neural network is applied to obtain the rotation angle information of the bounding box, the detection speed is fast, and it is suitable for application in the actual industrial production line.

Algorithm 1. Correction of abnormal number of small packages

Input: The upper-left and lower-right coordinates of each bounding box, like (x1,y1,x2,y2)
Output: N
1: **for** $i = 1 \rightarrow n$ **do** $//n$ is the number of detected objects
2: **Create** two empty lists xleft and xright
3: **xleft** stores x1
4: **xright** stores x2
5: **Calculate** w
6: **if** w > Experience Width Value **then**
7: flag = 1
8: **end if**
9: **end for**
10: **if** flag = 1 **then**
11: **Find** the minimum value in xleft
12: **Find** the maximum value in right
13: **Calculate** correct N
14: **end if**
15: **return** N

3.2 Correction of Abnormal Number of Small Packages

Due to the fast movement of small packages on the assembly line, the pictures taken by the depth camera are blurry, so it is easy to identify multiple objects as one. Therefor, it is necessary to correct the number of misjudgments so as not to affect the follow-up operation of the robot. Since the height of the depth camera and the platform where the small packages are placed is relatively fixed, the width of the detected object is similar or slightly different in normal circumstances. Therefore, the empirical object width value can be set. If the detected width is within the normal range, the reliability is high. If the detected width is much larger than the normal range, the reliability of results is low, the misjudgment occurs. It is necessary to correct the number of objects by recalculating where the misjudgment of the number occurs based on the empirical object width value. In the YOLO program, the pixel coordinate information of the upper left corner and lower right corner of each bounding box can be obtained, we create two empty lists, store the x coordinate information of the upper left corner and lower right corner of each bounding box, and find the minimum value and maximum value, divided by the normal value of the width to correct the count. The specific formula is:

$$n = \frac{x_{max} - x_{min}}{w_n} \tag{10}$$

where n represents the corrected number of objects, x_{max} represents the maximum value of the x-coordinate information in the lower right corner, x_{min} is the minimum value of the x-coordinate information in the upper left corner, w_n represents the normal value of the width information. After correction, the closing degree of the gripper can be controlled according to the corrected number.

When the number of small packages is recognized normally, we take the average of small packages' center pixel coordinates as the coordinate of multiple small packages. For example, the pixel coordinate of the first small package is (x_1, y_1) from left to right, the pixel coordinate of the second small package is (x_2, y_2), and the pixel coordinates of the nth small package are (x_n, y_n). So the coordinates that should be taken are as follows:

$$x = \frac{\sum_1^n x}{n} \tag{11}$$

$$y = \frac{\sum_1^n y}{n} \tag{12}$$

where n represents the number of identified small packages. When the number of objects is misjudged, the coordinate information obtained by using the above two formulas is not the midpoint in most cases, so the formulas need to be modified. The coordinates to be adopted are as follows:

$$x = \frac{x_{max1} + x_{min1}}{2} \tag{13}$$

$$y = \frac{y_{max} + y_{min}}{2} \tag{14}$$

where x_{max1} represents the maximum value of the x-coordinate of the bounding box, x_{min1} is the minimum value of the x-coordinate of the bounding box, y_{max} represents the maximum value of the y-coordinate of the bounding box, and y_{min} is the minimum value of the y-coordinate of the bounding box.

4 Experiments

Figure 5 shows our hardware equipment, including realsenseD435 camera, four-axis parallel robot arm and conveyor belt. The camera is placed at the starting position of the conveyor belt. It uses an infrared laser transmitter and an infrared camera to capture depth images. This camera can provide high-resolution (1280×720) depth images with a depth accuracy of ± 1 mm, which is suitable for the current work. The four-axis parallel robotic arm is a type of mechanical arm that consists of four parallel arms connected to a common base. The size of robotic arm used in the experiment is $670 * 650 * 830$ mm, its repeat positioning accuracy is ± 0.4 mm, and the speed beat is 60 times per minute. We use RS485 communication to control the robotic arm. The conveyor can be adjusted to ten different speeds, the fastest and slowest movement speed is 49.26 cm/s and 3.11 cm/s respectively.

Based on the method proposed above, the experiment is mainly divided into two parts. The first part is the abnormal detection of snack placement, the second part is the detection and correction of abnormal quantity.

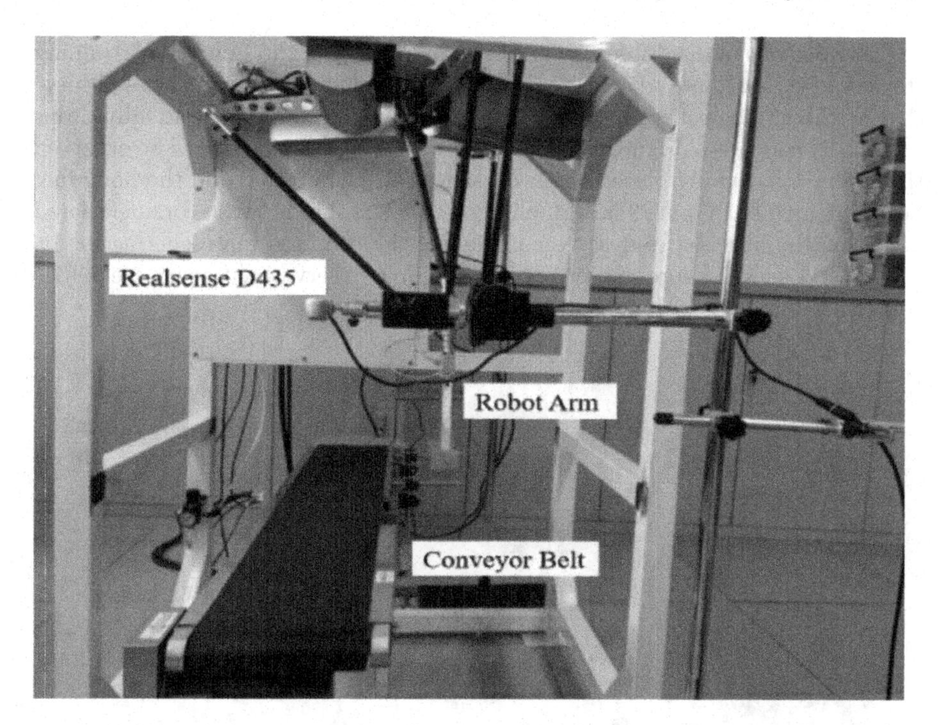

Fig. 5. The hardware equipment used in the experiment

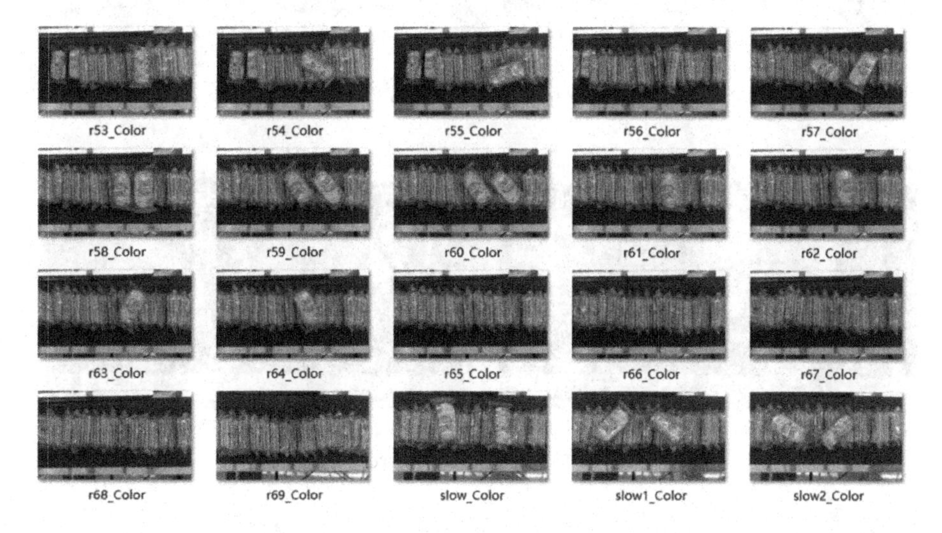

Fig. 6. Dataset collected by ourselves

For the abnormal detection of snacks, we applied improved YOLOv5 to train model and tested model on the dataset collected by ourselves. Part of the dataset are shown in Fig. 6. Figure 7 shows the relationship between the loss function and the number of training steps. The abscissa represents the number of training steps, and the ordinate represents the value of the loss function. We can observe that as the number of training steps increase, the loss function gradually drops and finally converges. Figure 8 is the result of object detection, it can be observed that every abnormally placed small packages can be detected and the bounding boxes are angular. Figure 9 is a graph of the relationship between confidence and accuracy, the abscissa represents confidence, the ordinate represents accuracy, snack0 and snack1 represent two categories. We observe accuracy increases as confidence increases in Fig. 10.

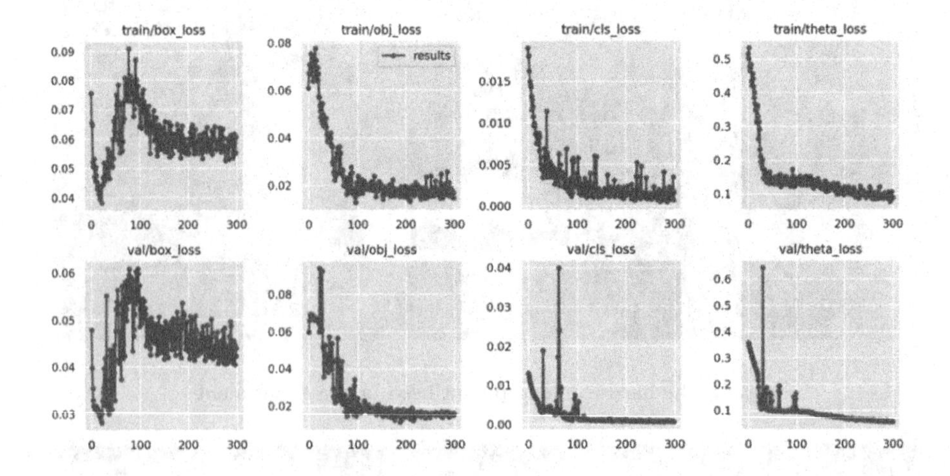

Fig. 7. Loss vs. training steps

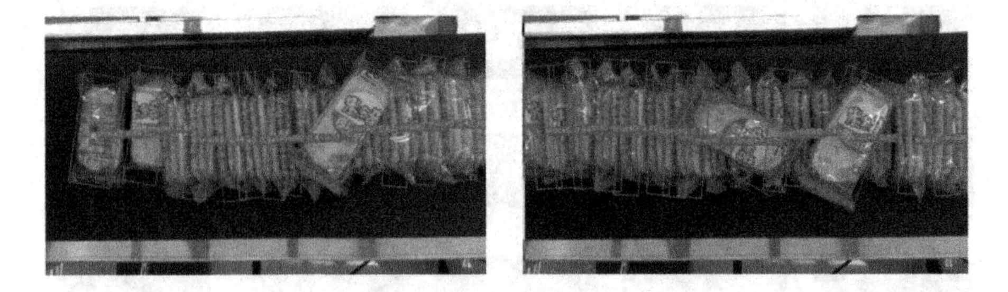

Fig. 8. Object detection results

For correcting the number of moving object, the normal value of width information can be set as 80 pixels in the experiment. We tested our algorithm and the results are shown in Fig. 10. It can be observed that there are several misjudgments of the small packages' number when the conveyor belt moves fast, and our algorithm based on above can accurately correct it.

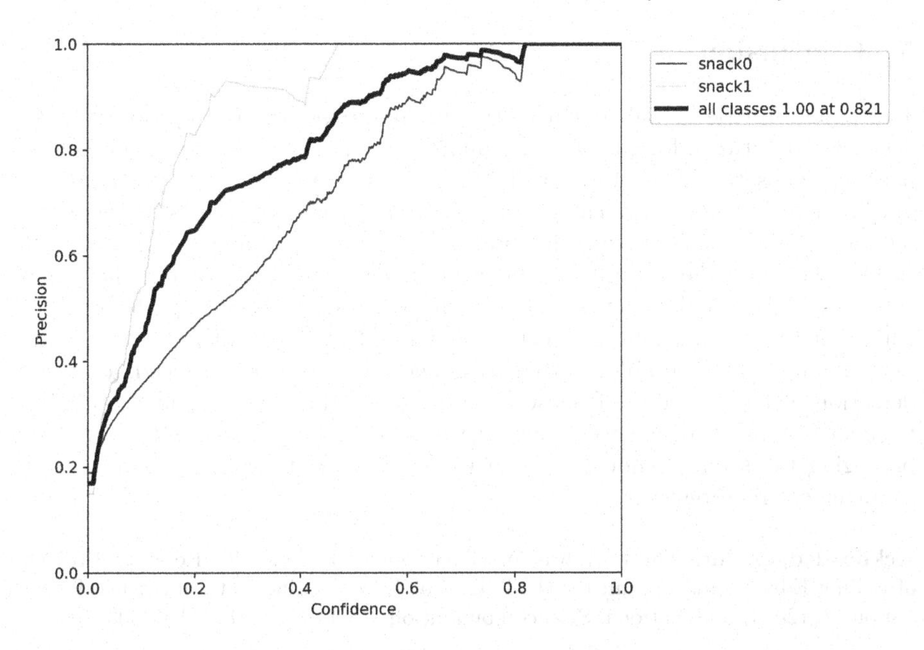

Fig. 9. The graph of the relationship between confidence and accuracy

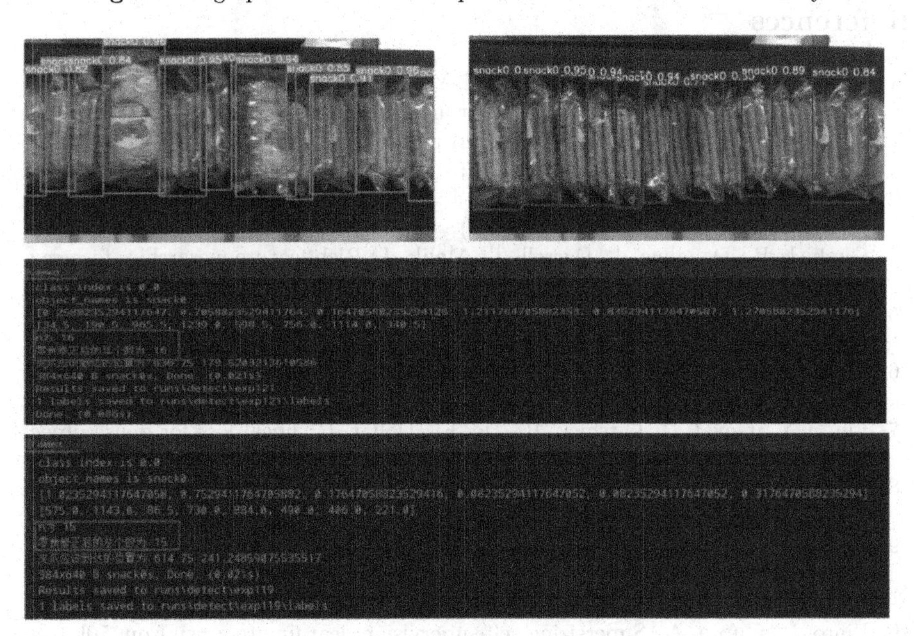

Fig. 10. Correction results for the number of small packages

5 Conclusion

This paper provides a real-time detection method based on the improved YOLO deep neural network for the accurate quantity and abnormal situation of small packages densely arranged vertically on the assembly line. The method can quickly detect the position, category, quantity and angle of objects. By combining the depth information with the rotation angle of the bounding box, abnormal snacks placed on the assembly line can be detected in time, so that the robot can remove abnormal objects in time. Then the wrong number of small packages which caused by image blur are corrected by rectifying the width of objects with low reliability. The Method has the advantages of fast detection speed, accurate detection of the quantity of densely arranged small packages, and the ability to identify abnormal objects, which is very important for robots to complete packaging tasks autonomously in real time according to various quantity and arrangement requirements.

Acknowledgement. The work was jointly supported by Scientific Research Project of Beijing Educational Committee (KM202110005023), Beijing Natural Science Foundation (4212033) and National Science Foundation of China (62273012, 62003010).

References

1. Bochkovskiy, A., Wang, C.-Y., Liao, H.-Y.M.: YOLOv4: optimal speed and accuracy of object detection. arXiv preprint arXiv:2004.10934 (2020)
2. Chu, F.-J., Xu, R., Vela, P.: Real-world multiobject, multigrasp detection. IEEE Robot. Autom. Lett. **3**, 3355–3362 (2018)
3. Ge, Z., Liu, S., Wang, F., Li, Z., Sun, J.: YOLOX: exceeding yolo series in 2021. arXiv preprint arXiv:2107.08430 (2021)
4. Girshick, R., Donahue, J., Darrell, T., Malik, J.: Rich feature hierarchies for accurate object detection and semantic segmentation. IEEE Computer Society (2014)
5. Glenn, J.: YOLOv5 release v6.1 (2022). https://github.com/ultralytics/yolov5/releases/tag/v6.1
6. Lenz, I., Lee, H., Saxena, A.: Deep learning for detecting robotic grasps. Int. J. Robot. Res. **34**(4–5), 705–724 (2015)
7. Lin, T.Y., Goyal, P., Girshick, R., He, K., Dollár, P.: Focal loss for dense object detection. arXiv e-prints (2017)
8. Mahler, J., et al.: Dex-Net 2.0: deep learning to plan robust grasps with synthetic point clouds and analytic grasp metrics. arXiv preprint arXiv:1703.09312 (2017)
9. Nguyen, A., Kanoulas, D., Caldwell, D.G., Tsagarakis, N.G.: Detecting object affordances with convolutional neural networks. In: 2016 IEEE/RSJ International Conference on Intelligent Robots and Systems (IROS), pp. 2765–2770. IEEE (2016)
10. Pinto, L., Gupta, A.: Supersizing self-supervision: learning to grasp from 50k tries and 700 robot hours. In: 2016 IEEE International Conference on Robotics and Automation (ICRA), pp. 3406–3413. IEEE (2016)
11. Qiao, S., Chen, L.C., Yuille, A.: Detectors: Detecting objects with recursive feature pyramid and switchable atrous convolution. arXiv (2020)
12. Redmon, J., Divvala, S., Girshick, R., Farhadi, A.: You only look once: unified, real-time object detection. In: Computer Vision & Pattern Recognition (2016)

13. Redmon, J., Farhadi, A.: YOLO9000: better, faster, stronger. In: IEEE Conference on Computer Vision & Pattern Recognition, pp. 6517–6525 (2017)
14. Redmon, J., Farhadi, A.: YOLOv3: an incremental improvement. arXiv e-prints (2018)
15. Saxena, A., Driemeyer, J., Ng, A.Y.: Robotic grasping of novel objects using vision. Int. J. Robot. Res. **27**(2), 157–173 (2008)
16. Tian, Z., Shen, C., Chen, H., He, T.: FCOS: fully convolutional one-stage object detection. In: Proceedings of the IEEE/CVF International Conference on Computer Vision, pp. 9627–9636 (2019)
17. Tian, Z., Shen, C., Chen, H., He, T.: FCOS: a simple and strong anchor-free object detector. IEEE Trans. Pattern Anal. Mach. Intell. **44**(4), 1922–1933 (2022)
18. Wang, C.Y., Bochkovskiy, A., Liao, H.Y.M.: Scaled-YOLOv4: scaling cross stage partial network. In: Computer Vision and Pattern Recognition (2021)
19. Wang, C.-Y., Yeh, I.-H., Liao, H.-Y.M.: You only learn one representation: unified network for multiple tasks. arXiv preprint arXiv:2105.04206 (2021)
20. Liu, W., et al.: SSD: single shot multibox detector. In: Leibe, B., Matas, J., Sebe, N., Welling, M. (eds.) ECCV 2016. LNCS, vol. 9905, pp. 21–37. Springer, Cham (2016). https://doi.org/10.1007/978-3-319-46448-0_2
21. Xu, S., et al.: PP-YOLOE: an evolved version of YOLO. arXiv preprint arXiv:2203.16250 (2022)
22. Yan, X., et al.: Data-efficient learning for sim-to-real robotic grasping using deep point cloud prediction networks (2019)

Robust License Plate Recognition Based on Pre-training Segmentation Model

Yanzhen Liao[1], Hanqing Yang[1], Ce Feng[2], Ruhai Jiang[3], Jingjing Wang[4], Feifan Huang[1], and Hongbo Gao[1,5](✉)

[1] University of Science and Technology of China, Hefei, China
liaoyanzhen@mail.ustc.edu.cn
[2] China Electronics Technology Group Co. Ltd., Beijing, China
[3] Hefei CAS Juche Technology Co. Ltd., Hefei, China
[4] Diankeyun Technology Co. Ltd., Beijing, China
[5] Institute of Advanced Technology of USTC, Hefei, China

Abstract. This work addresses the important task of recognizing license plates in outdoor environments, where many existing methods fail to achieve usable results under conditions such as rotation, deformation, extreme darkness, or brightness. We propose a license plate recognition method based on a pre-trained segmentation model, which has strong robustness. The algorithm consists of a license plate detection model and a license plate recognition model. The license plate detection model uses the pre-trained oCLIP-resnet model as the network backbone, which can obtain some text information during image encoding. In addition, we use a dataset synthesis tool to generate various license plate forms, which greatly increases the robustness of the recognition model. Our method achieves state-of-the-art performance on three public datasets, CCPD, ALOP, and PKUData, especially on difficult test samples, demonstrating the stability and robustness of the model.

Keywords: License plate recognition · Pre-training model · Image segmentation

1 Introduction

ALPR (Automatic License Plate Recognition) has a very wide range of applications in vehicle application scenarios, playing a role in vehicle identification, such as automatic gate machines at vehicle entrances and exits.

This work was supported in part by the National Natural Science Foundation of China (Grant Nos. U20A20225), in part by Anhui Province Natural Science Funds for Distinguished Young Scholar (Grant No. 2308085J02), in part by the Scientific and Technological Innovation 2030 - "New Generation Artificial Intelligence" Major Project (Grant 2022ZD0116305), in part by Innovation Leading Talent of Anhui Province TeZhi plan, in part by the Natural Science Foundation of Hefei, China (Grant No. 202321), and in part by the CAAI-Huawei Mind Spore Open Fund (Grant No. CAAIXSJLJJ-2022-011A).

F. Sun and J. Li (Eds.): ICCCS 2023, CCIS 2029, pp. 74–86, 2024.
https://doi.org/10.1007/978-981-97-0885-7_7

ALPR systems are deployed in various environments depending on their intended use, but the vast majority are used in outdoor settings. Outdoor environments pose significant challenges to the capture of license plates due to factors such as darkness, glare, obstruction, rain, snow, tilt, and blurriness. Detecting and recognizing license plates under constantly changing environmental and weather conditions is a challenging task. However, most existing algorithms can only perform well under testing conditions and have limitations in real-world applications.

In previous studies, license plate detection and recognition were often treated as separate tasks and solved using different methods. However, these two tasks are highly interrelated. Accurate detection of bounding boxes can improve recognition accuracy, while recognition results can be used to eliminate false positives.

Existing license plate recognition methods can be classified into two categories: traditional methods and deep learning-based methods. Traditional methods use hand-crafted features to detect or recognize license plates, but they are sensitive to interference. In contrast, deep learning-based methods can automatically learn lots of features from data, making them more robust. With the rapid development of computer vision technology, researchers typically divide the license plate recognition process into two stages: treating license plate detection as an object detection process and license plate recognition as an optical character recognition (OCR) process. In this work, we present a robust framework for license plate recognition that utilizes a pre-training segmentation model.

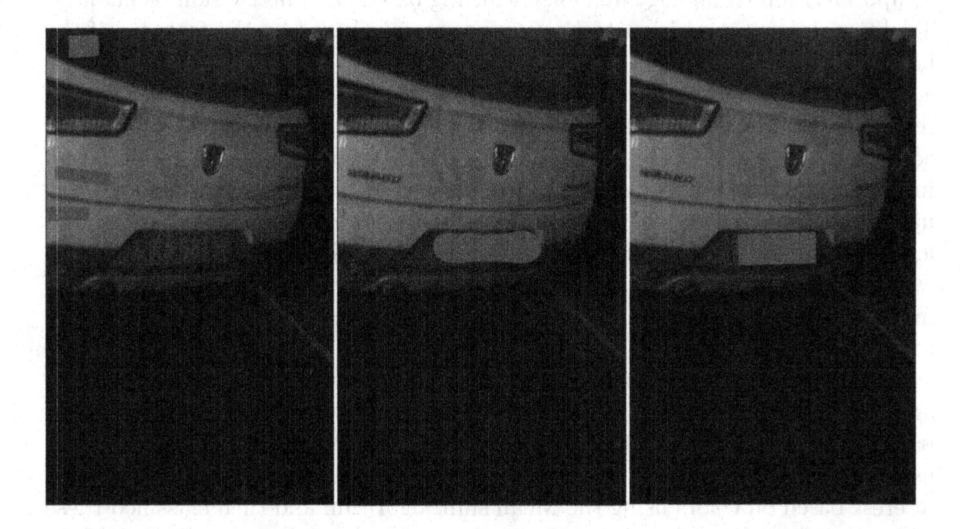

Fig. 1. The results of different methods. The image on the left shows the detection result of Mask R-CNN, which fails to produce correct results in low-light conditions. The image in the middle shows the detection result of TextSnake after fine-tuning, but some parts of the text edges are not included in the detection results. The image on the right shows our detection result, which accurately captures the license plate information.

Our framework is specifically designed to detect and recognize license plates in complex scenarios. The main contributions of this work are as follows (Fig. 1):

- A segmentation model based on visual text pre-training model is proposed for complex scene license plate detection.
- A license plate synthesis algorithm suitable for various scenarios has been proposed, including adding shadows, glare or darkness, perspective transformation, etc. By using this engine, license plate images with small data deviation can be generated, thereby obtaining models with good generalization ability.
- The proposed method is evaluated experimentally on AOLP, Media Lab, CCPD and PKUdata datasets, and the results show that our method is superior to the previous advanced methods.

2 Related Work

This section will briefly introduce the relevant work of license plate detection, license plate recognition, and visual language pre-training models.

2.1 License Plate Detection

License plate detection technology can be divided into two categories: traditional computer vision technology and deep learning based computer vision technology. Traditional computer vision techniques are mainly based on the features of the license plate such as shape, colour, texture, etc.

Shape-based approaches are commonly used to detect license plates, with the rectangular frame of the plate being a crucial feature. For instance, [1] utilized edge clustering to solve plate detection for the first time. Similarly, [2] employed image enhancement and Sobel operator to extract the vertical edges of a car image, followed by an effective algorithm to remove most of the background and noise edges. Finally, they searched for the plate region by using a rectangle window in the residual edge image and segmented the plate from the original car image. However, one of the major drawbacks of this approach is its susceptibility to noise.

Color-based methods have been developed to detect license plates by leveraging the fact that the color of a license plate is often distinct from that of the background. For instance, [3] proposed a robust license plate localization method using mean shift segmentation. They segmented vehicle images into regions of interest based on colors using the Mean shift algorithm and then classified these regions using a Mahalanobis classifier to detect license plates. Another approach, presented in [4], utilized fuzzy logic. In this method, the HSV color space was employed to extract color features. However, these methods are very sensitive to illumination variation and are not robust.

Texture-based methods detect license plates based on the unusual distribution of pixels on the plate. For example, [5] utilized Gabor filter, threshold, and connected component labeling (CCL) algorithms to obtain the license plate region. Another approach, proposed in [6], introduced a new segmentation technique called Sliding Concentric Windows (SCW) for license plate detection. By leveraging the irregularities in image textures, this method can detect plate regions more quickly and accurately. Nonetheless, these methods involve complex calculations and do not work well with complex backgrounds and varying lighting conditions.

In recent years, the development of Deep learning based computer vision techniques has been rapid. As mentioned earlier, license plate detection is often treated as an object detection task, with the YOLO series models [7–9] being the most popular in this field. However, directly applying the YOLO detector to license plate recognition often yields unsatisfactory results. Therefore, many scholars have made improvements to the YOLO model for this specific task. In [10], the YOLOv2 model was trained and fine-tuned to achieve 93.53% accuracy on the SSIG dataset, which is a significant improvement over previous models. Similarly, [11] customized a YOLO model for license plate recognition by modifying the grid size and bounding box parameters. However, it should be noted that object detection models are not capable of handling curved or distorted text, which can limit their effectiveness in certain scenarios.

2.2 License Plate Recognition

Previous work on license plate recognition usually involved character segmentation in the license plate followed by Optical Character Recognition (OCR) techniques for character recognition. However, character segmentation is a challenging task that can be influenced by uneven lighting, shadows, and noise in the image, which has an immediate impact on plate recognition. Even with a strong recognizer, if segmentation is improper, the plate cannot be recognized correctly. With the development of deep neural networks, approaches have been proposed to recognize the entire license plate directly without character separation. CRNN [12] is the most well-known method in scene text recognition and is currently widely used as a baseline. CRNN is a neural network structure that combines feature extraction, sequence modeling, and transcription. The current common practice is to introduce attention mechanisms in this task. ASTER [13] is an end-to-end neural network model that includes a rectification network and a recognition network. The rectification network adaptively transforms the input image into a new image and corrects the text in it. The recognition network is an attention-based sequence-to-sequence model that directly predicts character sequences from the corrected image.

[14] is a self-attention-based scene text recognizer that encodes input-output attention and learns self-attention to encode feature-feature and target-target relationships within the encoder and decoder. In addition, some work focuses on recognizing arbitrarily shaped text [15], including curved or rotated text. SVTR [16] abandons the sequence model and first decomposes the image text into feature components. Then, it repeatedly performs hierarchical stages through component-level mixing, merging, and/or combination. The design of global and

local mixing blocks perceives inter-character and intra-character patterns, resulting in the perception of multi-granularity character components.

2.3 Vision-Language Pre-training

Inspired by advanced Transformer-based pre-training techniques in the Natural Language Processing (NLP) community [17], many vision-language pre-training methods have been studied in recent years. These methods have greatly promoted the development of many multi-modal tasks in the computer vision community.

VilBERT [18] extends the popular BERT architecture to a multimodal two-stream model that handles visual and textual inputs separately through a shared attentional transformation layer. Model was pre-trained on two proxy tasks on a large automatically collected concept caption dataset and then transferred it to multiple established visual and language tasks - visual question answering, visual commonsense reasoning, referring expression, and image retrieval based on captions - with minimal additions to the underlying architecture. LXMERT [19] builds a large-scale transformer model and pre-trains it on a large number of image and sentence pairs through five different representative pre-training tasks: masked language modeling, masked object prediction (feature regression and label classification), cross-modal matching, and image question answering, learning relationships within and across modalities.

CLIP [20] collected 400 million image-text pairs without human annotation for model pre-training, demonstrating the potential for learning transferable knowledge and open-set visual concepts. Different neurons in the CLIP model can capture corresponding concepts literally, symbolically, and conceptually. For each text, the loss calculation is similar to each image. In this way, CLIP can be used for zero-shot image recognition. And a new method called TCM [21] is proposed, which focuses on using the CLIP model directly for text detection without a pre-training process.

3 Proposed Method

In this part, we first introduce the whole pipeline of the proposed license plate recognition model. Our advanced overview of the model is shown in Fig. 2. Then, the backbone network is introduced in detail, and how the algorithm detects and packages the license plate, and finally passes it to the recognition model for license plate recognition.

3.1 CLIP-ResNet Backbone

The CLIP model [20] has been popular for the past two years and has shown surprising performance in the image-text space. However, there are two prerequisites to leveraging the relevant information from this model: 1) an appropriate method to effectively request prior knowledge from CLIP, and 2) the original

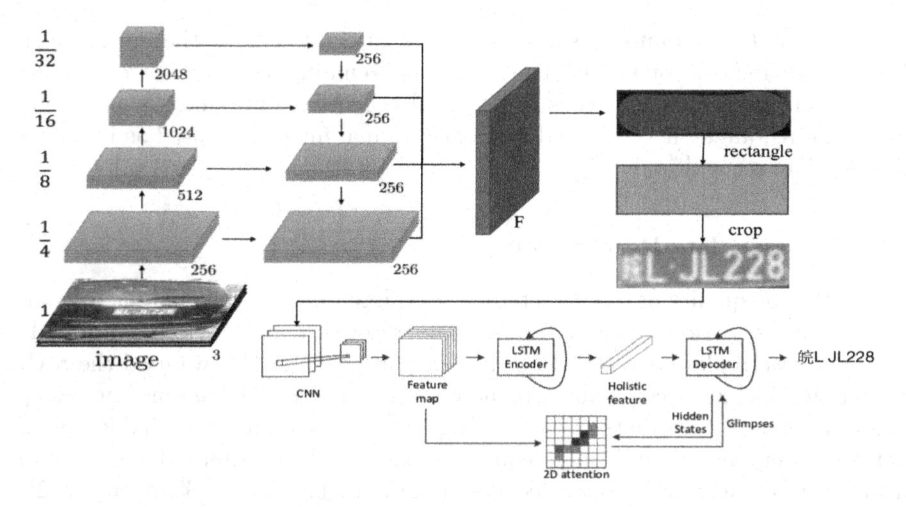

Fig. 2. The flowchart of license plate recognition system. It includes two parts: license plate detection and license plate recognition (Color figure online)

model can only measure similarity between one image and one word or sentence. For license plate detection, there are usually many text instances per image that need to be recalled equivalently.

Xue et al. [22] proposed a weakly supervised pre-training method that jointly learns and calibrates visual and partial textual information to learn effective visual-textual representations for scene text detection. oCLIP has three elements: an Image Encoder, a Character-Aware Text Encoder and a Visual-Textual Decoder. Given an input image, the image encoder (including the backbone with a subsequent multi-head attention layer) first extracts visual features. At the same time, each character in each text instance is converted into a character embedding, and the character-aware text encoder further extracts text instance embeddings from the character embeddings. The visual-textual decoder simulates the interaction between the text instance embeddings and the corresponding image embeddings. During training, one random character in each text instance is masked (highlighted with a red box), and the entire network is optimized by predicting the masked character.

3.2 License Plate Detection and Unwarping

For license plate detection, we chose to use an instance segmentation framework. We first use a ResNet-50 backbone network pre-trained with oCLIP and adopt an FPN architecture to extract multi-level features. We connect low-level texture features with high-level semantic features, and these feature maps are further fused in F to encode information with various receptive fields. Intuitively, this fusion is likely to help generate kernels of various scales. The feature map F is then projected to produce segmentation results.

However, the obtained results cannot be directly sent to the downstream license plate recognition model. The minimum bounding rectangle corresponding to the segmentation result needs to be calculated. After obtaining the minimum bounding rectangle, it is cropped from the original image and used as input for license plate recognition.

3.3 License Plate Recognition

Currently, the quality of the detected license plate image is the main factor that constrains the performance of license plate recognition. Therefore, we mainly focus on license plate detection. For license plate recognition, we used the SAR model suitable for recognizing irregular texts. The strong baseline for recognizing irregular scene texts is easy to implement, uses readily available neural network components, and only requires word-level annotations. It consists of a 31-layer ResNet, an encoder-decoder framework based on LSTM, and a 2D attention module. Although this method is simple, it has robustness and achieves state-of-the-art performance on both regular and irregular scene text recognition benchmarks.

4 Experiments

4.1 Datasets

CCPD Dataset. The CCPD dataset [23] is a large and diverse open-source dataset of Chinese city license plates that is extensively annotated. The dataset was primarily collected from parking lots in Hefei city and is divided into two versions based on release date: CCPD2019 and CCPD2020 (CCPD-Green). The CCPD2019 dataset only includes ordinary license plates (blue plates) and consists of 200,000 training images and 140,000 testing images. The CCPD2020 dataset only includes new energy license plates (green plates) and contains 10,000 images. Each image contains only one license plate, with a resolution of 720 (width) × 1160 (height) × 3 (channels).

AOLP Dataset. The AOLP dataset [1] comprises 2049 images of Taiwanese license plates that can be classified into three subsets based on shooting conditions: access control (AC), traffic law enforcement (LE), and road patrol (RP). AC represents the simplest case where a vehicle passes through a fixed channel at low speed or comes to a complete stop. These images were captured under varying lighting and weather conditions. LE, on the other hand, captures vehicles that violate traffic regulations and are captured by roadside cameras. The images often contain cluttered backgrounds, with road signs and multiple plates in one image. Finally, RP captures images from any angle or distance using cameras mounted on patrol vehicles.

PKU Dataset. The PKU dataset [24] includes 3977 Chinese license plate images from various scenes. The PKUdata dataset is divided into five subsets based on detection difficulty. The original PKUdata only provides bounding box annotations for license plate detection.

SynthLicense Dataset. We have developed a license plate generator based on Chinese license plate types to create the Synthlicense dataset. Using a provided corpus and font, we generate label text and paste it onto the background of the license plate style. We then apply various random transformations, such as adjusting size, brightness, and skew angle.

4.2 Implementation Details

We jointly trained license plate detection and recognition models on the CCPD dataset. For license plate detection, we pre-trained oCLIP on ICDAR2015 and fine-tuned it on CCPD2019 and CCPD2020. The test dataset included CCPD test set, AOLP and PKUdata datasets. For license plate recognition, we pre-trained on synthlicense and ICDAR2015 datasets and fine-tuned on CCPD2019, CCPD2020, and PKUdata datasets. The test dataset included CCPD2019 and CCPD2020 datasets. In order to improve accuracy, we stretched all images to (2240, 2240). The learning rate was set to 1e-04. All experiments were conducted on 8 NVIDIA Tesla V100 GPUs with 32GB of memory each.

4.3 Evaluation Metrics

We used Intersection over Union (IoU) to determine whether the model correctly detected the text region and calculated precision (P), recall (R), and F-measure (F) based on conventional practices for comparison.

$$\mathrm{IoU} = \frac{\mathrm{Area}(R_{\mathrm{det}} \cap R_{\mathrm{gt}})}{\mathrm{Area}(R_{\mathrm{det}} \cup R_{\mathrm{gt}})} \tag{1}$$

$$\text{F-measure} = \frac{2 \times (\mathrm{Precision} \times \mathrm{Recall})}{(\mathrm{Precision} + \mathrm{Recall})} \tag{2}$$

4.4 License Plate Detection Results

For the LPD task, we evaluated our method on the CCPD, AOLP and PKUdata. To ensure consistency in evaluation metrics, we set the IOU threshold to 0.7, which means we consider a detection correct only when the IOU between the ground truth and annotation is greater than 0.7. Based on the comparison we conducted between two models in the license plate detection task, as illustrated in Fig. 3, we present the smooth F1-score curves of our model and the PSENET model. It can be observed that our model outperforms PSENET in most recall rates. Additionally, we notice that our model achieves its peak F1-score and hmean score at a recall rate of 0.7. Our model exhibits better overall performance

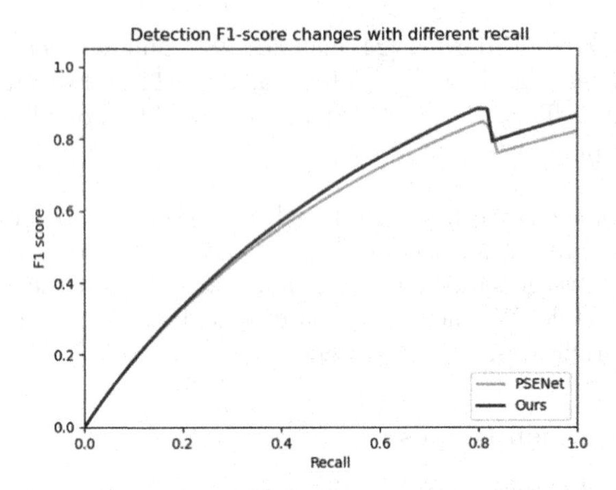

Fig. 3. The results of Detection F1-score changes with different recall

in both high and low recall rates. Therefore, we conclude that our model is superior to PSENET in terms of license plate detection.

It should be noted that although many datasets provide four-point annotations for license plates, such operations can cause trouble when calculating IOU. Considering that the original license plates are rectangular, we projected them as rectangles during actual computation.

Some detection results are shown in Fig. 4, indicating that our method can locate license plates accurately while ignoring the background. The experiment verifies that this model is capable of detecting and recognizing car license plates under various shooting conditions, such as daytime and nighttime, clear and snowy weather.

Tables 1, 2 and 3 demonstrate the results of our method and competing methods on different subsets of CCPD, AOLP, and PKUDATA datasets.

As shown in Table 1, our model exhibits the best performance on all test subsets of the CCPD dataset. Particularly, our model demonstrates strong robustness compared to other methods, especially on the weather subset.

Table 1. LPD Accuracy for CCPD

Method	Subset		
	DB	Rot	Weath
RPNet [23]	89.5	94.7	84.1
TE2E [25]	91.7	95.1	83.6
Ours	**92.4**	**95.4**	**88.7**

Fig. 4. The results of a car license plate detection and recognition instance based on a pre-trained segmentation model. The first row of images shows CCPD_blur, while the second row depicts CCPD_rotate.

As shown in Table 2, our model also shows the best performance on all test subsets of the AOLP. Even though the test environments vary greatly between subsets, our model handles this well.

Table 2. LPD Accuracy for AOLP

Method	Subset		
	AC	LE	RP
OpenALPR	92.4	86.1	93.1
HPA [26]	98.3	98.0	98.2
Ours	**99.6**	**99.2**	**99.4**

As shown in Table 3, PKUDATA has clear images, so various models perform well in its testing. However, on the most challenging G5 subset, our model still wins due to its robust performance.

4.5 License Plate Recognition Results

We evaluated our method on the CCPD2019 and CCPD2020 datasets for the LPR task. To overcome various lighting changes and blurriness, we created the

Table 3. LPD Accuracy for PKUDATA

Method	Subset					
	G1	G2	G3	G4	G5	Average
TE2E [25]	**99.9**	**99.7**	99.5	**99.8**	98.7	99.5
Baseline [24]	98.8	98.4	97.8	96.2	97.3	97.7
Ours	99.6	99.6	**99.7**	99.6	**99.4**	**99.6**

synthlicense dataset by setting different characters and license plate colors for the same license plate text. Compared to converting all images to grayscale before inference, using color images results in an average accuracy increase of 2.7%. During testing, for images with height greater than width after correction, we rotate the images 90° clockwise and counterclockwise and recognize them together with the original images.

Regarding the results on CCPD, Table 4 shows that our algorithm has achieved near-optimal performance on the DB, Rot, and Weather subsets.

Table 4. LPR Accuracy for CCPD

Method	Subset		
	DB	Rot	Weath
RPNet [23]	96.9	90.8	87.9
CLPD-baseline [27]	98.8	96.4	**98.5**
Ours	**99.2**	**98.4**	98.3

5 Conclusion and Future Work

With the transfer of transformer models to various tasks, target detection has received some attention. Recently, some transformer-based end-to-end OCR methods have achieved good performance on certain datasets. Intuitively, end-to-end OCR does not accumulate errors and has theoretically the highest accuracy, which should be the main direction of future research.

In addition, there is a trend of zero-shot learning in academia. Can zero-shot learning be achieved in license plate recognition? For example, can we recognize a type of license plate that has never been seen before in our country by only learning license plates from other countries? This performance is exciting, but we must not forget that currently, the annotation data for OCR is far less than that for image classification. Therefore, for a long time, how to improve the quality and quantity of datasets remains a challenge. Perhaps data generation tools can greatly alleviate this problem.

References

1. Hsu, G.-S., Chen, J.-C., Chung, Y.-Z.: Application-oriented license plate recognition. IEEE Trans. Veh. Technol. **62**(2), 552–561 (2013)
2. Luo, L., Sun, H., Zhou, W., Luo, L.: An efficient method of license plate location. In: 2009 First International Conference on Information Science and Engineering, pp. 770–773 (2009)
3. Jia, W., Zhang, H., He, X., Piccardi, M.: Mean shift for accurate license plate localization. In: Proceedings 2005 IEEE Intelligent Transportation Systems, pp. 566–571 (2005)
4. Wang, F., Man, L., Wang, B., Xiao, Y., Pan, W., Xiaochun, L.: Fuzzy-based algorithm for color recognition of license plates. Pattern Recogn. Lett. **29**(7), 1007–1020 (2008)
5. Caner, H., Gecim, H.S., Alkar, A.Z.: Efficient embedded neural-network-based license plate recognition system. IEEE Trans. Veh. Technol. **57**(5), 2675–2683 (2008)
6. Anagnostopoulos, C.N.E., Anagnostopoulos, I.E., Loumos, V., Kayafas, E.: A license plate-recognition algorithm for intelligent transportation system applications. IEEE Trans. Intell. Transp. Syst. **7**(3), 377–392 (2006)
7. Redmon, J., Divvala, S., Girshick, R., Farhadi, A.: You only look once: unified, real-time object detection. In: Proceedings of the IEEE Conference on Computer Vision and Pattern Recognition, pp. 779–788 (2016)
8. Redmon, J., Farhadi, A.: YOLO9000: better, faster, stronger. In: Proceedings of the IEEE Conference on Computer Vision and Pattern Recognition, pp. 7263–7271 (2017)
9. Li, C., et al.: YOLOv6: a single-stage object detection framework for industrial applications. arXiv preprint arXiv:2209.02976 (2022)
10. Laroca, R., et al.: A robust real-time automatic license plate recognition based on the YOLO detector. In: 2018 International Joint Conference on Neural Networks (IJCNN), pp. 1–10. IEEE (2018)
11. Hsu, G.-S., Ambikapathi, A.M., Chung, S.-L., Su, C.-P.: Robust license plate detection in the wild. In: 2017 14th IEEE International Conference on Advanced Video and Signal Based Surveillance (AVSS), pp. 1–6 (2017)
12. Shi, B., Bai, X., Yao, C.: An end-to-end trainable neural network for image-based sequence recognition and its application to scene text recognition. IEEE Trans. Pattern Anal. Mach. Intell. **39**(11), 2298–2304 (2016)
13. Shi, B., Yang, M., Wang, X., Lyu, P., Yao, C., Bai, X.: ASTER: an attentional scene text recognizer with flexible rectification. IEEE Trans. Pattern Anal. Mach. Intell. **41**(9), 2035–2048 (2019)
14. Ning, L., et al.: MASTER: multi-aspect non-local network for scene text recognition. Pattern Recogn. **117**, 107980 (2021)
15. Lee, J., Park, S., Baek, J., Oh, S.J., Kim, S., Lee, H.: On recognizing texts of arbitrary shapes with 2D self-attention. In: Proceedings of the IEEE/CVF Conference on Computer Vision and Pattern Recognition Workshops, pp. 546–547 (2020)
16. Du, Y., et al.: SVTR: scene text recognition with a single visual model. arXiv preprint arXiv:2205.00159 (2022)
17. Devlin, J., Chang, M.-W., Lee, K., Toutanova, K.: BERT: pre-training of deep bidirectional transformers for language understanding. arXiv preprint arXiv:1810.04805 (2018)

18. Lu, J., Batra, D., Parikh, D., Lee, S.: ViLBERT: pretraining task-agnostic visiolinguistic representations for vision-and-language tasks. In: Advances in Neural Information Processing Systems, vol. 32 (2019)
19. Tan, H., Bansal, M.: LXMERT: learning cross-modality encoder representations from transformers. arXiv preprint arXiv:1908.07490 (2019)
20. Radford, A., et al.: Learning transferable visual models from natural language supervision. In: International Conference on Machine Learning, pp. 8748–8763. PMLR (2021)
21. Yu, W., Liu, Y., Hua, W., Jiang, D., Ren, B., Bai, X.: Turning a clip model into a scene text detector. In: Proceedings of the IEEE/CVF Conference on Computer Vision and Pattern Recognition, pp. 6978–6988 (2023)
22. Xue, C., Hao, Y., Lu, S., Torr, P., Bai, S.: Language matters: a weakly supervised pre-training approach for scene text detection and spotting. arXiv preprint arXiv:2203.03911 (2022)
23. Xu, Z., et al.: Towards end-to-end license plate detection and recognition: a large dataset and baseline. In: Proceedings of the European Conference on Computer Vision (ECCV), pp. 255–271 (2018)
24. Yuan, Y., Zou, W., Zhao, Y., Wang, X., Xuefeng, H., Komodakis, N.: A robust and efficient approach to license plate detection. IEEE Trans. Image Process. **26**(3), 1102–1114 (2017)
25. Li, H., Wang, P., Shen, C.: Toward end-to-end car license plate detection and recognition with deep neural networks. IEEE Trans. Intell. Transp. Syst. **20**(3), 1126–1136 (2019)
26. Zhang, Y., Wang, Z., Zhuang, J.: Efficient license plate recognition via holistic position attention. In: Proceedings of the AAAI Conference on Artificial Intelligence, vol. 35, pp. 3438–3446 (2021)
27. Zhang, L., Wang, P., Li, H., Li, Z., Shen, C., Zhang, Y.: A robust attentional framework for license plate recognition in the wild. IEEE Trans. Intell. Transp. Syst. **22**(11), 6967–6976 (2021)

Predicting Cell Line-Specific Synergistic Drug Combinations Through Siamese Network with Attention Mechanism

Xin Bao[2], XiangYong Chen[2], JianLong Qiu[2], Donglin Wang[2], Xuewu Qian[3], and JianQiang Sun[1](✉)

[1] School of Information Science and Engineering, Linyi University, Linyi 276005, China
sjqyjs@sina.com
[2] School of Automation and Electrical Engineering, Linyi University, Linyi 276005, China
[3] Jinan Key Laboratory of 5G+ Advanced Control Technology, Jinan 250103, China

Abstract. Drug combination therapy holds great promise for curing cancer patients, as it can significantly enhance treatment effectiveness and reduce drug toxicity. However, efficiently identifying synergistic drug combinations from a vast pool of potential drug combinations remains challenging, mainly due to the expensive and time-consuming nature of traditional experimental methods. In this work, we introduce a novel deep learning method called SNAADC for discovering synergistic drug combinations. In SNAADC, we employ Siamese Network and attention mechanism to extract more effective drug features for better performance. The experimental results demonstrate that SNAADC outperforms traditional and popular methods.

Keywords: Combination therapy · Graph neural network · Deep learning · Siamese network architecture

1 Introduction

Drug combination therapy has become a prevalent treatment approach for complex human diseases [1, 2]. However, it is crucial to be mindful of the potential for adverse reactions (antagonism), which can arise from the combined pharmacological effects of these drugs [3]. While traditional wet lab is a reliable detection, conducting large-scale drug screening using the method can be both expensive and time-consuming [4]. In contrast, preliminary identification of drug combinations through computer can effectively narrow the search range for synergistic drug combinations, leading to cost reduction and improved efficiency [5]. In the last few years, computer assisted drug development has become a hot topic.

One major limitation in drug synergy modeling is the scarcity of available dataset. However, recent advancements in High-Throughput Screening (HTS) have significantly mitigated this challenge [6]. For instance, O'Neil et al. proposed an extensive collection of over 20,000 pairwise drug combinations, while DrugCombDB has amassed more than 6,000,000 quantitative drug dose responses [7, 8]. Furthermore, resources such

F. Sun and J. Li (Eds.): ICCCS 2023, CCIS 2029, pp. 87–95, 2024.
https://doi.org/10.1007/978-981-97-0885-7_8

as Genomics of Drug Sensitivity in Cancer (GDSC), Cancer Cell Line Encyclopedia (CCLE), Cancer Therapeutics Response Portal (CTRP) and NCI-60 provide comprehensive datasets on various cancer cell lines [9]. The publication of large-scale drug datasets and cancer cell line data has further accelerated the development of prediction algorithms. In specific studies, in 2014, Rahul et al. employed pharmacological screening data generated from Genomics of Drug Sensitivity in Cancer to develop quantitative structure activity relationship (QSAR) models [10]. These models were then employed to predict promiscuous inhibitors against 16 pancreatic cancer cell lines. In 2019, Pavle et al. proposed the utilization of Random Forest (RF) and Extreme Gradient Boosting (XGBoost) to predict synergistic combinations [11].

In addition to data, the advancements in computer machines have provided hardware support for the application of algorithms in synergistic drug combinations discovery. Especially, deep learning algorithms have developed rapidly, with their complex model structures and ability to handle large-scale data, surpassing the limitations of traditional machine learning methods. In 2018, Preuer proposed a deep learning model called DeepSynergy, which mapped drug structures and gene expression of cancer cell to a single the synergy score [12]. The performance of DeepSynergy outperformed traditional machine learning methods on the same dataset. In the same year, Ding et al. introduced the Ensemble Prediction framework of Synergistic Drug Combinations (CSS), a novel approach to improve accuracy [13]. The CSS framework integrated information from multiple-sources, including biological, chemical, pharmacological, and network knowledge, providing more opportunities for screening synergistic drug combinations. Besides, other deep learning methods such as the convolutional neural network (CNN) [14, 15], recurrent neural network (RNN) and Transformer have been successfully applied in screening synergistic drug combinations [16–19].

Recently, graph neural networks (GNNs) have gained significant attention in the field of biopharmaceutics due to their practicality in analyzing drug molecular structures [20]. By leveraging the graph structure of molecules, GNNs can effectively capture and learn the spatial and chemical relationships between atoms and their local environments. This enables them to extract meaningful features and representations from the molecular structure, which can be used for tasks such as drug property prediction, and drug-drug interaction analysis [21]. In 2018, Zitnik et al. developed a multimodal graph framework, known as Decagon, which effectively incorporated various types of biological relationships to predict drug combination of the exact side effect [22]. In 2020, Chen et al. introduced a novel end-to-end graph representation learning model known as GCN-BMP for the prediction of drug-drug interactions [23]. In the same year, Lin et al. presented KGNN, an end-to-end framework-based knowledge graph that effectively encodes multiple biological entities, including drugs, proteins, and genes, to construct a comprehensive knowledge graph. KGNN can capture the relationships between drugs and their potential neighborhoods, providing a robust solution for drug-drug interaction prediction [24]. In 2022, Zhang et al. developed a novel encoder-decoder relational graph convolutional network called SDCNet [25]. SDCNet learned and fused drug feature with multi-type cell line-specific features and employed attention to integrate drug feature, for identifying synergistic drug combinations.

Although many published studies have demonstrated improved performance, they still face certain limitations. Specifically, in terms of aggregating drug molecule features, many existing models fail to adequately explore the available molecule feature information, thereby hindering drug feature of overall effectiveness. Furthermore, in the fusion of features between drug pairs and cancer cells, a common practice among most models is to simply concatenate the features of the drug pair and cancer cell. However, this fusion method fails to fully capture the intricate interactions and dependencies that exist between these features. Therefore, we present a novel model named SNAADC (Siamese Network with Attention Mechanism for Anticancer Drug Combinations Prediction). In SNAADC, the Siamese network takes two drugs as input to extract their respective drug features. In the Siamese network, we develop a novel attention-based method for aggregating drug molecule features. Then the output results from the Siamese network are fused with cancer cell features using an element-wise addition function. Finally, the fused feature is fed into a classifier to predict the synergy of the drug combination. In this work, we use 5-fold-cross-validation to assess the performance of our model. The results demonstrate the promising potential of SNAADC compared to other published feature extraction methods.

2 Materials and Methods

2.1 Materials

In this work, we evaluate the performance of SNAADC using a large-scale cancer screening dataset published by O'Neil et al. [7]. In order to enable the model to better learn and predict synergy, and to effectively capture drug combinations that strong synergistic effects or pronounced antagonistic effects, we select data with Loewe scores greater than 10 or less than 0. There are a total of 12415 samples in the processed dataset.

2.2 Methods

The Overview of SNAADC is shown in Fig. 1. In SNAADC, the original data of drugs and cancer cells are processed and fed into the Siamese network to extract the drug feature vector. Next, the drug feature vector and cancer cell feature vector are input to predictor to predict the synergy value.

Cancer Cell Feature Extraction Module

According to the previous study [17], the cell gene expression data is utilized as the representation for the cancer cell vector. To facilitate the subsequent calculation between drugs and cancer cells, a multi-layer perceptron (MLP) is employed to balance the dimensions of the cell feature vectors.

$$X_C = MLP(X_{C-original}) \quad (1)$$

where X_C is the balanced cancer cell feature vector.

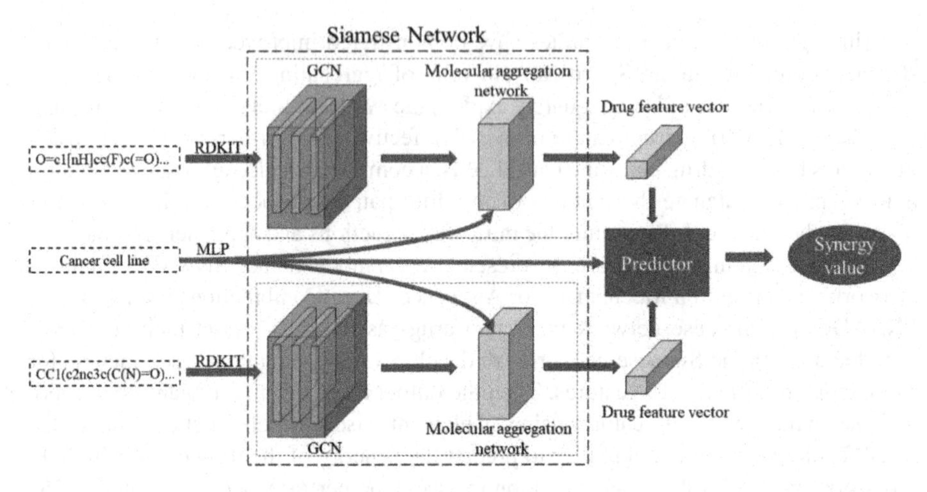

Fig. 1. Overview of SNAADC

Weight-Sharing-Based Drug Feature Embedding

The Siamese network is employed to extract features of two drugs. In this network, the weights are shared between the two drugs, leading the model to effectively capture the relationship and similarities between drug pairs.

The Siamese network is constructed two subnetworks: a multi-layer GCN and a molecular aggregation network based on attention mechanism. When two drug data is input into the Siamese network, the SMILES of two drugs are encoded as graph data, which including the adjacency matrix and atom feature matrix, using the RDKIT tool. Subsequently, the graph data is fed into multi-layer GCN, which iteratively update the atom feature vectors by aggregating information from neighboring atoms. After multi-layer GCN, we obtain the atom feature matrixes for two drugs, represented as X_A and X_B, respectively. Next, the feature matrixes of two drugs, along with cancer cell matrix, are passed into the molecular aggregation network. The molecular aggregation network employs an attention mechanism to effectively aggregate the atom features into a drug feature, culminating in a more comprehensive representation of the drug feature. In contrast to traditional self-attention mechanisms, SNAADC utilizes a mutual attention approach. In this approach, the cancer cell features are treated as the Q (Query), and the drug features are treated as the K (Key) and V (Value). The formula for calculating the drug feature vector for drug A and drug B are as follows:

$$Attn_A = \text{Softmax}(\frac{(W_Q X_C)(W_K X_A)^T}{\sqrt{d}})(W_V X_A) \quad (2)$$

$$Attn_B = \text{Softmax}(\frac{(W_Q X_C)(W_K X_B)^T}{\sqrt{d}})(W_V X_B) \quad (3)$$

where W_Q, W_K, W_V are the learning shared weight, d is the atom vector length. $Attn_A$ and $Attn_B$ represent the drug feature vector of drug A and drug B, respectively.

Predictor

After obtaining the representation vectors for the two drugs, we perform an element-wise addition of these two vectors to obtain the drug-drug combination vector X_{A-B}. Subsequently, we normalize both the drug-drug combination vector and the cancer cell feature vector. The normalized drug-drug combination vector is then concatenated with the normalized cancer cell feature vector, resulting in a new feature vector that captures the interactions between the drug combination and the cancer cells X_{A-B-C}. Finally, this new feature vector is fed into a MLP to predict drug combination synergy. The prediction process can be summarized by the following formulas:

$$X_{A-B} = Attn_A \oplus Attn_B \tag{4}$$

$$X_{A-B-C} = Concat(Norm(X_{A-B}), Norm(X_C)) \tag{5}$$

$$P = \sigma(MLP(X_{A-B-C})) \tag{6}$$

where $\oplus$ is element-wise addition function, and σ is softmax function.

3 Hyperparameter Setting

(See Table 1).

Table 1. The hyperparameter settings

Hyperparameter	Values
Optimizer	Adam
The number of input features of each atom in drug	78
Epoch	1000
Batch size	512
Learning rate	0.0005
Dropout	0.2
GCN layer	3
MLP layer for processing cancer cell line	3
MLP layer for classification	3

4 Results

4.1 Evaluation Metrics

In this work, synergistic drug combinations discovery is defined as a binary classification problem (i.e., the synergistic drug combination is 1 and the antagonistic drug combination is 0). Therefore, we use classify metrics such as ROC AUC, PR AUC, accuracy (ACC), precision (PREC), True Positive Rate (TPR) and f1-score (F1) to evaluate the performance of SNAADC. The formulas of these metrics mentioned above are as follows:

$$ACC = \frac{TP + TN}{TP + FN + FP + TN} \tag{7}$$

$$Precision = \frac{TP}{TP + FP} \tag{8}$$

$$TPR = \frac{TP}{TP + FN} \tag{9}$$

$$F1 = \frac{2 \times TP}{2 \times TP + FP + FN} \tag{10}$$

In the above formulas, TP and TN represent the number of synergistic and antagonistic drug combinations that are correctly predicted, respectively. FP and FN indicate the number of synergistic and antagonistic combinations that are wrongly predicted, respectively.

4.2 Comparison Results

We compared SNAADC with the state-of-the-art event prediction methods, namely DeepSynergy [12]. We also consider several other popular classification methods, including XGBoost, Random Forest, GBM, Adaboost, MLP and SVM. All these methods are subjected to 5-fold cross-validation using the same dataset. Through this rigorous comparison, we aim to demonstrate the superior performance of SNAADC in predicting drug combination synergy. The 5-fold mean of all methods is presented in the Table 2.

As a result, SNAADC achieved the best scores in terms of the ACC, PREC and TPR, which values of 0.94, 0.86, 0.87, 0.85, and 0.86, respectively. The results show that SNAADC can effectively identify synergistic drug combinations. In comprehensive metrics of ROC AUC, PR AUC and F1, SNAADC also achieves the best scores. Especially, SNAADC achieved a ROC AUC of 0.93, which is 5%, 1%, 7%, 8%, 10%, 28% and 35% higher than that of DeepSynergy, XGBoost, RF, GMB, AdaBoost, MLP and SVM, respectively. This indicates that SNAADC's overall performance surpasses that of other methods. In conclusion, the six performance metrics represent various aspects of the model's performance, and SNAADC demonstrates superior results across all of them.

4.3 Predicting New Drug Combinations

To further demonstrate the efficacy of the proposed SNAADC, we conducted tests on a new dataset to assess the generalization of the model. This new dataset, provided by AstraZeneca [27], comprises 57 drugs and 24 cell lines, all of which are distinct from the drugs and cells present in the training dataset. We utilized the train dataset to train all models, and subsequently, the new dataset was employed to test these models. The results are displayed in Table 3.

Table 2. Comparison results on dataset

Method	ROC AUC	PR AUC	ACC	PREC	TPR	F1
SNAADC	**0.93 ± 0.01**	**0.93 ± 0.01**	**0.86 ± 0.01**	**0.86 ± 0.01**	**0.85 ± 0.02**	**0.85 ± 0.02**
DeepSynergy	0.88 ± 0.01	0.87 ± 0.01	0.80 ± 0.01	0.81 ± 0.01	0.75 ± 0.01	0.69 ± 0.01
XGBoost	0.92 ± 0.01	0.92 ± 0.01	0.83 ± 0.01	0.83 ± 0.01	0.84 ± 0.01	0.83 ± 0.01
Random Forest	0.86 ± 0.02	0.85 ± 0.02	0.77 ± 0.01	0.77 ± 0.02	0.74 ± 0.01	0.75 ± 0.01
GBM	0.85 ± 0.02	0.85 ± 0.01	0.76 ± 0.02	0.77 ± 0.01	0.74 ± 0.01	0.75 ± 0.01
Adaboost	0.83 ± 0.01	0.83 ± 0.03	0.74 ± 0.01	0.74 ± 0.02	0.72 ± 0.01	0.72 ± 0.01
MLP	0.65 ± 0.02	0.63 ± 0.05	0.56 ± 0.06	0.54 ± 0.01	0.53 ± 0.22	0.53 ± 0.01
SVM	0.58 ± 0.01	0.56 ± 0.02	0.54 ± 0.01	0.54 ± 0.06	0.51 ± 0.12	0.52 ± 0.08

Table 3 demonstrates that SNAADC outperforms other models. Although the precision (PREC) value of SNAADC is 0.74, which is 1% lower than the highest value (Deep-Synergy), the F1 value of SNAADC is 27% higher than that of DeepSynergy. In addition, SNAADC achieves higher values in other metrics as well, with values of 0.65, 0.82, 0.67, 0.82 and 0.78, respectively. Notably, the RP AUC of SNAADC is 11%, 9%, 6%, 11%, 13%, 9% and 11% higher than that of DeepSynergy, XGBoost, RF, GMB, AdaBoost, MLP and SVM, respectively. These results demonstrate the superior generalization of SNAADC in predicting new drug combination synergy.

Table 3. Comparison results on new drug combinations

Method	ROC AUC	PR AUC	ACC	PREC	TPR	F1
SNAADC	**0.65 ± 0.02**	**0.82 ± 0.01**	**0.67 ± 0.02**	0.74 ± 0.02	**0.82 ± 0.07**	**0.78 ± 0.03**
DeepSynergy	0.55 ± 0.15	0.71 ± 0.13	0.47 ± 0.14	**0.75 ± 0.14**	0.39 ± 0.17	0.51 ± 0.2
XGBoost	0.52 ± 0.11	0.73 ± 0.12	0.49 ± 0.15	0.71 ± 0.09	0.38 ± 0.17	0.49 ± 0.1
Random Forest	0.53 ± 0.14	0.76 ± 0.16	0.50 ± 0.14	0.75 ± 0.14	0.49 ± 0.14	0.59 ± 0.1
GBM	0.51 ± 0.10	0.71 ± 0.09	0.45 ± 0.12	0.69 ± 0.14	0.43 ± 0.12	0.52 ± 0.1
Adaboost	0.49 ± 0.09	0.69 ± 0.14	0.47 ± 0.17	0.69 ± 0.14	0.46 ± 0.15	0.55 ± 0.1
MLP	0.53 ± 0.13	0.74 ± 0.12	0.53 ± 0.15	0.74 ± 0.13	0.53 ± 0.13	0.61 ± 0.1
SVM	0.47 ± 0.11	0.71 ± 0.13	0.47 ± 0.13	0.70 ± 0.13	0.63 ± 0.11	0.66 ± 0.1

5 Conclusion

In this work, we proposed SNAADC, a method based on a Siamese network with attention mechanism for predicting the synergy of drug combinations. Compared to other population methods, SNAADC achieve the best comprehensive performance. The ideal predictive ability of SNAADC mainly depends on the following factors. On one hand, the design architecture of SNAADC incorporates a Siamese network, which allows the model to share the same weights for both drugs. This shared weight scheme enables the model to be fully trained and effectively learn the relationships and similarities between drug pairs, enhancing its predictive capabilities for drug combination synergy [28]. On the other hand, the attention is used to aggregate the atom feature vector into drug feature vector. By focusing on atomic features that are more relevant to cancer cells, SNAADC can effectively extract a more effective drug feature representation.

References

1. Mokhtari, R.B., Homayouni, T.S., Baluch, N., et al.: Combination therapy in combating cancer. Oncotarget **8**(23), 38022 (2017)
2. Boyle, E.A., Li, Y.I., Pritchard, J.K.: An expanded view of complex traits: from polygenic to omnigenic. Cell **169**(7), 1177–1186 (2017)
3. Hecht, J.R., Mitchell, E., Chidiac, T., et al.: A randomized phase IIIB trial of chemotherapy, bevacizumab, and panitumumab compared with chemotherapy and bevacizumab alone for metastatic colorectal cancer. J. Clin. Oncol. **27**(5), 672–680 (2009)
4. Lewis, B.J., DeVita, V.T., Jr.: Combination chemotherapy of acute leukemia and lymphoma. Pharmacol. Ther. **7**(1), 91–121 (1979)
5. Ter-Levonian, A.S., Koshechkin, K.A.: Review of machine learning technologies and neural networks in drug synergy combination pharmacological research. Res. Results Pharmacol. **6**(3), 27–32 (2020)
6. Lehár, J., Krueger, A.S., Avery, W., et al.: Synergistic drug combinations tend to improve therapeutically relevant selectivity. Nat. Biotechnol. **27**(7), 659–666 (2009)
7. O'Neil, J., Benita, Y., Feldman, I., et al.: An unbiased oncology compound screen to identify novel combination strategies. Mol. Cancer Ther. **15**(6), 1155–1162 (2016)
8. Zheng, S., Aldahdooh, J., Shadbahr, T., et al.: DrugComb update: a more comprehensive drug sensitivity data repository and analysis portal. Nucleic Acids Res. **49**(W1), W174–W184 (2021)
9. Baptista, D., Ferreira, P.G., Rocha, M.: Deep learning for drug response prediction in cancer. Brief. Bioinform. **22**(1), 360–379 (2021)
10. Kumar, R., Chaudhary, K., Singla, D., et al.: Designing of promiscuous inhibitors against pancreatic cancer cell lines. Sci. Rep. **4**(1), 4668 (2014)
11. Sidorov, P., Naulaerts, S., Ariey-Bonnet, J., et al.: Predicting synergism of cancer drug combinations using NCI-ALMANAC data. Front. Chem. **7**, 509 (2019)
12. Preuer, K., Lewis, R.P., Hochreiter, S., et al.: DeepSynergy: predicting anti-cancer drug synergy with Deep Learning. Bioinformatics **34**(9), 1538–1546 (2018)
13. Ding, P., Yin, R., Luo, J., et al.: Ensemble prediction of synergistic drug combinations incorporating biological, chemical, pharmacological, and network knowledge. IEEE J. Biomed. **23**(3), 1336–1345 (2018)
14. Li, T.-H., Wang, C.-C., Zhang, L., et al.: SNRMPACDC: computational model focused on Siamese network and random matrix projection for anticancer synergistic drug combination prediction. Brief. Bioinform. **24**(1), bba503 (2023)

15. Zhang, C., Lu, Y., Zang, T.: CNN-DDI: a learning-based method for predicting drug–drug interactions using convolution neural networks. BMC Bioinform. **23**(1), 1–11 (2022)
16. Santiso, S., Perez, A., Casillas, A., et al.: Exploring joint AB-LSTM with embedded lemmas for adverse drug reaction discovery. IEEE J. Biomed. **23**(5), 2148–2155 (2018)
17. Liu, Q., Xie, L.: TranSynergy: mechanism-driven interpretable deep neural network for the synergistic prediction and pathway deconvolution of drug combinations. PLoS Comput. Biol. **17**(2), e1008653 (2021)
18. Cocos, A., Fiks, A.G., Masino, A.J.: Deep learning for pharmacovigilance: recurrent neural network architectures for labeling adverse drug reactions in Twitter posts. J. Am. Med. Inform. Assoc. **24**(4), 813–821 (2017)
19. Lin, S., Wang, Y., Zhang, L., et al.: MDF-SA-DDI: predicting drug–drug interaction events based on multi-source drug fusion, multi-source feature fusion and transformer self-attention mechanism. Brief. Bioinform. **23**(1), bbab421 (2022)
20. Al_Rabeah, M.H., Lakizadeh, A.: GNN-DDI: a new data integration framework for predicting drug-drug interaction events based on graph neural networks. BMC Bioinformat. (2022)
21. Han, K., Lakshminarayanan, B., Liu, J.: Reliable graph neural networks for drug discovery under distributional shift. arXiv preprint arXiv:2111.12951 (2021)
22. Zitnik, M., Agrawal, M., Leskovec, J.: Modeling polypharmacy side effects with graph convolutional networks. Bioinformatics **34**(13), i457–i466 (2018)
23. Chen, X., Liu, X., Wu, J.: GCN-BMP: investigating graph representation learning for DDI prediction task. Methods **179**, 47–54 (2020)
24. Lin, X., Quan, Z., Wang, Z.-J., et al.: KGNN: knowledge graph neural network for drug-drug interaction prediction. In: IJCAI 2020, pp. 2739–2745 (2020)
25. Zhang, P., Tu, S., Zhang, W., et al.: Predicting cell line-specific synergistic drug combinations through a relational graph convolutional network with attention mechanism. Brief. Bioinform. **23**(6), bbac403 (2022)
26. Landrum, G.: RDKit: a software suite for cheminformatics, computational chemistry, and predictive modeling. Greg Landrum **8**, 31 (2013)
27. Menden, M.P., Wang, D., Mason, M.J., et al.: Community assessment to advance computational prediction of cancer drug combinations in a pharmacogenomic screen. Nat. Commun. **10**(1), 1–17 (2019)
28. Zhang, X., Wang, G., Meng, X., et al.: Molormer: a lightweight self-attention-based method focused on spatial structure of molecular graph for drug–drug interactions prediction. Brief. Bioinform. **23**(5), bbac296 (2022)

Pressure Pain Recognition for Lower Limb Exoskeleton Robot with Physiological Signals

Yue Ma[1,2], Xinyu Wu[1,2,3(✉)], Xiangyang Wang[1,2], Jinke Li[1,2], Pengjie Qin[1,2], Meng Yin[1,2], Wujing Cao[1,2], and Zhengkun Yi[1,2]

[1] Guangdong Provincial Key Lab of Robotics and Intelligent System, Shenzhen Institute of Advanced Technology, Chinese Academy of Sciences, Shenzhen, Guangdong, China

[2] SIAT-CUHK Joint Laboratory of Robotics and Intelligent Systems, Shenzhen, Guangdong, China

xy.wu@siat.ac.cn

[3] Shandong Institute of Advanced Technology, Chinese Academy of Sciences, Jining, Shandong, China

Abstract. Pain is a feedback mechanism the body uses to protect itself. The perception of human pain is still missing in lower limb exoskeleton robots, such as the pain caused by the pressure between the knee joint and the baffle when the lower limb exoskeleton stands up, or the pain caused by the mismatch between the gait and the body function of the wearer when walking. Aiming at the pain perception problem of the exoskeleton robot, this study designed a pressure pain data acquisition paradigm combined first considering the movement mode of the exoskeleton robot, and collected facial electromyography and electrocardiography signals, as well as five levels of pain. Then a deep learning neural Network based on Temporal Convolutional Network and Long Short-Term Memory is designed. The composite Loss of Center Loss, InfoNEC Loss and Softmax Loss is designed. The designed deep neural network model was evaluated using the thermal pain dataset BioVid, and the average recognition rate using only electrocardiography signals reached 86.44%. By Using the fusion signals of electrocardiography, electrocardiogram and electromyogram signals, the average recognition accuracy can reach 87.16%, which is superior to other similar methods, and verifies the effectiveness of the proposed method. The proposed network is also used to verify the collected pressure pain dataset, and the recognition accuracy can reach 87.5%, which proves the validity of the collected data set.

Keywords: Pain Recognition · Deep Learning Method · TCN · LSTM

1 Introduction

Acute pain is an unpleasant sensory and emotional experience, which helps to identify harmful situations and to avoid tissue damage.

F. Sun and J. Li (Eds.): ICCCS 2023, CCIS 2029, pp. 96–106, 2024.
https://doi.org/10.1007/978-981-97-0885-7_9

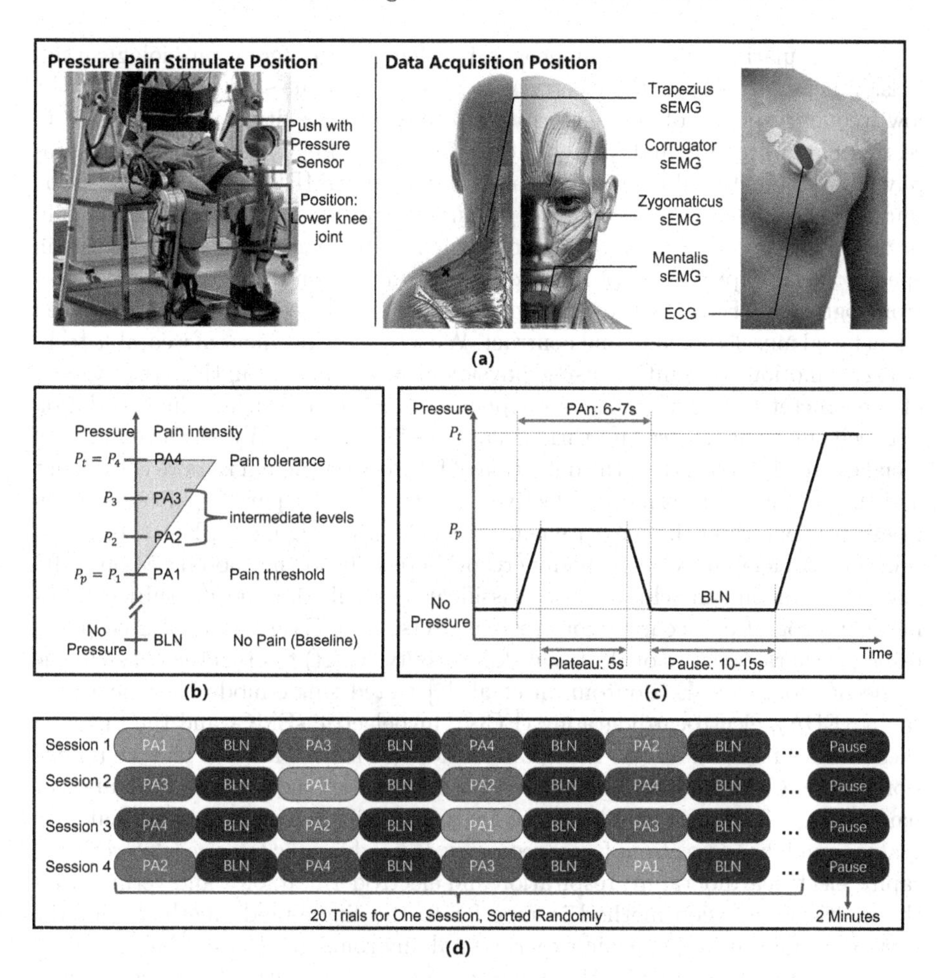

Fig. 1. The pressure pain acquisition paradigm. a) the pressure pain stimulation location and data acquisition location; b) the pain grading rules; c) data acquisition methods for a single trial; d) data collection rules for individual subjects.

Therefore, acute pain is a necessary sensory mechanism for human to protect their own safety.

However, for people with impaired motor and sensory abilities, such as paraplegics, not only cannot control the movement of their limbs, but also cannot feel pain to protect themselves.

With the development of intelligent rehabilitation assist devices, many types of robotic devices such as rehabilitation exoskeletons have been used to assist patients walking [1,2], but there are still few studies on the perception of patients' pain by this kind of robots. Therefore, how to effectively and stably evaluate pain can provide a necessary theoretical basis for designing the next generation of human-machine deeply integrated service robots.

Pain regulation will result in measurable changes in different physiological signals. When pain occurs, the electrical conductivity of the skin will increase, the low-frequency power of the heart rate will increase, resulting in an increase in resting blood pressure and pupil dilation. Correspondingly, electroencephalography (EEG), functional magnetic resonance imaging (fMRI) and functional near-infrared spectroscopy (fNIRS) also produce corresponding changes when pain occurs. Pain also produces a range of behavioral responses, such as changes in specific facial expressions, protective reflexes, and sounds. In turn, facial electromyography (sEMG) senses and amplifies the small electrical pulses produced when facial muscles expand and contract. With the advancement of wearable technology, emotion recognition using physiological signals strengthens the theory of the concretization of emotional expression, and also confirms the correlation between emotional states and various patterns [3]. Compared with the method of facial expression detection, the detection of bioelectrical signal is more convenient and the measurement is more objective, so a considerable part of pain evaluation research is based on the measurement of bioelectrical signal. A. Kelati et al. [4] proposed A facial myoelectric pain perception recognition method based on EMG and KNN, which can achieve good classification results for specific subjects, but has low recognition accuracy for cross-subjects. H. T. Tran et al. [5] used phase response characteristics of electrical skin activity (GSR) to objectively assess the grade of toothache. E. Pouromran et al. [6] tested single-mode electrocardiography (EDA), electrocardiography (ECG), myoelectric sEMG, and the fusion of these modes to evaluate the heat pain grade, and found that the feature regression accuracy of the fusion multi-mode was higher, but the error was still about 1 pain grade. P. Thiam et al. [7] proposed a pain rating estimation method that integrates sound, head position, facial geometry, facial dynamic texture, electromyography, electrocardiography, respiration and electrodermatology, and verified that the multi-mode fusion method is superior to the single-mode method with the best performance in a specific experimental environment. K. N. Phan et al. [8] proposed an end-to-end neural network method using multi-level convolution to achieve pain level estimation using only electrodermal and electrocardiogram signals, which achieved better results than traditional methods using manual feature extraction. Although many methods have been proposed, the effectiveness of pain level recognition still needs to be improved.

The existing pain database consists of four main modes of stimulation: thermal, electrical, mechanical, and chemical. Thermal and electrical stimulation on the skin are the most commonly used types in pain recognition studies. Other stimuli include cold pressure tasks in which the arm or forearm is immersed in cold water, mechanical stimulation of the skin using an electronic handheld manometer, and ischemic stimulation of muscle pain using a tourniquet on the arm. Ramin Irani et al. [9] collected a dataset of shoulder mechanical tenderness for RGB, depth, and thermal facial images. S. Gruss et al. [10] created a thermal pain biologic potential database. They collected the myoelectric, skin conductance levels and ECG signals of 85 subjects under controlled thermal stimulation conditions, extracted 159 features from the mathematical grouping of amplitude, frequency, stationarity, entropy, linearity, variability and similarity and used SVM to classify them. Finally, a good recognition effect is obtained. E. K. Naeini et al. [11] specially collected a set of pain data set of electrical stimulation with ECG alone for heart rate variability (HRV) caused by pain, such as changes in time interval between heartbeats, and verified the feasibility of achieving pain grade estimation with ECG alone. For wearable robots, tenderness is the most common approach, but no data set of physiological electrical signals for pressure to stimulate pain has been seen.

Aiming at the problem that the accuracy of pain classification methods based on nerve electrical signals still needs to be improved, and the lack of physiological electrical signal data set for pain assessment of wearable robots such as exoskeletons, we first designed a collection paradigm of tenderness data set based on physiological electrical signals according to existing studies, and collected two types of data, namely sEMG and ECG. Then a deep learn-based tenderness classification algorithm is designed to improve the classification accuracy.

The main contributions of this work are as follows:

1) A pressure pain data acquisition paradigm and data acquisition system were designed, and a 5-level pressure pain data set containing 5 subjects was established. The effectiveness of the collected data set was verified by deep learning model.
2) A deep learning network model based on tCN+LSTM was designed, and a composite form of loss function combining multiple indicators was designed. The effect of the proposed model was verified through the BioVid PartA dataset. The recognition accuracy of level 0 pain vs level 4 pain was 87.16%, higher than that of similar studies.

The organization and structure of this paper are as follows: the Sect. 2 introduces the designed data collection paradigm and data set form; Sect. 3 introduces the data preprocessing method and the designed classification algorithm based on deep learning; Sect. 4 verifies the classification effect of the data set through experiments; Sect. 5 summarizes and discusses the research content and results of this paper.

2 Paradigm Design and Data Collection

This study aims to study the relationship between tenderness and bioelectrical signals. In order to ensure a reasonable data collection paradigm, the collection methods of similar data sets can be referred to. The BioVid heat pain dataset is one of the reference paradigms for pain datasets, which collects multi-modal physiological signal expression. It collected data from 90 people in three ages, using a camera and a bioelectrical signal sensor to record the pain induced by thermal stimulation of the right arm. In the image data acquisition part, three synchronous cameras were used to capture the image data of the participants' right, left and right sides, respectively. When the participants' heads were turned 45° to the left or right, the cameras on the side captured the front faces respectively, and the participants were explicitly allowed to move their heads freely. At the same time, a Microsoft Kinect camera was also used to obtain depth information. In terms of participants' bioelectrical information, SCL, ECG, EMG of three pain-related muscles, and EEG were recorded [12]. It can be seen that video data acquisition is relatively complicated, and observation information such as observation Angle and depth need to be taken into account, which is not suitable for wearable systems. Therefore, this study refers to the collection paradigm of BioVid dataset and only collects bioelectrical signal information.

2.1 Paradigm Design

The experimental paradigm design of this study is shown in Fig. 1. The rigid lower limb exoskeleton robot has multiple connections with the wearer to ensure that the exoskeleton can effectively support the wearer's limbs and its own weight. Especially for the paraplegia walking assist lower limb exoskeleton robot, as shown in the left figure of Fig. 1(a), since paraplegia patients have almost no active movement ability, when they stand up, a baffle under the knee joint is needed to install to ensure that the patient will not fall to the front. Therefore, the baffle will exert greater pressure on the wearer's lower leg during the standing up process. Paraplegic patients cannot feel pain and cannot protect themselves well. As can be seen from the picture, soft pink protective towels are added to the knee joint to prevent skin damage due to excessive pressure. In order to realize the objective perception of pain in this situation, this study applied pain stimulation under the knee joint. In terms of data collection, referring to the collection process of BioVid dataset, we mainly collect facial electromyography, shoulder electromyography and electrocardiogram signals. For facial electromyography, this study added the sEMG collection of mental muscle to capture facial pain expression changes in a more comprehensive way. In terms of ECG, in order to ensure the synchronization of signals, the ECG acquisition equipment is not directly used for acquisition, but the sEMG equipment is used instead. After testing, it is found that the sEMG equipment can also capture the ECG pulse well.

Before data collection, the pain thresholds of different subjects needs to be determined, which is generated by interpolating the two thresholds of pain perception and maximum pain tolerance. We will increase the pressure on the shank of subjects, and record a pressure level P_p when the subject feels pain, and record another pressure level P_t when the subject feels the maximum tolerance level. Two pressure values are then inserted evenly between each other to indicate the intermediate level of pain, as shown in Fig. 1(b). As shown in Fig. 1(c), the pressure value corresponding to any pain level will be applied for a total of 7–8 s, in which the time to reach the pain level is not less than 5 s, and then a short rest of 10–15 s will be carried out before the next random pain stimulus. It must not be possible to produce pain when it is not touched, so we here consider zero pressure to be equivalent to no contact.

The process of data collection was divided into 4 sessions, with 20 random pain level stimuli performed in each session. Since there were 4 levels, each level of stimulation was performed 5 times, and each trial was randomly distributed. Take a 2 min break after each session.

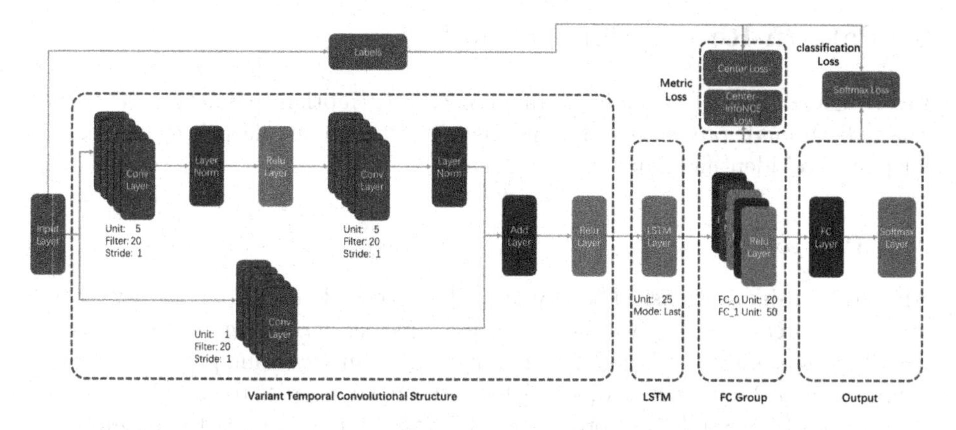

Fig. 2. The proposed deep learning network based on the TCN and LSTM.

2.2 Dataset Description

A total of 5 subjects' data were collected in this study (Age 31 ± 3, male, weight 68 ± 5 Kg). Among them, each subject experienced different degrees of pressure pain after 4 sessions of 20 trials in each group. 20 No pain trials were randomly selected by rest time of each subject and no pain trials are also added to the dataset, so there were 100 trials and data of 5 pain levels for each subject. Therefore, there are a total of 500 trial pain data in this dataset. The above pain perception experiment data collection experiment has been approved by the Ethics Committee, and the ethics number is SIAT-IRB-200715-H0512.

The PartA of the BioVid dataset is the primary pain assessment dataset used in the previous studies. The bioelectrical signals in this dataset include EDA (measured at index and ring fingers), ECG (two electrodes measured in the upper right and lower left of the body), and EMG (measured at trapezius muscle behind the shoulder). The database included data on 90 participants in three age groups of 30 subjects in each of 18–35, 36–50 and 51–65 years. Each age group was divided into 15 male and 15 female subjects on average. Due to technical problems in the recording process, data was missing for three of the subjects. As a result, the available data set contained data for only 87 subjects. The dataset is balanced for the classification task and contains both the original signal and the preprocessed signal. In this paper, only the pre-processed electrical signal is used for recognition. For the preprocessed signals, the dataset uses 20–250 Hz and 0.1–250 Hz Butterworth filters to filter the EMG and ECG signals, respectively. EMG signals are also filtered through empirical mode decomposition. The sequence length of each signal is 2816 sample points containing 5.5 s signal (sample rate is 512 Hz). For each subject, there were 20 random number of pain stimuli for each level. The dataset consisted of 87 subjects * 20 times * 5 levels = 8700 samples.

3 Pattern Recognition Method

Before using the self collected data, unnecessary motion artifacts need to be removed through pre-processing, and then input into the deep learning model for pain level identification.

3.1 Preprocess

Although subjects are asked to stay as still as possible in the laboratory, there is no guarantee that they will not move for some reason during the experiment. Relatively speaking, the frequency of expression changes during pain stimulation is higher than the frequency of restricted motion, so in order to contain as much information as possible, a 4-order Butterworth filter is used, with the filtering frequency set to 10–200 Hz for EMG and 0.1–200 Hz for ECG [13]. The pain perception data for each subject was segmented into data segments of the same length, the length of which was 7 s back from the label position to cover the time of the longest pain stimulus in the paradigm.

3.2 Deep Learning Method

The deep learning network structure designed in this paper is shown in Fig. 2. The network consists of four main components: a variant time domain Convolutional network (V-tCN), a short term memory network (LSTM), a set of fully connected layers (FC), and a Softmax output layer. The V-tCN mainly refers to the residual network structure model, and the main change is that the Relu layer of the series one-dimensional convolutional filter bank is placed outside the Addition layer. Then, a set of LSTM layers are used to extract the context association of the features after tCN filtering. The LSTM layer adopts the mode of the final output of the time series. Then, the features of the LSTM layer output are nonlinear transformed using a set of two-layer fully connected networks, and the pattern probabilities are identified using softmax output finally. In the part of Loss design, Center Loss is adopted to increase the tightness of features of the same category. The loss function L_{ct} is defined as follows:

$$\begin{aligned} L_{ct} &= \sum_{i=1}^{m} \|\boldsymbol{x}_i - \boldsymbol{c}_{y_i}\|_2^2, \\ \boldsymbol{c}_j^{t+1} &= \boldsymbol{c}_j^t + \alpha \cdot \Delta \boldsymbol{c}_j^t, \\ \Delta \boldsymbol{c}_j^t &= \frac{\sum_{i=1}^{m} \delta(y_i = j) \cdot (\boldsymbol{c}_j - \boldsymbol{x}_i)}{1 + \sum_{i=1}^{m} \delta(y_i = j)}, \end{aligned} \tag{1}$$

where, m is the number of samples in the batch, $\boldsymbol{x}_i$ is the sample feature of the network output, $\boldsymbol{c}_{y_i}$ is the central position of the features of different categories, α is the scalar parameter that adjusts the update distance of the central position, and $\delta(\cdot)$ is the 0–1 conditional function.

At the same time, the addition of contrast learning infoNEC loss is used to learn the contrast difference between different classes, which only needs to compare the center point of different classes with the mean point of the maximum and minimum features in the batch, which can effectively reduce the memory consumption and calculation speed (infoNEC loss requires the calculation of a batch correlation matrix), The loss function L_{cr} is defined as follows:

$$L_{cr} = \frac{\sum_{j=1}^{p} \sum_{i=1}^{q} L_{i,j}}{p \cdot q},$$
$$L_{i,j} = -log \frac{e^{sim(\boldsymbol{z}_i, \boldsymbol{z}_j)/\tau}}{\sum_{k=1}^{2 \cdot m} \delta(k \neq i) \cdot e^{sim(\boldsymbol{z}_i, \boldsymbol{z}_k)/\tau}}, \tag{2}$$
$$sim(\boldsymbol{u}, \boldsymbol{v}) = \frac{\boldsymbol{u} \cdot \boldsymbol{v}^T}{\|\boldsymbol{u}\| \cdot \|\boldsymbol{v}\|},$$

in which, m and $\delta(\cdot)$ are defined the same as before, $sim(\cdot, \cdot)$ is the cosine distance of the positive and negative samples, and τ is the temperature coefficient. i and j are the different characteristics of the comparison tasks, and p and q are the maximum number of samples of different types of tasks, respectively. The characteristic loss for positive and negative samples is $L_{i,j}$, and the final loss is the mean of all positive and negative sample pairings. As can be seen from the above formula, when the distance between positive sample pairs is smaller and the distance between negative samples is larger, the loss will be smaller.

Finally, we use softmax loss in the classification part of the output to obtain better classification accuracy. The loss function L_{cl} is defined as follows:

$$L_{cl} = -\sum_{k=1}^{m} y_k \cdot log(f(z_k)), \tag{3}$$

where, m is the same as the previous definition, y_k is the true value of the k sample, and $f(z_k)$ is the output of the k sample of the softmax layer. The final loss L is defined as:

$$L = L_{cl} + \beta \cdot (L_{ct} + L_{cr}). \tag{4}$$

4 Experiments

We designed two experiments to compare our proposed deep learning methods and validate our collected dataset. In the first experiment, the commonly employed BioVid PartA dataset was used to test the recognition accuracy. Similar to other research methods, we also used 86 subjects' data from 87 subjects as the training set, left 1 subject's data as the test set, and left a cross-verified average recognition accuracy as the final data set recognition result. The comparison between the proposed method and other methods is shown in Table 1. The designed deep learning method is carried out by single-mode EDA and multi-mode fusion EDA+ECG+EMG tests. It can be seen that the single-mode recognition accuracy of the proposed method can reach 86.44%, and the multi-mode recognition accuracy can reach 87.16%, which is higher than the recognition results of other comparison methods, proving the effectiveness of the proposed method.

We used the designed network to test the collected pressure pain dataset. Since the amount of data was less than the BioVid data set, we increased the number of iterations by 4 times to train the network. The obtained recognition results of grade 4 pain of 5 subjects were shown in Fig. 3, which showed that the pain recognition accuracy was similar to that of the BioVid data set. The recognition accuracy of the collected data set is up to 87.5%, which proves the validity of the designed pressure pain data acquisition paradigm. In some subjects, the discriminating accuracy of pain grade 1 and grade 2 was opposite to that of the BioVid dataset, which might be due to the low degree of pressure pain differentiation.

5 Discussion and Conclusion

Table 1. Performance comparison based on Part A of the BioVid heat pain.

Approach	Modality	0 vs 1	0 vs 2	0 vs 3	0 vs 4
Werner et al. [12]	Video	53.3%	56.0%	64.0%	72.4%
Lopez et al. [14]	EDA	56.44%	59.40%	66.00%	74.21%
Wang et al. [15]	EDA, ECG, EMG	58.5%	64.2%	75.1%	83.3%
Thiam et al. [16]	EDA	61.67%	66.93%	76.38%	84.57%
Phan et al. [8]	EDA, ECG	59.5%	65.7%	75.2%	84.8%
Ours	EDA	**62.39%**	**71.06%**	**78.79%**	**86.44%**
Ours	EDA, ECG, EMG	**63.12%**	**71.96%**	**79.51%**	**87.16%**

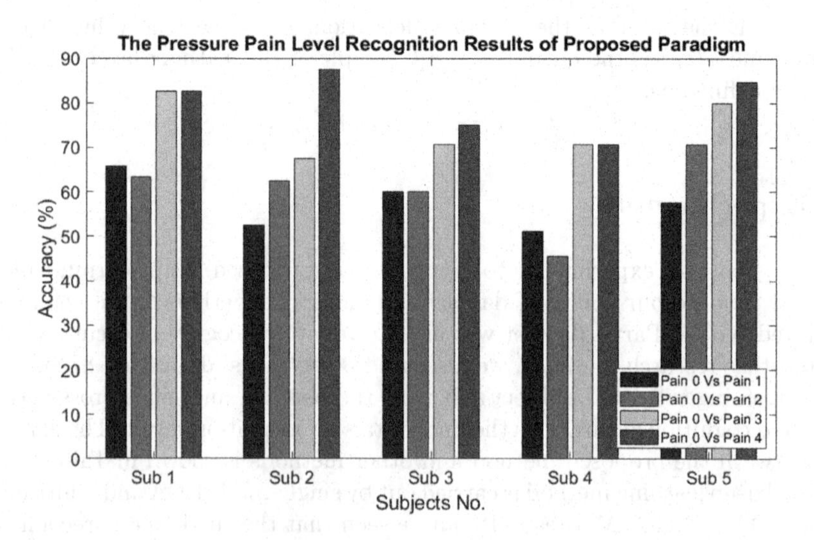

Fig. 3. The pressure pain recognition results of the proposed paradigm.

In view of the lack of research on the pain perception of the wearer by the exoskeleton robot, this paper designed a pressure-based pain data acquisition paradigm, taking into account the possible causes of pain caused by the lower limb exoskeleton robot, which collected the facial electromyography and electrocardiogram signals, graded the pain into 5 levels, and collected the data of 5 subjects. Then, in order to identify the pain threshold, we designed a deep learning network model based on V-tCN+LSTM, and designed the complex Loss function of Center Loss, infoNEC Loss and softmax Loss. The network model is tested and verified by using BioVid public data set. Compared with similar methods, it can be found that the proposed method has better recognition accuracy, and the effectiveness of the proposed method is verified. Then the proposed network is used to test the recognition results of the collected data set, and the validity of the collected data set is proved.

However, the pressure pain dataset collected in this paper only collected the data of 5 subjects, and did not take into account gender and age grouping. In addition, the pain feedback data collected in this study mainly came from the action response of pain, but no sensor data directly related to skin deformation. Future studies will consider using flexible pressure sensors to collect skin deformation and directly map pain and pressure to achieve a more objective and more suitable for the evaluation of robotic auxiliary systems such as exoskeleton robots.

Acknowledgments. This research was partially supported by National Natural Science Foundation of China (Grants No. 62125307), the National Natural Science Foundation of China (Grant No. 62103401), Guangdong Basic and Applied Basic Research Foundation (2023A1515011321), Shenzhen Basic Research Project under Grand No. JCYJ20220818101416035 and the Shenzhen Science and Technology Program under Grant RCBS20210706092252054.

Disclosure of Interests. The authors have no competing interests to declare that are relevant to the content of this article.

References

1. Ma, Y., et al.: Online gait planning of lower-limb exoskeleton robot for paraplegic rehabilitation considering weight transfer process. IEEE Trans. Autom. Sci. Eng. **18**(2), 414–425 (2021)
2. Ma, Y., Wu, X., Yi, J., Wang, C., Chen, C.: A review on human-exoskeleton coordination towards lower limb robotic exoskeleton systems. Int. J. Robot. Autom. **34**(4), 431–451 (2019)
3. Campbell, E., Phinyomark, A., Scheme, E.: Feature extraction and selection for pain recognition using peripheral physiological signals. Front. Neurosci. **13**, 437 (2019)
4. Kelati, A., Nigussie, E., Dhaou, I.B., Plosila, J., Tenhunen, H.: Real-time classification of pain level using zygomaticus and corrugator EMG features. Electronics **11**(11), 1671 (2022)

5. Tran, H.T., Kong, Y., Talati, A., Posada-Quintero, H., Chon, K.H., Chen, I.-P.: The use of electrodermal activity in pulpal diagnosis and dental pain assessment. Int. Endod. J. **56**(3), 356–368 (2022)
6. Pouromran, E., Radhakrishnan, S., Kamarthi, S.: Exploration of physiological sensors, features, and machine learning models for pain intensity estimation. PLoS ONE **16**(7), e0254108 (2020)
7. Thiam, P., et al.: Multi-modal pain intensity recognition based on the *SenseEmotion* database. IEEE Trans. Affect. Comput. **12**(3), 743–760 (2021)
8. Phan, K.N., Iyortsuun, N.K., Pant, S., Yang, H.-J., Kim, S.-H.: Pain recognition with physiological signals, using multi-level context information. IEEE Access **11**, 20114 (2023)
9. Irani, R., et al.: Spatiotemporal analysis of RGB-D-T facial images for multimodal pain level recognition. In: 2015 IEEE Conference on Computer Vision and Pattern Recognition Workshops (CVPRW), Boston, MA, USA, pp. 88–95 (2015). https://doi.org/10.1109/CVPRW.2015.7301341
10. Gruss, S., et al.: Pain intensity recognition rates via biopotential feature patterns with support vector machines. PLoS ONE **10**(10), e0140330 (2015)
11. Naeini, E.K., et al.: Pain recognition with electrocardiographic features in post-operative patients: method validation study. J. Med. Internet Res. **23**(5), e25079 (2021)
12. Werner, P., Al-Hamadi, A., Limbrecht-Ecklundt, K., Walter, S., Gruss, S., Traue, H.C.: Automatic pain assessment with facial activity descriptors. IEEE Trans. Affect. Comput. **8**(3), 286–299 (2017)
13. Walter, S., Gruss, S., Limbrecht-Ecklundt, K., Traue, H., Werner, P., Al-Hamadi, A., et al.: Automatic pain quantification using autonomic parameters. Front. Neurosci. **7**, 363–380 (2014)
14. Lopez-Martinez, D., Picard, R.: Continuous pain intensity estimation from autonomic signals with recurrent neural networks. In: Proceedings of the 40th Annual International Conference of the IEEE Engineering in Medicine and Biology Society (EMBC), pp. 5624–5627 (2018)
15. Wang, R., Xu, K., Feng, H., Chen, W.: Hybrid RNN-ANN based deep physiological network for pain recognition. In: Proceedings of the 42nd Annual International Conference of the IEEE Engineering in Medicine and Biology Society (EMBC), pp. 5584–5587 (2020)
16. Thiam, P., Bellmann, P., Kestler, H.A., Schwenker, F.: Exploring deep physiological models for nociceptive pain recognition. Sensors **19**(20), 4503 (2019)

Research on UAV Target Location Algorithm of Linear Frequency Modulated Continuous Wave Laser Ranging Method

Yanqin Su and Jiaqi Liu(✉)

Naval Aviation University, Yantai 264001, Shandong, China
460869693@qq.com

Abstract. The improved multi-track ranging target location algorithm is analyzed to achieve accurate location of typical obstacles in landing operations. On the basis of analyzing the requirements of typical obstacles on positioning accuracy and efficiency, the target positioning algorithm of multi-track ranging was improved to multi-machine positioning, and further improved to linear frequency modulated continuous wave laser ranging method. After in-depth study, the number of UAVs required at different altitude levels was given, and the positioning errors were constantly approaching the requirements of the positioning accuracy at the meter level of obstacles, and simulation verification and comparative analysis were carried out.

Keywords: UAV positioning · Multi-track ranging target location method · Linear frequency modulated CW laser ranging

With the arrival of the era of unmanned intelligent warfare, UAVs have become a new field of competition in modern warfare. Rapid and Accurate target positioning of UAV is an inevitable requirement for adapting to battlefield development. By virtue of UAV position algorithms, as far as possible to reduce positioning errors, and improve positioning accuracy, is the inevitable trend of UAV target location.

In the case of high of real-time requirements, the positioning accuracy of active positioning is higher than that of passive positioning. Therefore, the main methods of methods of active location include range-angle location, location based on range value of multiple track points, location based on state estimation, etc.

1 Improvement of Target Location Algorithm for Multi-track Ranging

1.1 Multi-track Ranging Target Location Algorithm

The target location algorithm based on the ranging value of multi-track points refers to the random laser ranging of the target during the flight of the UAV to obtain the laser ranging value $l_i (i = 1, 2, 3)$, as shown in Fig. 1. The observation equation can be

F. Sun and J. Li (Eds.): ICCCS 2023, CCIS 2029, pp. 107–122, 2024.
https://doi.org/10.1007/978-981-97-0885-7_10

constructed by combining the coordinate value of the UAV at different times with the ranging value and the coordinate position of the target point [1].

$$\begin{cases} l_1^2 = (x_1 - x')^2 + (y_1 - y')^2 + (z_1 - z')^2 \\ l_2^2 = (x_2 - x')^2 + (y_2 - y')^2 + (z_2 - z')^2 \\ \vdots \\ l_i^2 = (x_i - x')^2 + (y_i - y')^2 + (z_i - z')^2 \end{cases} \tag{1}$$

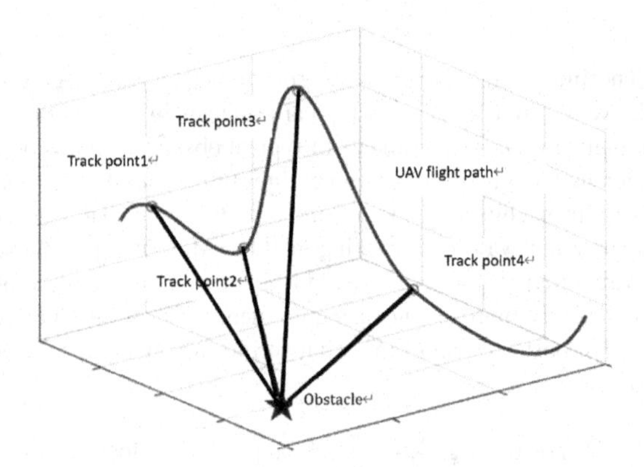

Fig. 1. Multi-track ranging target location.

1.2 Multi-track Ranging Target Positioning Improved Multi-UAV Ranging Target Positioning

The target location algorithm of multi-track ranging has some disadvantages such as long time, requirement on flight path and weak survivability of single machine. In the actual battlefield, the flight path of UAV should consider the enemy fire attack threat and electronic interference and other factors. The more reciprocating flight times, the more unfavorable the survival detection of UAV. Therefore, it is necessary to improve the target location algorithm of multi-track ranging, so that the improved positioning method does not have too high requirements on the flight path of UAV. It makes up for the disadvantages of long time, weak survivability and low measurement efficiency of multi-track ranging target location algorithm.

Let the coordinates of the target be $Q(a, b, c)$, the coordinates of n UAVs in the monitoring area are (x_1, y_1, z_1), (x_2, y_2, z_1), ... (x_n, y_n, z_1), the distance from the target to the observation station is respectively $l_1, l_2, \ldots l_n$

$$\begin{cases} (x_1 - a)^2 + (y_1 - b)^2 + (z_1 - c)^2 = l_1^2 \\ (x_2 - a)^2 + (y_2 - b)^2 + (z_2 - c)^2 = l_2^2 \\ \vdots \\ (x_n - a)^2 + (y_n - b)^2 + (z_n - c)^2 = l_n^2 \end{cases} \tag{2}$$

By subtracting from formula (2) from top to bottom, the formula can be turned into:

$$\begin{cases} x_1^2 - x_n^2 - 2(x_1 - x_n)a + y_1^2 - y_n^2 - 2(y_1 - y_n)b + z_1^2 - z_n^2 - 2(z_1 - z_n)c = l_1^2 - l_n^2 \\ \vdots \\ x_{n-1}^2 - x_n^2 - 2(x_{n-1} - x_n)a + y_{n-1}^2 - y_n^2 - 2(y_{n-1} - y_n)b + z_{n-1}^2 - z_n^2 - 2(z_1 - z_n)c = l_{n-1}^2 - l_n^2 \end{cases} \tag{3}$$

To estimate the target position, the least squares estimation of the target position can be obtained by using the least squares principle, namely:

$$A = -2\begin{bmatrix} x_1 - x_n & y_1 - y_n & z_1 - z_n \\ x_2 - x_n & y_2 - y_n & z_2 - z_n \\ & \vdots & \\ x_{n-1} - x_n & y_{n-1} - y_n & z_{n-1} - z_n \end{bmatrix}$$

$$B = \begin{bmatrix} l_1^2 - l_n^2 + x_n^2 + y_n^2 + z_n^2 - x_1^2 - y_1^2 - z_1^2 \\ l_2^2 - l_n^2 + x_n^2 + y_n^2 + z_n^2 - x_2^2 - y_2^2 - z_1^2 \\ \vdots \\ l_{n-1}^2 - l_n^2 + x_n^2 + y_n^2 + z_n^2 - x_{n-1}^2 - y_1^2 - z_1^2 \end{bmatrix} \tag{4}$$

$$X = \begin{bmatrix} a \\ b \\ c \end{bmatrix}$$

Formula (4) is the matrix representation of Formula (3), which is the most commonly used positioning algorithm in target positioning system based on observation distance. Here, to make the solution of Eq. (4) unique, at least four UAV observation stations need to be deployed in non-coplanar space.

1.3 Improvement of Ranging Method Based on Linear Frequency Modulate Continuous Wave Laser Ranging Principle

In the target location algorithm model of multi-UAV ranging, the error mainly comes from the ranging error. MATLAB is used, Gaussian white noise with variance of 5 is used as the disturbance error, Monte Carlo method is used to establish the error disturbance model [2], repeated experiments are conducted, and it is found that when the UAV is measured within 100 m, 300 UAVs are needed to meet the requirement that the average positioning accuracy reaches the meter level, and the number of UAVs is too large. Therefore, the ranging method needs to be improved to reduce the ranging error [3].

Principle of Linear Frequency Modulated CW Laser Ranging

Linear frequency modulation signal frequency changes with time in a zigzag shape. The frequency increases linearly with the increase of time. When it reaches the maximum value, it changes abruptly to the initial frequency, and then changes linearly with the increase of time, repeating the previous cycle [4] (Fig. 2).

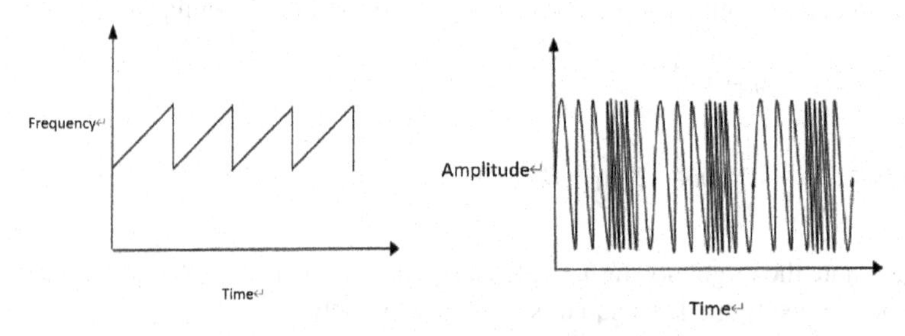

Fig. 2. Schematic diagram of linear frequency modulation signal

The principle of linear FM CW laser ranging is to use the frequency expression of the intermediate frequency signal and the time delay expression of the transmitted signal and the echo signal to solve the distance between the signal transmitting position and the target [5]. As shown in Fig. 3, the solid line is the change of the frequency of the transmitted signal with time. Because it takes time for the signal to travel back and forth to the target, there is a frequency delay between the echo signal and the transmitted signal. The initial frequency of the linear frequency modulation signal is F_0, the sweep frequency band width is b, the sweep frequency period is T, and the time required for the signal to travel to and from the target and ranging is

$$\tau = \frac{2L}{c} \tag{5}$$

The expression of the transmitted signal is

$$S_{\text{launch}}(t) = A\cos(2\pi F_0 t + \pi K t^2 + \varphi'), 0 \le t \le T\sum_{i=1}^{n} X_i \tag{6}$$

In this formula, A is the amplitude of the transmitted signal, K = b/T is the sweep slope, φ' is the initial phase. The expression of echo signal is

$$S_{\text{echo}}(t) = \eta A\cos\left[2\pi F_0(t-\tau) + \pi K(t-\tau)^2 + \varphi'\right], \tau \le t \le T + \tau \tag{7}$$

In the formula, η is the attenuation coefficient, and the instantaneous frequency expressions of transmitted signal and echo signal are respectively

$$f_{\text{launch}}(t) = F_0 + Kt, 0 \le t \le T \tag{8}$$

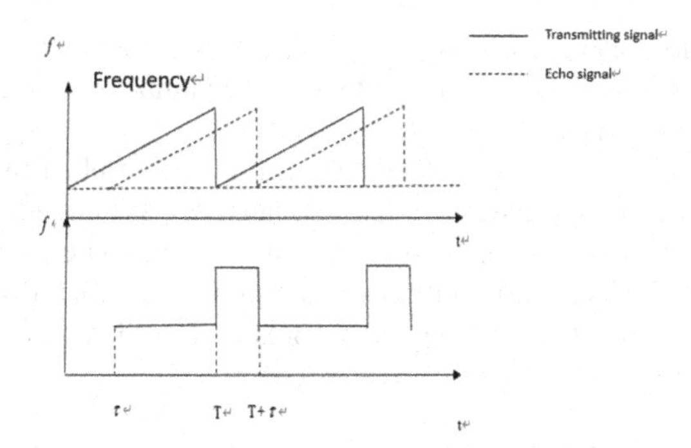

Fig. 3. Frequency delay diagram of transmitted signal and echo signal

$$f_{\text{instantaneous}}(t) = F_0 + K(t - \tau), \tau \le t \le T + \tau \tag{9}$$

After the low pass filter mixes the transmitted signal with the echo signal, the expression of the signal can be written as

$$S_{\text{mix}}(t) = \frac{1}{2}\eta A^2 \cos(2\pi K\tau t + 2\pi F_0\tau), \tau \le t \le T \tag{10}$$

It can be seen from formula (10) that the frequency of IF signal is

$$f_{if} = K\tau \tag{11}$$

By substituting K = b/T and formula (5) into formula (11), the principle expression of ranging is

$$L = \frac{Tc}{2b} f_{if} \tag{12}$$

Error Analysis of the Improved Ranging

It can be seen from formula (12) that the measurement error is related to T, b, c and f_{if} Linear frequency modulation signal is generated by DDS programming supplemented by necessary filtering circuit, so the sweep frequency band width b and sweep frequency period T are basically determined. The propagation rate c of light has little difference in values in different propagation media. The error of propagation rate in atmospheric environment and vacuum environment is very small and can be ignored. Therefore, only the influence of measurement error of IF signal on distance measurement error is emphasized. Δf_{if} is the error value of IF signal, and δL is the distance error value. Then, the expression of the influence of IF signal error value on distance measurement error is the analysis of distance error

$$\delta L = \frac{Tc}{2b} \cdot \Delta f_{if} \tag{13}$$

Linear frequency modulation continuous wave ranging using fast Fourier algorithm, fast Fourier algorithm can in the noise ratio is not conducive to measurement, through the signal correlation operation, make useful signals accumulation, the premise is to increase the number of sampling points, as long as the number is enough, the frequency of the signal can be estimated.

The frequency estimation error of this algorithm is controlled by the actual total sampling time of the analyzed signal T_{actual}. Because the system adopts the principle of proximity for judgment, the maximum frequency estimation error occurs when the frequency of IF signal falls in the middle of two sampling points. The distance expression of adjacent frequency points in the frequency domain of fast Fourier algorithm is

$$l = \frac{1}{T_{\text{actual}}} \tag{14}$$

The expression of frequency measurement limit error is

$$\Delta f_{if} = \pm \frac{1}{2T_{\text{actual}}} \tag{15}$$

If you substitute formula (15) into formula (13), you get

$$\delta L = \pm \frac{c}{4KT_{\text{actual}}} \tag{16}$$

As the error of the speed of light can be ignored, the size of δL is only related to the denominator. When the product of the sweep frequency slope of the sweep frequency signal and the sampling time of the if frequency is larger, δL is smaller; when the product of the sweep frequency slope of the sweep frequency signal and the sampling time is smaller, δL is larger. Therefore, the sampling time is fixed. To make δL as small as possible, the sweep frequency slope needs to be increased. Similarly, when the sweep slope is determined, the sampling time needs to be increased. For sawtooth linear frequency modulation signal, where the limit value of frequency sampling time is $T - \tau$, $\delta L = \pm \frac{c}{4K(T-\tau)}$, in the case of low and medium altitude measurement, $\tau \ll T$, then

$$\delta L = \pm \frac{c}{4K(T-\tau)} \approx \pm \frac{c}{4KT} = \pm \frac{c}{4B} \tag{17}$$

As can be seen from the above formula, the distance error measured by linear frequency modulated CW laser ranging is inversely proportional to the sweep band width of the sweep signal. In the algorithm, it is stipulated that the sweep frequency band width B = 5 GHz is used, and the measurement error in the model is ±1.5 cm.

2 Matlab Simulation Analysis

In order to verify the effect of the improved algorithm, MATLAB is used to simulate the algorithm before and after the improvement, and analyze and compare.

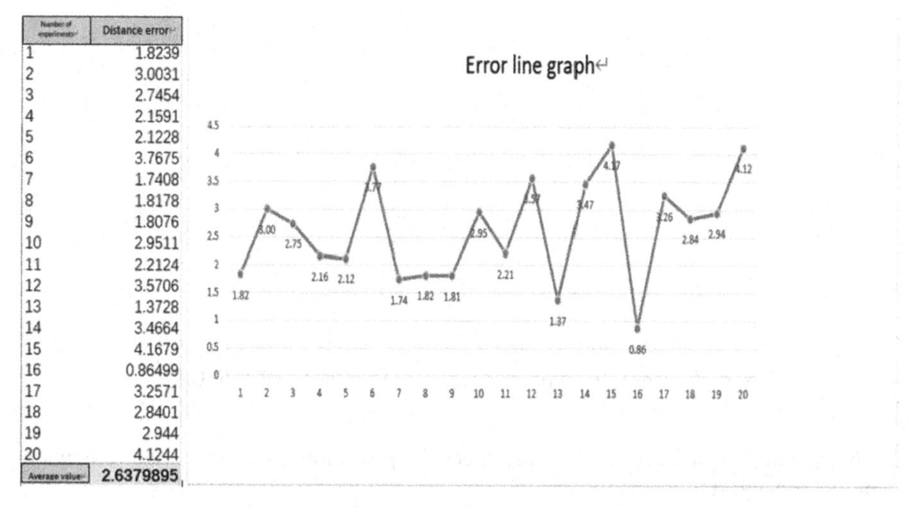

Fig. 4. 10 UAV 20 measurement error analysis

2.1 MATLAB Simulation and Analysis Before Improvement

The observation coordinates and target coordinates of 10 UAVs were randomly generated by repeated experiments in matlab, and the disturbance sequence with a variance of 5 Gaussian white noise was added. The results are shown in the figure below (Fig. 4).

Continue to change the number of UAVs for experiments. This paper conducts experiments on 15, 20, 25, 30, 35, 40, 45, 50 and 100 UAVs respectively. Table 1 and Fig. 5 are obtained by synthesizing data.

Table 1. Average values of errors corresponding to different numbers of UAVs

Number of UAVs	Distance error (m)
10	2.63799
15	2.418807
20	2.470117
25	1.863135
30	1.859808
35	1.720524
40	1.76495
45	1.686897
50	1.534554
100	1.381518

It is found that the positioning error corresponding to 100 UAVs is still more than 1 m, which does not meet the target positioning accuracy requirements. According to

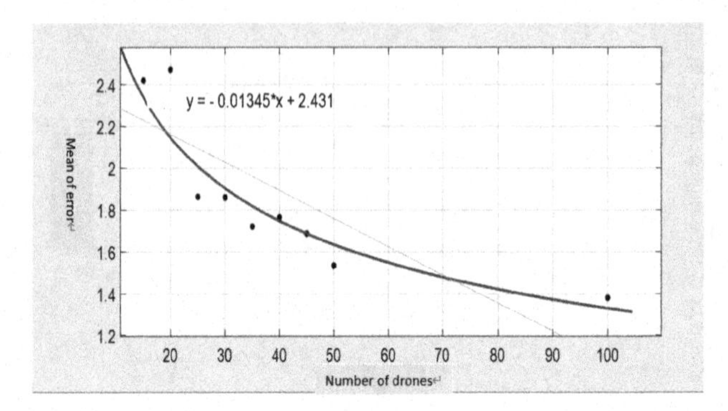

Fig. 5. Fitting diagram of error variation with the number of UAVs

Fig. 6, the number of UAVs required to meet the positioning accuracy requirements is predicted to be 300.

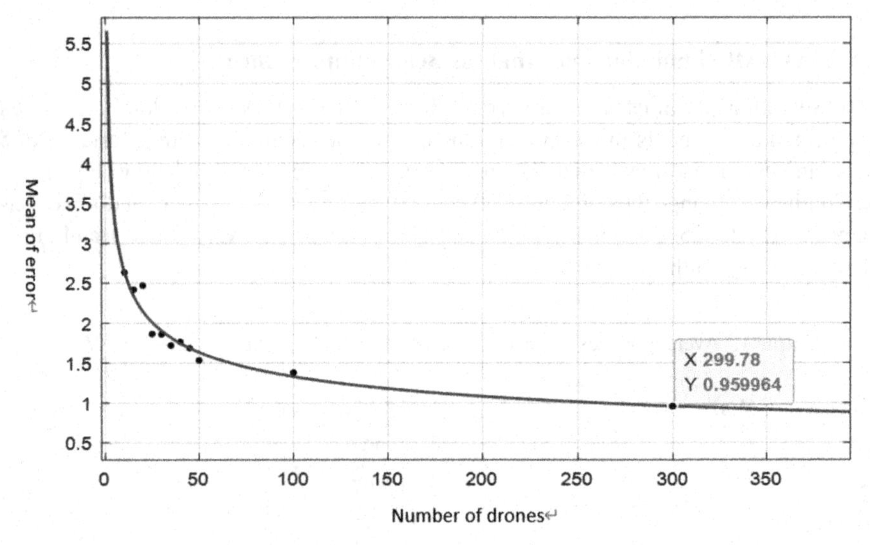

Fig. 6. Satisfies the trend prediction of positioning accuracy

2.2 Improved MATLAB Simulation and Analysis

This section uses MATLAB simulation to improve the algorithm model of the ranging method, and the ranging error analysis shows that the ranging error is ± 1.5 cm. The following is a simulation experiment based on two cases of ranging error distribution.

When the number of simulated UAVs is 20, 15, 10 and 4, the error of target positioning is 20 times. When the error is normally distributed, the most unfavorable case, the standard deviation $\sigma = 1.5$, is considered, so the proportion of error falling within the range of ± 1.5 cm is 68% of the total value.

20 UAVs		15 UAVs		10 UAVs		4 UAVs	
Number of experiments	Distance error	Number of experiments	Distance error	Number of experiments	Distance error	Number of experiments	Distance error
1	0.0077578	1	0.019279	1	0.0039376	1	0.027599
2	0.010574	2	0.018429	2	0.0092386	2	0.12964
3	0.0072243	3	0.030886	3	0.010447	3	0.037596
4	0.0065729	4	0.011838	4	0.014354	4	0.066584
5	0.0042605	5	0.017636	5	0.015546	5	0.44683
6	0.014666	6	0.018916	6	0.01657	6	0.15917
7	0.020975	7	0.017264	7	0.005847	7	0.0062254
8	0.0053108	8	0.02386	8	0.0072199	8	0.083613657
9	0.011582	9	0.007459	9	0.025539	9	0.037359
10	0.007031	10	0.046061	10	0.029535	10	0.026667
11	0.0057216	11	0.0023867	11	0.01821	11	0.16415
12	0.01129	12	0.0082344	12	0.013592	12	0.022476
13	0.010043	13	0.0074473	13	0.003489	13	0.069353
14	0.0087599	14	0.016799	14	0.011617	14	0.016787
15	0.0091314	15	0.017567	15	0.010952	15	0.036073
16	0.016357	16	0.015389	16	0.0290564	16	0.016917
17	0.023186	17	0.020732	17	0.011127	17	0.007494
18	0.0077066	18	0.013245	18	0.011738	18	0.10383
19	0.012252	19	0.018249	19	0.0065521	19	0.20818
20	0.0045687	20	0.0060911	20	0.02735	20	0.00572909
Average value	0.010248525	Average value	0.016888425	Average value	0.01409588	Average value	0.083613657

Fig. 7. Measurement data of normal distribution of error

20 UAVs		15 UAVs		10 UAVs		4 UAVs	
Number of experiments	Distance error	Number of experiments	Distance error	Number of experiments	Distance error	Number of experiments	Distance error
1	0.0058191	1	0.01636	1	0.02315	1	0.2014
2	0.0073362	2	0.0050456	2	0.020684	2	0.12744
3	0.0030511	3	0.012216	3	0.023151	3	0.17125
4	0.00700556	4	0.0098983	4	0.011175	4	0.0066601
5	0.014657	5	0.0064805	5	0.0027985	5	0.064815
6	0.015517	6	0.0097537	6	0.01616	6	0.038067
7	0.0075689	7	0.0026865	7	0.0070392	7	0.012812
8	0.01024	8	0.011686	8	0.016134	8	0.059619
9	0.014147	9	0.0048364	9	0.017622	9	0.070064
10	0.0095808	10	0.012504	10	0.01461	10	0.11015
11	0.006125	11	0.0090652	11	0.01763	11	0.0073499
12	0.0047805	12	0.0090914	12	0.018845	12	0.057025542
13	0.22224	13	0.015429	13	0.0074844	13	0.03126
14	0.006733	14	0.011869	14	0.010291	14	0.012213
15	0.11063	15	0.0046112	15	0.010291	15	0.029139
16	0.005503	16	0.024047	16	0.00858814	16	0.02524
17	0.0025884	17	0.015536	17	0.016318	17	0.015263
18	0.10329	18	0.0303	18	0.0094489	18	0.0094853
19	0.015667	19	0.0060599	19	0.0033833	19	0.039397
20	0.010735	20	0.017169	20	0.012427	20	0.051861
Average value	0.029160728	Average value	0.011732235	Average value	0.013361522	Average value	0.057025542

Fig. 8. Measurement data with uniform distribution of errors

It can be seen that the improved experimental data all meet the requirements that the positioning accuracy reaches the meter level. As long as the ranging error is small enough, the number of UAVs will not affect the positioning accuracy and has nothing to do with the distribution of ranging error. By comparing the average positioning errors of different number of UAVs, it is found that the ranging error is more evenly distributed than the ranging error is normally distributed, and with the increase of the number of UAVs, the average error of both decreases gradually.

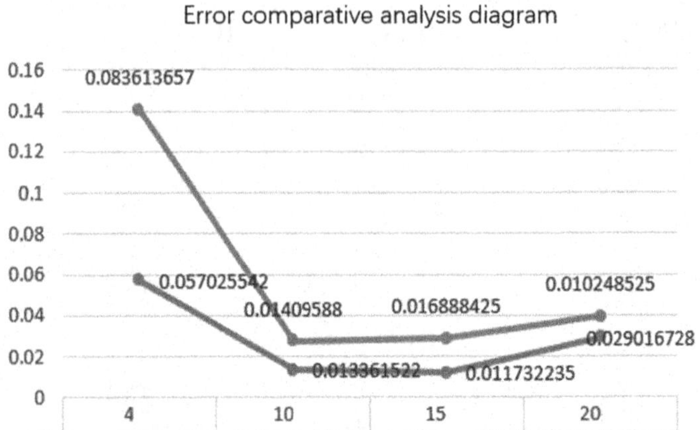

Error comparative analysis diagram

	4	10	15	20
Normal distribution	0.083613657	0.01409588	0.016888425	0.010248525
Uniform distribution	0.057025542	0.013361522	0.011732235	0.029016728

Fig. 9. Comparison between normal distribution and uniform distribution of errors

3 Analyze the Influence of UAV Flight Height on Positioning Accuracy After the Improved Algorithm

3.1 Height Limitation Range of Multi-UAV Ranging Target Positioning Algorithm

The sampling period (unit: s) is mostly the order of 10^{-3}. Therefore, if $T - \tau \approx T$ in $\delta L = \pm\frac{c}{4K(T-\tau)}$ is to be established, the order of τ must be smaller than 10^{-5}. Therefore, if $\tau = \frac{2L}{c}$ is substituted, L must be less than 1500 m. Therefore, based on the principle of linear frequency modulated CW laser ranging, the height setting range of the target positioning model should be 0–1500 m.

3.2 The UAV has No Low Altitude Flight Restrictions

Using MATLAB simulation, with 100 m as the independent variable interval, in the case of positioning of 4 UAVs, the maximum height change of UAVs flight area and positioning error data were analyzed. At this time, the lowest altitude of aircraft distribution was 0 m.

The maximum restricted height of the flight area was taken as the X-axis, and the average value of positioning errors was measured for several times as the Y-axis. The point plot of positioning errors changing with the height was drawn, and it could be seen that each point presented discrete distribution (Figs. 10, 11 and 12).

The area where the UAV measures the target (the maximum height is no more than 1500 m under the conditions allowed by the model) determines the flight height range of the UAV. The larger the area, the more discrete the distribution of the UAV. According to the data graph, the following is obtained: Under the premise of unlimited low-altitude flight, no matter how the ranging error is distributed, the size of the flight area does

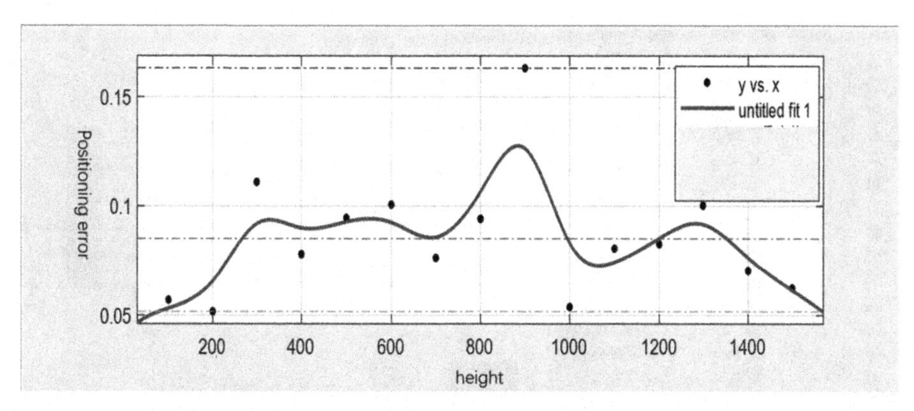

Fig. 10. Data chart of positioning error changing with the maximum restricted height of the flight area when the error is uniformly distributed

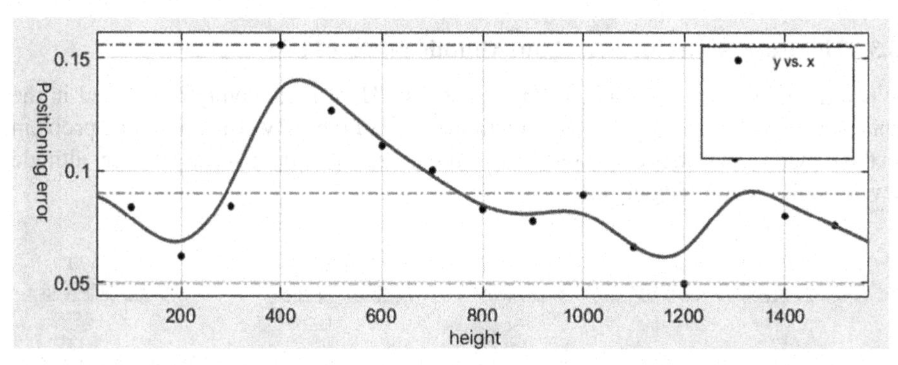

Fig. 11. Data chart of positioning error changing with the maximum restricted height of the flight area when the error is normally distributed

not affect the positioning error, and the positioning error is distributed in a discrete dot pattern with the change of the maximum restricted altitude. The influence of normal distribution and uniform distribution of error on positioning error can not be judged, and they are discrete distribution. When there is no minimum flight height restriction, the UAV is randomly distributed in the hemispheres, and the maximum flight height is limited, which only expands the possible flight area of the UAV. Because the probability of the UAV randomly appearing in any position within the region is the same, so the above experiments do not mean that the UAV falls near the maximum height exactly, although the flight height is specified. Nor does it represent the positioning error measured by the UAV at the corresponding height.

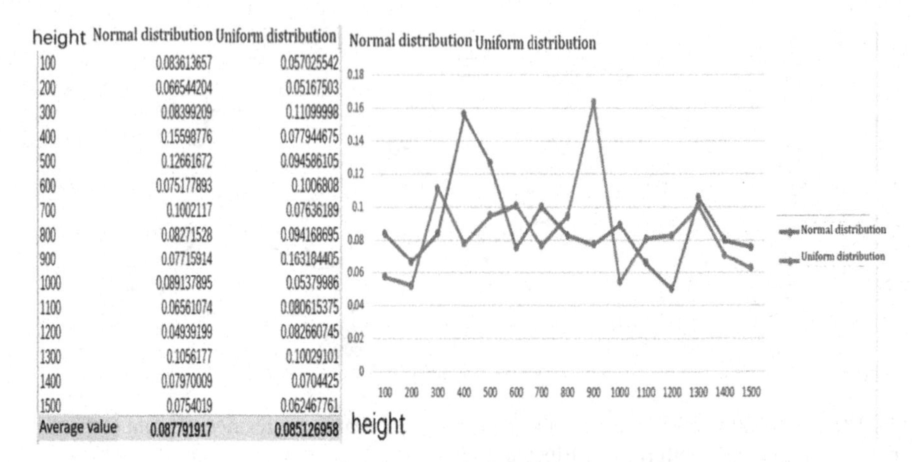

height	Normal distribution	Uniform distribution
100	0.083613657	0.057025542
200	0.066544204	0.05167503
300	0.08399209	0.11099998
400	0.15598776	0.077944675
500	0.12661672	0.094586105
600	0.075177893	0.1006808
700	0.1002117	0.07636189
800	0.08271528	0.094168695
900	0.07715914	0.163184405
1000	0.089137895	0.05379986
1100	0.06561074	0.080615375
1200	0.04939199	0.082660745
1300	0.1056177	0.10029101
1400	0.07970009	0.0704425
1500	0.0754019	0.062467761
Average value	0.087791917	0.085126958

Fig. 12. Error comparison of the two distributions of error at the restricted height

3.3 There are Restrictions on Low-Altitude Flight of UAV

When the minimum flight height limit exists, the UAV is randomly distributed in the sphere ring, which can limit the flight altitude level of the UAV. The remaining problem in the previous section is discussed, that is, the influence of the UAV at different altitude levels on the positioning error.

0-100		100-200		200-300		300-400	
Number of experiments	Distance error	Number of experiments	Distance error	Number of experiments	Distance error	Number of experiments	Distance error
1	0.2014	1	0.31686	1	0.11687	1	0.7762
2	0.12744	2	0.071852	2	0.36693	2	0.13021
3	0.17125	3	0.093215	3	0.072973	3	0.60347
4	0.0066601	4	0.02948	4	0.39966	4	0.21483
5	0.064815	5	0.70127	5	0.3789	5	0.10584
6	0.038067	6	0.18147	6	0.076366	6	0.16411
7	0.012812	7	0.13517	7	0.23807	7	0.31645
8	0.059619	8	0.21491	8	0.091441	8	0.38867
9	0.070064	9	0.1067	9	0.1714	9	0.11652
10	0.11015	10	0.060209	10	0.056261	10	0.24339
11	0.0073499	11	0.048928	11	0.057618	11	0.13539
12	0.05702554	12	0.013242	12	0.11421	12	0.70857
13	0.03126	13	0.38989	13	0.44652	13	0.19434
14	0.012213	14	0.043204	14	0.16939	14	0.15699
15	0.029139	15	0.042066	15	0.29631	15	0.33351
16	0.02524	16	0.12891	16	0.18827	16	1.0586
17	0.015263	17	0.086248	17	0.090754	17	0.64486
18	0.0094853	18	0.053995	18	0.12674	18	0.62531
19	0.039397	19	0.025256	19	0.1645	19	0.53787
20	0.051861	20	0.041145	20	0.29608	20	0.402
Average value	0.05702554	Average value	0.139201	Average value	0.19596315	Average value	0.3928565

Fig. 13. 4 UAVs data at different altitude levels

Four UAVs were selected and their flight height was tested by MATLAB for several times, as shown in Figs. 5, 6, 7, 8 and 9. Due to limited space, only 20 times of results were listed, and multiple experiments were actually conducted (more than 200 times to ensure sufficient number of experiments). The positioning error of the 16th time exceeded 1 m, which did not meet the target positioning accuracy requirements. That is to say, within the flying altitude of 300–400 m, the measurement of four UAVs may lead to invalid positioning. See Fig. 13, the average positioning error increases with the elevation of the altitude layer. Therefore, it can be concluded that there may be invalid positioning situation when four UAVs are used at the altitude level above 300 (Fig. 14).

300-400		400-500		500-600		600-700		700-800	
Number of experiments	Distance error	Number of experiments	Distance error	Number of experiments	Distance error	Number of experiments	Distance error	Number of experiments	Distance error
1	0.11583	1	0.32158	1	0.30478	1	0.24713	1	0.2933
2	0.23504	2	0.032835	2	0.22312	2	0.12157	2	0.3923
3	0.074166	3	0.18603	3	0.086422	3	0.32622	3	0.31075
4	0.10561	4	0.1033	4	0.061816	4	0.12472	4	0.46913
5	0.048607	5	0.17025	5	0.15674	5	0.26	5	0.36109
6	0.14311	6	0.22355	6	0.10572	6	0.6846	6	0.26591
7	0.1512	7	0.10297	7	0.2109	7	0.29345	7	1.106268
8	0.13945	8	0.079489	8	0.28007	8	0.18327	8	0.22878
9	0.090506	9	0.12775	9	0..235	9	0.16547	9	0.6737
10	0.1952	10	0.3573	10	0.19035	10	0.17882	10	0.17769
11	0.25207	11	0.12789	11	0.13197	11	0.0754	11	0.15591
12	0.20081	12	0.12966	12	0.60004	12	0.21078	12	0.19933
13	0.38191	13	0.088317	13	0.13844	13	0.10549	13	0.18386
14	0.200083	14	0.19055	14	0.12065	14	0.35186	14	0.10992
15	0.059308	15	0.099475	15	0.16357	15	0.093279	15	0.26025
16	0.1045	16	0.077557	16	0.07182	16	0.8863	16	0.37021
17	0.038752	17	0.18028	17	0.16567	17	0.47	17	1.12376
18	0.10793	18	0.32547	18	0.17313	18	0.28183	18	0.20371
19	0.01354	19	0.063971	19	0.1277	19	0.28377	19	0.16924
20	0.13677	20	0.099378	20	0.125552	20	0.03458	20	0.29371
Average value	0.13972	Average value	0.15438	Average value	0.180972	Average value	0.268927	Average value	0.367441

Fig. 14. 6 UAVs at different altitude levels

In the experiment of 700–800 m, special data was found. The positioning error of the 7th time was more than 1 m, which did not meet the requirement of target positioning accuracy. Therefore, when the minimum altitude of flight altitude layer was greater than 700 m, the required number of UAV was more than 6 (Fig. 15).

In the experiment of 1200–1300 m, special data was found. The positioning error of the 14th time was more than 1 m, which did not meet the requirement of target positioning accuracy. Therefore, when the minimum altitude of flight altitude layer was greater than 1200 m, the required number of UAV was more than 10 (Fig. 16).

No positioning error of more than 1 m was found, so 15 UAVs met the positioning effectiveness of 100% in the existing experimental conditions.

It is concluded that when the UAV locates the target at different altitude levels, it meets the condition that the positioning effectiveness is 100%, and the range of the UAV is required. In combat, by querying scope Table 2, the corresponding number of UAVs can be selected on the premise of determining combat altitude, so as to ensure that fewer UAVs are used to achieve combat objectives, complete target positioning and reduce resource consumption.

700-800		800-900		900-1000		1000-1100		1100-1200		1200-1300	
Number of experiments	Distance error	Number of experiments	Distance error	Number of experiments	Distance error	Number of experiments	Distance error	Number of experiments	Distance error	Number of experiments	Distance error
1	0.26589	1	0.17037	1	0.2059	1	0.4608	1	0.37536	1	0.44076
2	0.043378	2	0.11609	2	0.21809	2	0.14397	2	0.42498	2	0.2213
3	0.21866	3	0.18239	3	0.23787	3	0.21925	3	0.099382	3	0.15926
4	0.28468	4	0.16686	4	0.25065	4	0.279	4	0.2096	4	0.39889
5	0.21124	5	0.21236	5	0.47371	5	0.18153	5	0.34586	5	0.35788
6	0.13081	6	0.058338	6	0.55725	6	0.069556	6	0.089096	6	0.30437
7	0.47515	7	0.34949	7	0.26843	7	0.26728	7	0.42401	7	0.16534
8	0.25539	8	0.26092	8	0.16234	8	0.28265	8	0.16549	8	0.34576
9	0.10147	9	0.18577	9	0.22407	9	0.32336	9	0.31191	9	0.42228
10	0.22154	10	0.16446	10	0.3274	10	0.072037	10	0.40322	10	0.21704
11	0.099474	11	0.33247	11	0.15474	11	0.28731	11	0.089315	11	0.25201
12	0.18904	12	0.28153	12	0.63028	12	0.27663	12	0.23358	12	0.29869
13	0.10934	13	0.23144	13	0.12282	13	0.30391	13	0.33548	13	0.15901
14	0.14797	14	0.1745	14	0.26331	14	0.13104	14	0.28646	14	1.2556
15	0.12337	15	0.16733	15	0.18055	15	0.12835	15	0.26666	15	0.16855
16	0.2552	16	0.28394	16	0.17354	16	0.45033	16	0.14675	16	0.41877
17	0.21368	17	0.18834	17	0.10774	17	0.18216	17	0.22287	17	0.19913
18	0.26861	18	0.32008	18	0.25456	18	0.39872	18	0.18788	18	0.23569
19	0.23191	19	0.3005	19	0.12692	19	0.195745	19	0.20477	19	0.15862
20	0.26944	20	0.38337	20	0.1989	20	0.29325	20	0.57847	20	0.39658
Average value	0.205812	Average value	0.226527	Average value	0.256954	Average value	0.247344	Average value	0.270057	Average value	0.328777

Fig. 15. 10 UAVs at different altitude levels

1200-1300		1300-1400		1400-1500	
Number of experiments	Distance error	Number of experiments	Distance error	Number of experiments	Distance error
1	0.1791	1	0.38645	1	0.34401
2	0.2062	2	0.29458	2	0.28251
3	0.21006	3	0.31248	3	0.18278
4	0.16376	4	0.21312	4	0.4215
5	0.21485	5	0.30249	5	0.3285
6	0.17483	6	0.2375	6	0.145896
7	0.242598	7	0.46215	7	0.31254
8	0.10483	8	0.23458	8	0.3124
9	0.15265	9	0.31058	9	0.366261
10	0.12456	10	0.44598	10	0.5466
11	0.26548	11	0.38459	11	0.2707
12	0.052696	12	0.13458	12	0.19596
13	0.245204	13	0.18459	13	0.31258
14	0.13546	14	0.16489	14	0.16548
15	0.165485	15	0.24841	15	0.18457
16	0.254958	16	0.19443	16	0.23549
17	0.151456	17	0.29458	17	0.47892
18	0.07448	18	0.29845	18	0.214598
19	0.11988	19	0.45981	19	0.31258
20	0.53326	20	0.12548	20	0.5468
Average value	0.18859	Average value	0.284486	Average value	0.308034

Fig. 16. 15 UAVs at different altitude levels

Table 2. Relationship between the number of UAVs and the altitude level

Height layer	Number of drones
0–100	4–6
100–200	
200–300	
300–400	6–10
400–500	
500–600	
600–700	
700–800	10–15
800–900	
900–1000	
1000–1100	
1100–1200	
1200–1300	>=15
1300–1400	
1400–1500	

4 Closing Remarks

In this paper, the target positioning method based on the ranging value of multi-track points has shortcomings such as low measurement efficiency, long time and limited survival ability of a single UAV. The multi-track point ranging positioning method of a single UAV is changed to multi-UAV multi-point target positioning method. This improvement not only improves the positioning speed of the UAV, but also does not consider the flight path problem of the UAV. MATLAB was used for simulation, white Gaussian noise was added to increase the authenticity of the simulation environment, and the feasibility of multi-UAV ranging target positioning method was verified. Aiming at the problem that the ranging error caused by noise interference was too large, so that the measurement error could not meet the requirements of target positioning accuracy, the linear frequency modulation continuous wave laser ranging principle was used for ranging. Two kinds of distribution of error, uniform distribution and uniform distribution, were respectively simulated by MATLAB, qualitative and quantitative analysis of the multi-UAV ranging target positioning method after the improved ranging principle is in line with the accuracy requirements of target positioning, and the applicable height range of the algorithm was determined. The impact of the UAV flight height on the positioning error was discussed respectively when there is no low altitude limitation. The height range is divided into sections, the number of UAVs required by the corresponding height level to meet the positioning requirements is discussed, the advantages and disadvantages of target positioning method are summarized, and the problems that need to be solved in the follow-up research are put forward.

References

1. Huang, X., Wang, Z.: Accuracy analysis and improvement method of photoelectric active positioning method. In: China Academic Journal Electronic Publishing House. Electronic Technology and Software Engineering 2021, vol.1, pp. 109–114. Changchun, Jilin Province (2021)
2. Zhang, J.: Enhanced Accuracy of Optical Target Positioning of UAV Based on Error Identification and Compensation. Nanjing University of Aeronautics and Astronautics, Nanjing (2019)
3. Ma, Z., Gong, Q., Chen, Y., Wang, H.: Analysis of influencing factors on target positioning accuracy of photoelectric reconnaissance system, In: China Academic Journal Electronic Publishing House. Journal of Applied Optics 2018, vol.1, pp.1–6. Xi'an, Shaanxi Province (2018)
4. Sharma, K.K., Joshi, S.D.: Time delay estimation using fractional fourier transform. Signal Process.rocess. **87**(5), 853–865 (2007)
5. Tsuchida, H.: Regression analysis of FMCW-LiDAR beat signals for non-linear chirp mitigation. Electron. Lett. **55**(16), 914–916 (2019)

A Flexible Tactile Sensor for Robots Based on Electrical Impedance Tomography

Zhiqiang Duan[1], Lekang Liu[2], Jun Zhu[3], Ruilin Wu[3], Yan Wang[2], Xiaohu Yuan[4(✉)], and Longlong Liao[1(✉)]

[1] College of Computer and Data Science/College of Software, FuZhou University, Fuzhou, China
t20104@fzu.edu.cn
[2] School of Physics and Electronic Information, Yantai University, Yantai, China
[3] School of Automation, Nanjing University of Information Science and Technology, Nanjing, China
20211249095@nuist.edu.cn
[4] Beijing Embodied Intelligent Technology Co., Ltd., Beijing, China
874839874@qq.com

Abstract. During social interactions, people can obtain a great deal of important information from their tactile senses to improve their relationship with their surroundings. The development of similar capabilities in robots will contribute to the success of intuitive human-robot interaction in the future. In this paper, a tactile sensing method based on the principle of electrical impedance tomography (EIT) is introduced, which with the help of EIT technology and combined with the flexible piezoresistive material Velostat, thin, lightweight, stretchable, and flexible skin can be designed for robots, and at the same time, information about the touch position, duration, and intensity can be acquired, and the image reconstruction is carried out using a dual finite element model, and the experimental results show that based on the flexible The experimental results show that the EIT tactile sensing technology based on the flexible material Velostat can be applied to robotic flexible skin applications.

Keywords: Velostat · Electrical Impedance Tomography · Bi-model

1 Introduction

Tactile perception, as a key technology in artificial intelligence research, can empower robots to perceive changes in surrounding objects and environments The research on robotic skin occupies an indispensable position in terms of tactile perception of anthropomorphic human beings [1]. Traditional rigid sensors are greatly limited in adapting to the irregular surfaces of robots, while array sensors suffer from the disadvantages of too many wires and clutter, as well as the inability to completely cover the surface, which has prompted researchers to seek new solutions. In recent years, the rapid development of flexible materials has provided new possibilities for robotic skin sensing and has become

F. Sun and J. Li (Eds.): ICCCS 2023, CCIS 2029, pp. 123–131, 2024.
https://doi.org/10.1007/978-981-97-0885-7_11

a hot research topic in robotic tactile sensing by virtue of its highly stretchable and customizable capabilities [2, 3]. Velostat is one of the compressible polymer composites, whose special structure gives it excellent piezoresistive properties [4]. When pressure is applied externally to Velostat, its conductivity changes significantly. This makes Velostat an ideal flexible piezoresistive material, which is promising in applications to robotic tactile sensing.

The initial application area of EIT is in medicine, which can be used to detect the impedance distribution in the human brain. In recent years, more and more researchers have applied EIT technology to the field of tactile sensing. Haptic information is mainly obtained through the electrodes discretely distributed on the sensor, a constant excitation current is applied to these electrodes, at this time, pressing the sensor brings about changes in the electrical impedance at the same time, it will also change the magnitude of the measured voltage, and through the voltage difference before and after the pressure is applied to reconstruct an image of the change in electrical conductivity, so as to obtain the information of the load contact such as the pressure, position, and shape of the load [5, 6]. The EIT technology can be Velostat piezoresistive material into the flexible skin of robots, thus endowing robots with more natural and intelligent tactile sensing ability and opening up new possibilities for the innovation and development of robotics.

2 EIT-BASED Flexible Sensor

Flexible The flexible skin for robotic surfaces studied in this paper was fabricated based on the principles of electrical impedance imaging (EIT), a non-invasive imaging technique that allows conductivity image reconstruction through the measurement of voltage changes across electrodes. For a typical electrical impedance system, if an adjacent current excitation mode is used, a constant current signal is injected into adjacent electrode pairs that are uniformly distributed in the sensing area, and the information of the voltage change between adjacent electrode pairs is measured at the same time, a total of $n(n - 3)$ measurements can be obtained, where n is the number of electrodes, usually 16 or 32 electrodes, and a 16-electrode model is used in the sensing system in this paper (Fig. 1).

2.1 Forward Problem

The forward problem of EIT is the prerequisite and basis for solving the inverse problem, i.e., solving the electromagnetic field boundary value problem. For a sensing region Ω with a smooth boundary $\partial\Omega$, the forward problem of EIT aims to predict the boundary potential ϕ on the haptic sensor boundary $\partial\Omega$ by using the known conductivity distribution σ and DC current injected into the boundary electrodes I_{inj}. Based on Kirchhoff's law, it can be deduced that the boundary potential ϕ with the conductivity distribution σ:

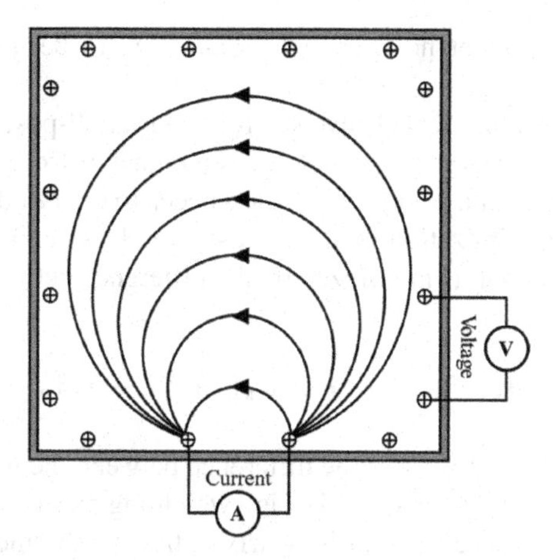

Fig. 1. The principle of the EIT-based tactile sensor

$$\nabla \cdot \left[\sigma(x, y)\nabla\phi(x, y)\right] = 0, (x, y) \in \Omega \quad (1)$$

For a fixed conductivity σ, we can calculate the boundary current density j on $\partial\Omega$ based on the boundary conditions this current density j can be expressed by the following equation:

$$j = \sigma \nabla\phi \cdot \boldsymbol{n} \text{ on } \partial\Omega \quad (2)$$

In this formulation, $\boldsymbol{j}$ represents the current density, while $\boldsymbol{n}$ denotes the unit normal vector of the boundary $\partial\Omega$. This constitutes a Dirichlet-Neumann boundary value problem. The potential ϕ is uniquely determined given a particular conductivity σ and current density j. The numerical solution of this potential value can be obtained by the Finite Element Method (FEM). The FEM divides the sensing area into n cells, and the potential distribution is then calculated from the values at the nodes between each cell by a simulation tool.

2.2 Inverse Problem

Differential imaging is a fast, non-iterative imaging method that reduces the disturbances associated with unknown contact impedances and inaccurate electrode positions [7]. The essence of the method is to first calculate the initial voltage $\boldsymbol{v}_0$ of a homogeneous sensing domain with a substrate conductivity $\boldsymbol{\sigma}_0$, and then calculate the difference $\Delta\boldsymbol{v}$ between the current voltage $\boldsymbol{v}_1$ and the initial voltage $\boldsymbol{v}_0$ The discrete model is solved linearly and this linear approximation is used to calculate the difference conductivity $\Delta\boldsymbol{\sigma}$ between the inhomogeneous sensing domain $\boldsymbol{\sigma}_1$ and the homogeneous sensing domain $\boldsymbol{\sigma}_0$. After calculating the Jacobi matrix $\boldsymbol{J}$ between the boundary potential and the variation of the internal conductivity, the discrete form of the linearized problem becomes:

$$\Delta v = J \cdot \Delta\sigma + N \quad (3)$$

where $\boldsymbol{N}$ is the measurement noise, which is assumed to be uncorrelated Gaussian measurement noise.

Image reconstruction for the EIT inverse problem is an ill-posed problem because it is not possible to estimate a large number of image parameters from limited measurement data. In this case, additional a priori information needs to be introduced to complement the available data. A minimization method is used to obtain an approximate solution, where the objective function is minimized by the difference between the measured and predicted data.

$$arg\ \underset{\Delta\sigma}{min}\|\Delta\boldsymbol{v}-\boldsymbol{J}\cdot\Delta\boldsymbol{\sigma}\|_{\boldsymbol{W}}^{2}+\|\hat{\boldsymbol{\sigma}}-\boldsymbol{\sigma}_{*}\|_{\boldsymbol{R}}^{2} \tag{4}$$

The first of these $\Delta\boldsymbol{v}-\boldsymbol{J}\cdot\Delta\boldsymbol{\sigma}$ is the difference between the measured data and the data estimated by the forward model, and $\boldsymbol{W}$ is a weighting matrix for the measurements, representing the inverse noise covariance matrix of the measurements, representing the measurement accuracy. The second term $\hat{\boldsymbol{\sigma}}-\boldsymbol{\sigma}_{*}$ is the difference between the reconstructed estimate $\hat{\boldsymbol{\sigma}}$ and the a priori estimate of the image $\boldsymbol{\sigma}_{*}$, and $\boldsymbol{R}$ is a regularization matrix representing the image's certain a priori assumptions. By solving Eq. (5), a linearized inverse solution can be obtained, denoted as:

$$\Delta\sigma=\left(\boldsymbol{J}^{T}\boldsymbol{W}\boldsymbol{J}+\lambda^{2}\boldsymbol{R}\right)^{-1}\boldsymbol{J}^{T}W\cdot\Delta v=H\cdot\Delta v \tag{5}$$

where $\boldsymbol{H}=\left(\boldsymbol{J}^{T}\boldsymbol{W}\boldsymbol{J}+\lambda^{2}\boldsymbol{R}\right)^{-1}\boldsymbol{J}^{T}\boldsymbol{W}$ is the linear reconstruction matrix, and the hyper-parameter λ controls the relative weights of the a priori solutions and those based on measured data-based relative weights of the exact solution. Multiple iterations are computed until the iteration value that minimizes the error value is reached. For small changes in conductivity, a single-step calculation is usually used and an iterative approach is not required.

2.3 Flexible Sensor

Velostat is a composite polymer material consisting of carbon-impregnated polyethylene, which has a surface resistance of about 31,000 ohms/sq.cm. Our earliest models for making flexible sensors had a size of 30 cm * 30 cm, and used copper foil as electrodes and hand-soldered wires. Since the temperature range of Velostat is around 45 °C to 65 °C, it is not possible to directly by welding the copper foil on the surface of Velostat, instead, a double-sided conductive copper foil with adhesive backing was pasted on the surface of Velostat (Fig. 2). The electrodes were connected to the data acquisition system via wires, and by pressing on the Velostat surface, a change in conductivity was caused, which resulted in a voltage difference signal that was transmitted to the host computer via the data acquisition system and the conductivity change was imaged using a reconstruction algorithm.

By simple fabrication of the flexible sensor, we can verify our feasible experimental scheme (Fig. 3). We also found that the instability of our EIT reconstructed image signals is mainly due to the floating wires interfering with the running flexible sensors, in addition to the fact that similar to most piezoresistive materials, its main drawback is that its conductive properties can change as the material ages. Piezoresistive materials deform over time and do not return to their initial state quickly. Piezoresistive sensors are typically nonlinear and exhibit significant hysteresis. Based on our preliminary experiments, we have updated the design of the flexible sensor, optimized the reconstruction of the garlic peddler, and enriched the structure of the flexible sensor so that it can display touch information more stably.

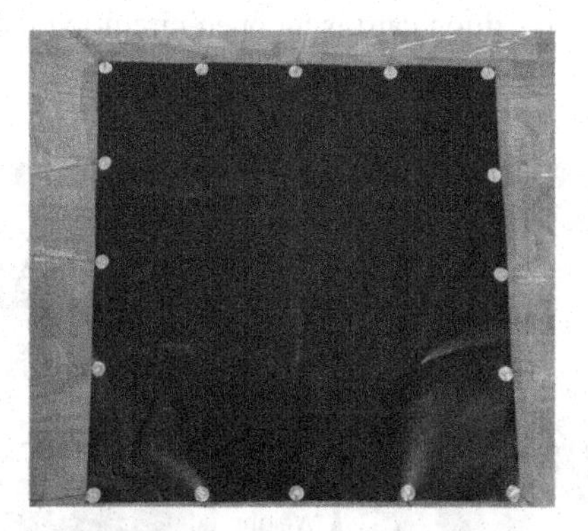

Fig. 2. Flexible Material: Velostat

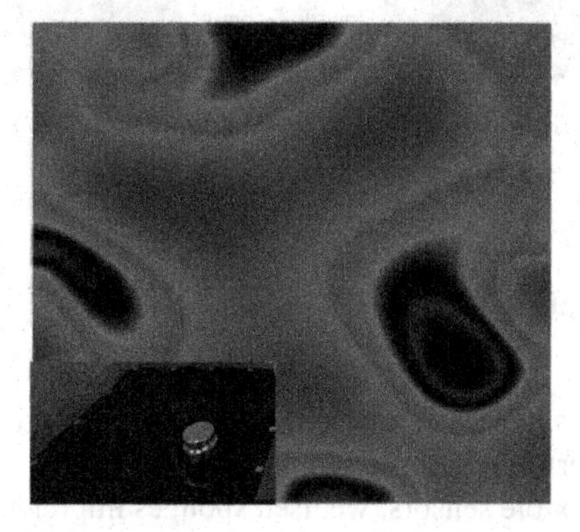

Fig. 3. Preliminary validation of imaging using flexible sensors

3 Experimental Studies

3.1 Data Acquisition System

EIT data acquisition system EIT sensor, multiplexer, constant current source, data acquisition card composition, data acquisition system acquisition rate of 20 kHz, data acquisition card using 12 V power supply, the input multiplexing switch current is 25 mA. Multiplexer selection of two 16-to-1 multiplexing switches MAX306, the data acquisition card, and the host computer using the network port connection between the data acquisition card and the host computer. The host computer sends commands to control the switching of the analog switch through the network port, and the constant current source is injected into the electrode pairs of the sensor through the analog switch, and then the voltage values of the 16 electrodes are measured in parallel, and then the dark loss is given to the software of the host computer through the network port. In order to ensure that the analog data measurement is not interfered with, the constant current source and the digital circuit are powered by batteries separately (Fig. 4). The digital output part of the data acquisition card is an open-circuit output, and pull-up resistors are added to the peripheral circuit.

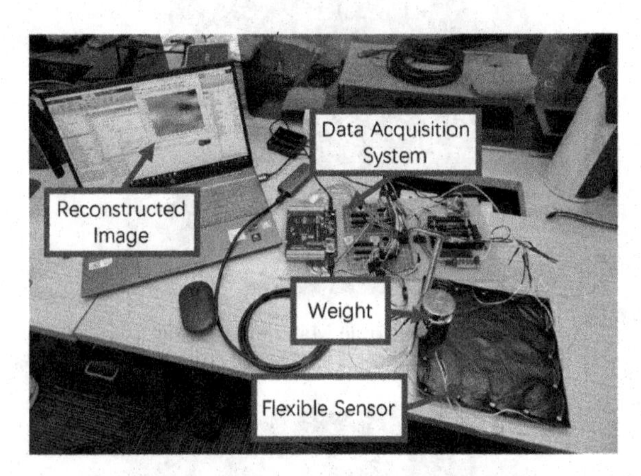

Fig. 4. Preliminary validation of imaging using flexible sensors

3.2 Flexible Sensor Fabrication

As mentioned in Sect. 2.3, excessive wire wrapping can lead to poor imaging performance of flexible sensors fabricated using Velostat, we optimized the wire arrangement to reduce the signal interference caused by the wires; in order to enhance the acquired signal strength of the flexible sensors, we used sponges impregnated with graphite conductive spray paint to increase the electrical conductivity of the surface of the Velostat as well as to improve the piezoresistive characteristics of the Velostat surface. To simulate the rigid surface of the robot, we fixed the optimized Velostat sensors together with the conductive sponge on a 20 cm * 20 cm epoxy resin plate. The epoxy resin plate provides a stable substrate to ensure that the flexible sensor can maintain a stable and reliable working condition when it comes into contact with the load. The electrodes are metal electrodes, uniformly distributed at the boundary of the sensing domain (Fig. 5).

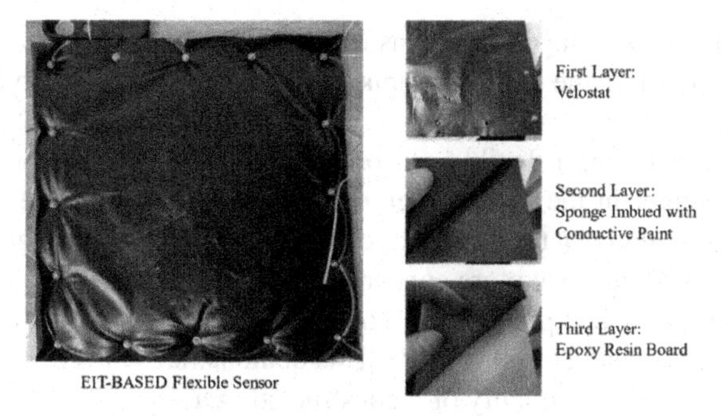

Fig. 5. Structure of flexible sensor

3.3 Bi-model Structure

When performing EIT image reconstruction, the number of conductivity elements is greater than the number of measurement data, which results in an underdetermined EIT system. In an underdetermined system, there are more unknowns (conductivity elements) and fewer measurement data points. This means that there are an infinite number of possibilities to generate multiple conductivity distributions that all match the measured data.

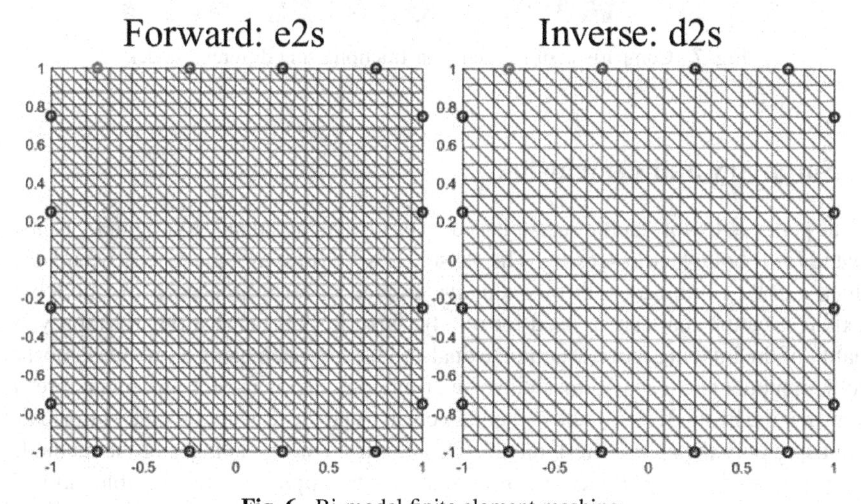

Fig. 6. Bi-model finite element meshing

We use the Eidors toolkit [8], and for the inverse problem, the model is dissected with a sparse mesh and for the positive problem solution, the model is dissected with a fine mesh (Fig. 6). The fine finite element model is used to perform the positive problem to compute the measurements from the parameters, the number of pixels is higher and the size of the pixels is smaller, so the resolution and accuracy of the reconstructed image is relatively high and the parameters of the flexible sensor can be computed more accurately. The sparse mesh is used to model the sensor using the sparse finite element model, given the measured boundary voltages, to reconstruct the internal conductivity distribution of the sensor, which reduces the number of unknown parameters and simplifies the inverse problem solving.

We experimented with a flexible sensor optimized using a bi-model (e2s-d2s) by placing a 100g weight in the central region of the sensor to simulate pressing on the robot's skin. It was found that the optimization of the reconstructed images using the dual model resulted in significantly more concentrated pixels, clearer boundaries, and more accurate locations of contact points in the contact region compared to the reconstruction using only the sparse model (d2s) (Fig. 7). This demonstrates the effectiveness of dual-model EIT in improving the quality of reconstructed images.

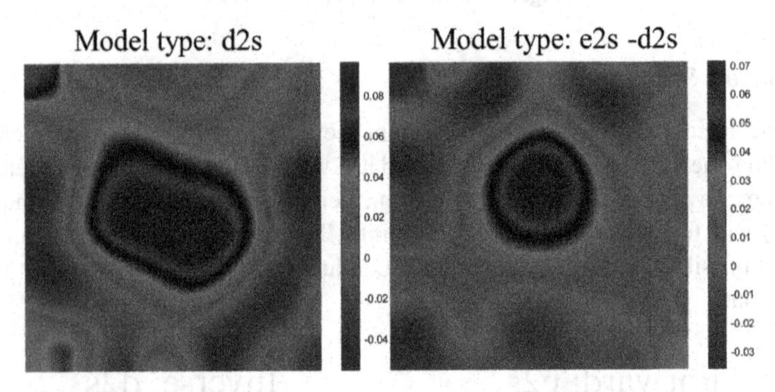

Fig. 7. Comparison of image reconstruction using different models

4 Summary and Conclusion

In this paper, a flexible tactile sensor based on electrical impedance tomography is investigated for robotic tactile sensing. The sensor uses Velostat as a piezoresistive material, which can be combined with a conductive sponge to form an elastic and stretchable flexible structure that can be used for robotic flexible skin fabrication. To improve the quality of the imaging reconstruction, a dual-model approach using a fine-mesh positive problem model and a sparse-mesh inverse problem model is used. Experimental results show that the bi-model structure can obtain clearer and more accurate touch information than using only the sparse mesh model. This confirms the advantages of the method.

In this study, an EIT tactile sensor that can be applied to the flexible skin of a robot was realized, which provides a new method for tactile sensing in robots. Future work can further optimize the sensor performance, improve the sensitivity, stability and resolution, and expand the sensing range. The effect of different material parameters on the imaging quality can also be investigated. This technology has a broad application prospect in empowering robots with tactile interaction capabilities.

Acknowledgments. The work was partially supported by the Natural Science Foundation of Fujian Province of China (Grant No. 2021J01617) and the Program of National Natural Science Foundation of China (Grant Nos. 83321016, U21A20471).

References

1. Liatsis, P.: Learning to "see" touch. In: 2020 International Conference on Systems, Signals and Image Processing (IWSSIP), Niteroi, Brazil, pp. 9–10 (2020)
2. Yang, Y., Zhou, W., Chen, X., Ye, J., Wu, H.: A flexible touching sensor with the variation of electrical impedance distribution. Measurement **183**, 109778 (2021)
3. Zhang, Y., Harrison, C.: Pulp nonfiction: low-cost touch tracking for paper. In: Proceedings of the 2018 CHI Conference on Human Factors in Computing Systems (CHI 2018), Paper 117, pp. 1–11. Association for Computing Machinery, New York (2018)
4. Chen, H., Yang, X., Geng, J., Ma, G., Wang, X.: A skin-like hydrogel for distributed force sensing using an electrical impedance tomography-based pseudo-array method. ACS Appl. Electron. Mater. **5**(3), 1451–1460 (2023)
5. Dzedzickis, A., et al.: Polyethylene-carbon composite (Velostat®) based tactile sensor. Polymers **12**, 2905 (2020)
6. Silvera-Tawil, D., Rye, D., Velonaki, M.: Artificial skin and tactile sensing for socially interactive robots: a review. Robot. Auton. Syst. **63**(Part 3), 230–243 (2015)
7. Silvera Tawil, D., Rye, D., Velonaki, M.: Touch modality interpretation for an EIT-based sensitive skin. In: 2011 IEEE International Conference on Robotics and Automation, Shanghai, China, pp. 3770–3776 (2011). https://doi.org/10.1109/ICRA.2011.5979697
8. Adler, A., Lionheart, W.R.: Uses and abuses of EIDORS: an xtensible software base for EIT. Physiol. Meas. **27**(5), S25 (2006)

Joint Domain Alignment and Adversarial Learning for Domain Generalization

Shanshan Li[1(✉)], Qingjie Zhao[1(✉)], Lei Wang[2], Wangwang Liu[2], Changchun Zhang[3], and Yuanbing Zou[1]

[1] School of Computer Science and Technology, Beijing Institute of Technology, Beijing 100081, China
{3120185489,Zhaoqj}@bit.edu.cn

[2] Beijing Institute of Control Engineering, Beijing 100190, China

[3] School of Technology, Beijing Forestry University, Beijing 100083, China

Abstract. Domain generalization aims to extract a classifier model from multiple observed source domains, and then can be applied to unseen target domains. The primary challenge in domain generalization lies in how to extract a domain-invariant representation. To tackle this challenge, we propose a multi-source domain generalization network called Joint Domain Alignment and Adversarial Learning (JDAAL), which learns a universal domain-invariant representation by aligning the feature distribution of multiple observed source domains based on multi-kernel maximum mean discrepancy. We adopt an optimal multi-kernel selection strategy that further enhances the effectiveness of embedding matching and approximates different distributions in the domain-invariant feature space. Additionally, we use an adversarial auto-encoder to bound the multi-kernel maximum mean discrepancy for rendering the feature distribution of all observed source domains more indistinguishable. In this way, the domain-invariant representation generated by JDAAL can improve the adaptability to unseen target domains. Extensive experiments on benchmark cross-domain datasets demonstrate the superiority of the proposed method.

Keywords: Domain generalization · Domain alignment · Adversarial learning

1 Introduction

With the rapid development of deep learning technology in recent years, various computer vision tasks achieve impressive success [1,2]. However, most deep models rely on large-scale labeled datasets for supervised learning, which may result in poor performance when the training domains and testing ones follow different distributions. To address this issues, domain generalization (DG) studies [3,4] have been conducted to extract a classifier model from multiple source

This work was supported by Pre-research Project on Civil Aerospace Technologies of China National Space Administration (D010301).

F. Sun and J. Li (Eds.): ICCCS 2023, CCIS 2029, pp. 132–146, 2024.
https://doi.org/10.1007/978-981-97-0885-7_12

domains and apply it to a target domain, with the aim of minimizing the impact of inconsistent distributions.

Previous domain generalization methods have focused on reducing domain shift by learning a universal domain invariant feature across different source domains. For example, Mohammad *et al.* [5] proposed a framework for correlation-aware adversarial domain adaption and domain generalization, which minimizes the domain discrepancy among source domains by integrating correlation alignment with adversarial training. The Meta-Variational Information Bottleneck (MetaVIB) [6] proposed a novel approach to extract domain-invariant representations by leveraging the meta-learning setting of domain generalization and deriving variational bounds of mutual information.

Recent works [7,8] aligned the source domain distribution to generate domain-invariant representations by minimizing the domain discrepancy among multiple source domains. Li *et al.* [9] extended the Adversarial Auto-Encoder (AAE) by imposing the Maximum Mean Discrepancy (MMD) metric to approximate different distributions across domains for domain generalization. However, using a single or fixed kernel to map various distributions of domain invariant feature space may lead to inaccurate and inadequate outcomes. Therefore, the selection of an optimal kernel for mapping various distributions to a domain invariant feature space is necessary. In this paper, we propose a novel model called Joint Domain Alignment and Adversarial Learning (JDAAL) to acquire a universal feature among multi-domains. Our main work can be summarized as follows:

1. We propose a multi-source domain generalization network called Joint Domain Alignment and Adversarial Learning (JDAAL), which learns a universal domain invariant representation by aligning the domain feature distribution of multiple seen source domains based on the multi-kernel maximum mean discrepancy. An optimal multi-kernel selection strategy is adopted to further improve embedding matching effectiveness and approximate different distributions in the domain invariant feature space.
2. The adversarial auto-encoder is employed to bound multi-kernel maximum mean discrepancy, making the domain feature distribution of all seen source domains more indistinguishable and reducing the risk of overfitting.
3. The experiments show that our model effectively addressed overfitting in the source domain data and achieved good classification accuracy compared to other approaches.

2 Proposed Method

Notations and Definitions. We are given K domains. $\mathbf{X}= [x_1, ..., x_n]$ is the feature space, $\mathbf{Y} = [y_1, ..., y_n]$ is the label space of each domain, and $\mathbf{X}$ corresponding to $\mathbf{Y}$ one to one.

2.1 Network Architecture

Considering only the source domains observed during the training process, we assume that both source and target domains can be effectively mapped onto a domain-invariant feature space. In this work, we propose a novel approach joint domain alignment and adversarial learning. Figure 1 shows the overall architecture of our proposed method for DG. This network introduces a priori distribution to regulate the training data distributions, enabling it to be mapped to the domain invariant feature space. Specifically, the encoder $Q(X)$ is adopted to map the train data distribution into the hidden space, generating hidden codes $\mathbf{H}$. Then, a set of characteristic kernels maps the hidden codes to a reproducing kernel Hilbert space (RKHS), and the multi-kernel maximum mean discrepancy is used to measure the distribution discrepancy between the two hidden codes. Finally, decoder $P(h)$ reconstructs the feature vector from the hidden space. In this process, our model employs adversarial auto-encoder to match the prior distribution with train data distribution, making the domain feature distribution of all seen source domains more indistinguishable and reducing the risk of overfitting. The encoder and decoder share weights among all training domains in the training stage. The hidden codes generated by the encoder serves as compressed low-dimensional feature vectors, which are utilized for feature training and domain discrimination. The universal domain invariant representation produced by the proposed model improves the adaptability to unseen domains.

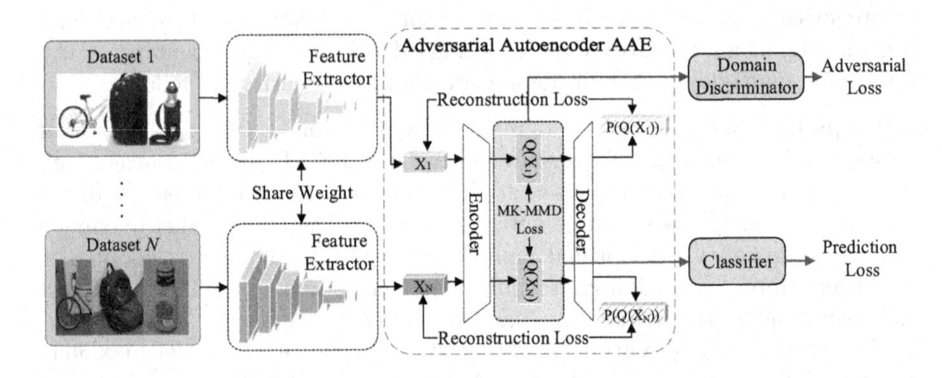

Fig. 1. An overview of JDAAL framework for domain generalization

2.2 Domain Alignment Module

In our study, we utilize the multiple kernel maximum mean discrepancy (MK-MMD) to align the hidden sub-space distributions among several source datasets. MK-MMD is an extension of the MMD metric. The feature vectors from various source domains are mapped to a reproducing kernel Hilbert space using a characteristic kernel, and the MMD metric then measures the distribution discrepancy after mapping. The characteristic kernel of MMD is predetermined and fixed; however, relying solely on a single kernel mapping may lead to inadequate and imprecise measurements. By incorporating multiple kernels that utilize different characteristic kernels, the MK-MMD test can be enhanced to yield more precise and reliable results.

Let $\mathcal{H}_k$ denote the Reproducing Kernel Hilbert Space equipped with a characteristic kernel k. Using the encoder in the AAE, the feature vectors $\mathbf{X}_l$ and $\mathbf{X}_t$ $(l, t \in K)$, characterized by probability distributions $\mathcal{P}_l$ and $\mathcal{P}_t$ respectively, are mapped to the hidden space and generate the compressed feature vectors $\mathbf{H}_l$ and $\mathbf{H}_t$, also referred to as hidden codes. The compression feature vectors $\mathbf{H}_l$ and $\mathbf{H}_t$ obtained from the encoder are then projected onto a shared RKHS $\mathcal{H}_k$, where $\phi(\cdot)$ represents the mapping operation function. Subsequently, we define a mean map $\mu_k(p)$, which represents the expected value of each hidden feature vector obtained from the same probability distribution and mapped to a common RKHS $\mathcal{H}_k$. Let $\mathbf{X}_l = \{\mathbf{x}_{l,i}\}_i^{n_l}$ and $\mathbf{X}_t = \{\mathbf{x}_{t,j}\}_j^{n_t}$ denote the sample sets of the distributions $\mathcal{P}_l$ and $\mathcal{P}_t$, where n_l and n_t represent the same sizes for each batch-size. Utilizing the MMD, we compute the discrepancy in domain feature distribution between domains l and t (or $\mathcal{P}_l$ and $\mathcal{P}_t$) as follow:

$$\mathrm{MMD}(\mathcal{P}_l, \mathcal{P}_t)^2 = \|E_{X_l\sim\mathcal{P}_l}[\phi(\mathbf{X}_l)] - E_{X_t\sim\mathcal{P}_t}[\phi(\mathbf{X}_t)]\|^2_{\mathcal{H}_k}. \tag{1}$$

The high-dimensional feature vectors $\mathbf{X}_l = \{\mathbf{x}_{l,i}\}_i^{n_l}$ and $\mathbf{X}_t = \{\mathbf{x}_{t,j}\}_j^{n_t}$, generated by backbone network, are mapped to the hidden layer to obtain the hidden codes $\mathbf{H}_l = \{\mathbf{h}_{l,i}\}_i^{n_l}$ and $\mathbf{H}_t = \{\mathbf{h}_{t,j}\}_j^{n_t}$. This process involves mapping high-dimensional feature vectors to low-dimensional feature vectors, resulting in a significant reduction in subsequent computational requirements. In summary, it effectively captures the domain-specific distribution differences.

$$\mathrm{MMD}(\mathbf{H}_l, \mathbf{H}_t)^2 = \left\| \frac{1}{n_l}\sum_{i=1}^{n_l}\phi(\mathbf{h}_{l,i}) - \frac{1}{n_t}\sum_{j=1}^{n_t}\phi(\mathbf{h}_{t,j}) \right\|^2_{\mathcal{H}_k}. \tag{2}$$

It has been proven that the mapping to the RKHS $\mathcal{H}_k$ is injective if the kernel $k(\cdot,\cdot)$is characteristic. This injectivity indicates that an element in RKHS uniquely represents any arbitrary distribution $\mathcal{P}$. The feature map $\phi(\cdot)$ is determined by the characteristic kernel k, such that the kernel function can be expressed as $k(\mathbf{h}_{l,i}, \mathbf{h}_{t,j}) = \langle\phi(\mathbf{h}_{l,i}), \phi(\mathbf{h}_{t,j})\rangle$. According to the polynomial expansion rule, the MMD metric formulated in Eq. 2 can be written as follows:

$$\begin{aligned}\mathrm{MMD}(\mathbf{H}_l, \mathbf{H}_t)^2 = &\frac{1}{(n_l)^2}\sum_{i=1}^{n_l}\sum_{i'=1}^{n_l} k(\mathbf{h}_{l,i}, \mathbf{h}_{l,i'}) \\ &- \frac{1}{(n_t)^2}\sum_{j=1}^{n_t}\sum_{j'=1}^{n_t} k(\mathbf{h}_{t,j}, \mathbf{h}_{t,j'}) \\ &- \frac{2}{n_l \cdot n_t}\sum_{i=1}^{n_l}\sum_{j=1}^{n_t} k(\mathbf{h}_{l,i}, \mathbf{h}_{t,j}).\end{aligned} \tag{3}$$

The MK-MMD method bears resemblance to MMD, yet it distinguishes itself by employing a diverse set of characteristic kernels for distance calculation. Unlike the conventional MMD approach that relies on a single predefined kernel, the MK-MMD framework encompasses a family of kernels, each yielding distinct MMD values. This family of characteristic kernels can be represented as follows:

$$\kappa = \left\{ k = \sum_{u=1}^{m} \beta_u k_u : \sum_{u=1}^{m} \beta_u = 1, \beta_u \geq 0, \forall u \right\}, \tag{4}$$

where m denotes the number of characteristic kernels. Each kernel $k_u \in \kappa$ is uniquely associated with an RKHS $\mathcal{H}_k$. The weight of a linear k_u is denoted by β_u. Combining Eq. 3 and Eq. 4, we can express the MK-MMD metric as follows:

$$\begin{aligned} \mathrm{MKMMD}(\mathbf{H}_l, \mathbf{H}_t)^2 = & \frac{1}{(n_l)^2} \sum_{i=1}^{n_l} \sum_{i'=1}^{n_l} \kappa\left(\mathbf{h}_{l,i}, \mathbf{h}_{l,i'}\right) \\ & - \frac{1}{(n_t)^2} \sum_{j=1}^{n_t} \sum_{j'=1}^{n_t} \kappa\left(\mathbf{h}_{t,j}, \mathbf{h}_{t,j'}\right) \\ & - \frac{2}{n_l \cdot n_t} \sum_{i=1}^{n_l} \sum_{j=1}^{n_t} \kappa\left(\mathbf{h}_{l,i}, \mathbf{h}_{t,j}\right). \end{aligned} \tag{5}$$

To further heighten the domain invariance of the latent feature, we employed a principled method to obtain an optimal kernel by linearly combining multiple kernels, thereby seeking more efficient kernels. Hence, we seek to learn optimal kernel parameter β for MK-MMD by jointly maxmizing the test power and minimizing the Type **II** error [10], leading to the optimization,

$$\max_{k \in \kappa} d_k^2\left(\mathbf{H}_l, \mathbf{H}_t\right) \sigma_k^{-2}, \tag{6}$$

where $\sigma_k^2 = \mathbf{E}_{\mathrm{z}} g_k^2(\mathrm{z}) - [\mathbf{E}_{\mathrm{z}} g_k(\mathrm{z})]^2$ denotes the estimation variance by the k-th characteristic kernel in a family kernels. More specifically, we denote quad-tuple $\mathrm{z}_i \triangleq (x_{2i-1}^l, x_{2i}^l, x_{2i-1}^t, x_{2i}^t)$, and evaluate multi-kernel function k on each quad-tuple z_i by $g_k(\mathrm{z}_i) \triangleq k(x_{2i-1}^l, x_{2i}^l) + k(x_{2i-1}^t, x_{2i}^t) - k(x_{2i-1}^l, x_{2i}^t) - k(x_{2i}^l, x_{2i-1}^t)$. Let $\mathbf{d} = (d_1, d_2, ..., d_m)^T$, where each d_u is evaluated using MMD via kernel k_u. Therefore, in this paper, we formulate the kernel selection strategy as a minimax problem to address the entire objective.

The JDAAL model leverages multiple domains (K domains) to learn domain-invariant representations for domain generalization, and the MK-MMD loss L_{MKMMD} is calculated as follows:

$$L_{\mathrm{MKMMD}}(\mathbf{H}_1, \mathbf{H}_2, ...\mathbf{H}_K) = \frac{1}{K^2} \sum_{i \leq 1, j \leq K} \mathrm{MKMMD}(\mathbf{H}_i, \mathbf{H}_j). \tag{7}$$

Adversarial Learning. Our model is designed to prioritize domain-invariant features over domain-specific visual information, in order to extract a representation that is robust across different data distributions. To further obfuscate the label information of each domain, we employ an adversarial autoencoder. Let x

denote the input and z denote the latent code vector (hidden units) of an autoencoder with a deep encoder and decoder. Let $p(\mathbf{h})$ denotes the prior distribution imposed on the codes, $q(\mathbf{h}|\mathbf{x})$ denotes the encoding distribution, and $p(\mathbf{x}|\mathbf{h})$ represents the decoding distribution. Additionally, let $p_d(\mathbf{x})$ be indicative of the data distribution while $p(\mathbf{x})$ denotes the model distribution. The encoding function $q(\mathbf{h}|\mathbf{x})$ defines an aggregated posterior distribution $q(\mathbf{h})$ on the hidden code vector, which represents a fundamental aspect of the autoencoder as follows:

$$q(\mathbf{h}) = \int_{\mathbf{x}} q(\mathbf{h}|\mathbf{x}) p_d(\mathbf{x}) d\mathbf{x}. \tag{8}$$

The adversarial autoencoder is an autoencoder that is regularized by aligning the aggregated posterior, $q(\mathbf{h})$, with an arbitrary prior, $p(\mathbf{h})$. The AAE consists of three basic components: Encoder $Q(\mathbf{X})$, Decoder $P(\mathbf{H})$ and Discriminator D. The feature vector $\mathbf{X}$ is encoded by the encoder $Q(\mathbf{X})$ to obtain hidden codes $\mathbf{H}$ (i.e., $\mathbf{H} = Q(\mathbf{X})$), which are then used by thr decoder $P(\mathbf{H})$ to reconstruct inputs. Meanwhile, discriminator D is trained to discriminate between the prior distribution and real data distribution. During training, the AAE is optimized using two loss functions, namely reconstruction loss and adversarial loss.

As is widely acknowledged, images features extracted by the convolutional neural networks are high-dimensional data. An encoder can compress this set of features into a hidden space to obtain a set of low-dimensional features, which significantly reduces computational complexity for subsequent calculations. However, there will inevitably be some errors in the reconstructed features generated by the encoder and decoder. The reconstruction error of AAEs can be expressed as follows:

$$error_{rec} = \left(X - P\left(Q\left(X\right)\right)\right)^2. \tag{9}$$

The reconstruction loss of all the training domains can be expressed as:

$$L_{rec} = E_{X \sim \mathcal{P}(X)}\left[\left(X - P\left(Q\left(X\right)\right)\right)^2\right]. \tag{10}$$

Although autoencoder are capable of acquiring low-dimensional features from convolutional features, the distribution of these features in the latent space cannot be specified. To address this issue, we train encoder and discriminator networks through adversarial learning to approximate a hidden code distribution to an arbitrary distribution. In our work, we adopt a Normal distribution $\mathcal{N}$ with the hyperparameters μ, σ^2 as the prior distribution. The encoder and discriminator are trained with an adversarial loss, which compares the latent vector sampled from the prior distribution with the latent vectors encoded from the real data distribution. Hence, a discriminator D is trained to distinguish whether the input is sampled from the encoder generated or prior Normal distribution. In contrast, the encoder Q is trained to align its generated output with the prior Normal distribution. The adversarial loss can be defined as follows:

$$L_{adv} = E_{\mathbf{h} \sim \mathcal{P}(\mathbf{h})}\left[\log(D(\mathbf{h}))\right] + E_{X \sim \mathcal{P}(X)}\left[\log\left(1 - D\left(Q\left(X\right)\right)\right)\right]. \tag{11}$$

During adversarial training, minimizing the adversarial loss L_{adv} can facilitate features obtained from each seen source domain embedding in a latent space generated by the prior distribution and confuse the domain label information for each seen source domain. This operation makes it challenging for the domain discriminator to accurately identify corresponding domain labels. After the training of AAE, the encoder generates hidden codes that exhibit a higher proximity to the hidden codes sampled from the prior distribution. Consequently, we are able to reconstruct the feature vector obtained from each source domain without relying on domain label information, thereby enabling generation of domain invariant features through the decoder.

2.3 Model Training

In the training process, the classes label information $\mathbf{Y} = [y_1, ..., y_n]$ is embedded into the latent variable to generate new feature vectors for a particular type in supervised domain generalization. Therefore, Cross-Entropy loss is applied to address this prediction task:

$$L_{pre} = \frac{1}{n}\sum_{i=1}^{n} \log\left(\mathbf{h}_i, y_i\right) = \frac{1}{n}\sum_{i=1}^{n} \log\left(Q(x_i), y_i\right). \tag{12}$$

The JDAAL model is trained using multiple domain data, and its training procedure is considered a multi-objection optimization process. Hence, the overall training losses can be formulated as follows:

$$L = L_{pre} + \alpha_0 L_{rec} + \alpha_1 L_{\mathrm{MKMMD}} + \alpha_2 L_{adv}, \tag{13}$$

where α_0, α_1 and α_2 are the hyperparameters that control the trade-off among several lossed. They are assigned different values in various experiments.

Our JDAAL network aims to minimize training loss L, and jointly minimize prediction, reconstruction, adversarial and MK-MMD metric on the latent variable. The AAE training procedure is based on that of Generative Adversarial Networks (GAN). Similar to other adversarial learning models, the JDAAL method's training procedures are divided into three stages. These training stages require multiple iterations until the model achieves stability. Let C, Q, P and D represent parameters of the classifier, encoder, decoder and discriminator, respectively. The training procedure is described as follow:

1) **Update Encoder and Decoder for reconstruction.** The encoder and decoder are trained to reconstruct the feature vector obtained from the backbone network. During this stage, the reconstruction loss is utilized for optimizing and updating encoder Q and decoder P through minimizing L_{rec}.
2) **Update Discriminator.** The discriminator is trained to distinguish between the hidden codes derived from the real feature vector and those sampled from the prior distribution. During this stage, we optimize and update the discriminator D using adversarial loss, specifically by maximizing L_{adv}.

3) **Update Encoder for Generation.** The discriminator D is capable of distinguishing the hidden codes generated from either the prior distribution or real feature vector more effectively. The encoder Q is trained to generate hidden codes that can be predicted by the discriminator as being generated from the prior distribution. At this stage, the encoder is updated to confuse the discriminator by jointly minimizing prediction loss L_{pre} and MKMMD loss L_{MKMMD}, i.e., learning Q and C through minimizing joint loss.

Algorithm 1 summarizes detailed procedures of the JDAAL model.

Algorithm 1. JDAAL Training

Input: Multiple Dataset domains $\mathbf{X} = \{X_1, \ldots, X_K\}, \mathbf{Y} = \{Y_1, \ldots, Y_K\}$, initialized parameters Q, P, C and D.
Output: Learned Parameters Q^*, P^*, C^* and D^*.
1: **for** $t = 1$ to max iteration **do**
2: **for** get a batch data $\mathbf{X}_d$ and $\mathbf{Y}_d$ from the $\mathbf{X}$ and $\mathbf{Y}$, respectively **do**
3: Calculate: $\mathbf{H}_d = Q(\mathbf{X}_d)$;
4: Calculate: $\hat{\mathbf{X}}_d = P(\mathbf{H}_d)$;
5: Calculate: $minL_{rec} = \mathbf{X}_d - \hat{\mathbf{X}}_d \rightarrow minL_{rec} = (\mathbf{X}_d - P(Q(\mathbf{X}_d)))^2$;
6: Update the Encoder Q and Decoder P;
7: Sample $\mathbf{h}$ from prior distribution;
8: Calculate: $maxL_{adv} = E_{\mathbf{h}\sim\mathcal{P}(\mathbf{h})}[\log(D(\mathbf{h}))] + E_{\mathbf{X}_d\sim\mathcal{P}(\mathbf{X}_d)}[\log(1 - D(Q(\mathbf{X}_d)))]$;
9: Update the Discriminator D;
10: Calculate: $minL_{\text{MKMMD}}(\mathbf{H}_1, \mathbf{H}_2, ...\mathbf{H}_K) = \frac{1}{K^2}\sum_{i\leq 1, j\leq K} \text{MKMMD}(\mathbf{H}_i, \mathbf{H}_j)$;
11: Calculate: $minL_{pre} = \frac{1}{n}\sum_{i=1}^{n} \log(\mathbf{h}_i, y_i) = \frac{1}{n}\sum_{i=1}^{n} \log(Q(x_i), y_i)$;
12: Update the Encoder Q and Classifier C;
13: **end for**
14: **end for**
return E_n;

Accordingly, the objective of JDAAL can be written as following,

$$\min_{C,Q,P} \max_{D} L_{pre} + \alpha_0 L_{rec} + \alpha_1 L_{\text{MKMMD}} + \alpha_2 L_{adv}. \quad (14)$$

3 Experiments

3.1 Datasets and Baseline

To evaluate the effectiveness of the proposed JDAAL model, we conduct a seried of experiments on two benchmark datasets for domain generalization: VLCS dataset [11], Office-Home dataset [12], as shown in Fig. 2.

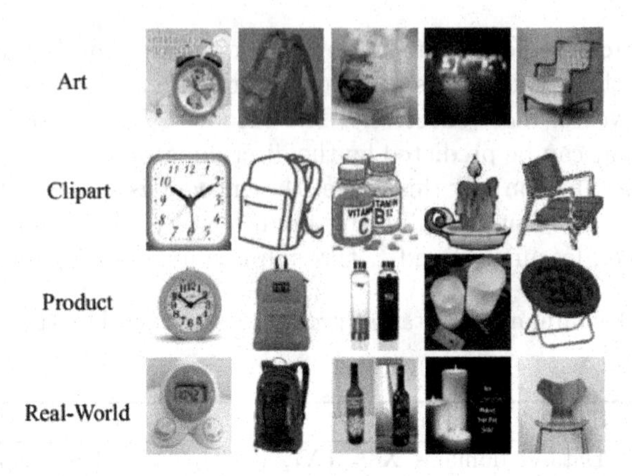

Fig. 2. Example image from Office-Home dataset

3.2 Implementation Details

Following previous literature [13], we adopt the strategy of selecting one domain as testing domain, and the remaining domains as training domains. The encoder Q and decoder P share weights and have an identical network architecture. We evaluate classification accuracy of each generalization task, as well as an overall average accuracy, as our performance metric. The input of the adversarial subnetwork corresponds to the implicit code generated by the encoder, while the input of the classification subnetwork refers to the deep convolutional feature. To train the AAE and whole model, we utilized Adam optimizer with an initial learning rate 10^{-4} for the AAE and an initial learning rate 10^{-5} for the whole model. The batch size is set to 64. Both the classification sub-network and adversarial sub-networks employed ReLU activation function as their non-linear activation function. All experiments were implemented using PyTorch framework on a single GeForce RTX 2080 Ti GPU for training.

3.3 Experiment on Office-Home Dataset

We followed the experimental setup in Sect. 3.5 to train and evaluate our model. The backbone network was choosn as the ImageNet pre-trained ResNet50 with its last layer removed. The source domains dataset was splitted into 9 (train): 1 (val). The entire model is trained for 5K iterations. The encoder Q selects the tanh activation and an output size of 2000 for hidden feature space. The decoder P employs a linear activation function, with an input size of 2000. We report the average classification accuracy of ten independent runs. The parameters of the Eq. 14 are set as $\alpha_0 = 0.5$, $\alpha_1 = 1$, and $\alpha_2 = 2$. Table 1 shows the experimental results in the Office-Home dataset. Five state-of-the-art methods were chosen for comparison with our approach, and their respective scores were derived from the original code implementation using ResNet50 as the underlying architecture.

Table 1. Classification accuracy (%) on the Office-Home dataset with backbone of ResNet-50 architecture.

Method	Art	Clipart	Product	Real world	Average
MMD-AAE [9]	72.92	69.25	84.89	85.74	78.20
JiGen [13]	68.35	56.51	79.53	80.88	71.32
MMLD [14]	67.78	56.33	79.84	82.10	71.51
EISNet [15]	68.36	56.19	79.08	80.61	71.06
DADG [16]	70.63	60.43	81.08	81.65	73.44
JDAAL(ours)	**73.35**	**70.04**	**85.67**	**85.99**	**78.76**

The JDAAL model achieves the best performance in all domain generalization tasks, as shown in Table 1. In comparison to MMD-AAE, JDAAL outperforms it significantly across all four tasks, with improvements of 0.43%, 0.79%, 0.78% and 0.25%, respectively. These superior results can be attributed to the selection of optimal kernels in JDAAL, which effectively leverage diverse characteristic kernels to enhance the MK-MMD test and reduce domain distribution discrepancy more effectively than MMD-AAD. Notably, our approach outperforms MMLD, which also employs an adversarial training strategy for domain generalization model training. This superiority can be attributed to JDAAL's distinctive method of learning domain invariant features, which simultaneously optimizes an autoencoder, a discriminator, and a classifier by jointly leveraging MK-MMD and adversarial training to align distributions across multiple domains.

3.4 Experiment on VLCS Dataset

VLCS dataset is an image classification dataset. We randomly split the data from each domain into 70% for training and 30% for evaluating. And the DeCAF model is adopted to obtain FC6 features (DeCAF6) for overall experiments. We report the average classification accuracy over 20 independent runs. We set $\alpha_0 = 6$, $\alpha_1 = 2$ and $\alpha_2 = 0.2$ of Eq. 14 for VLCS dataset. Other experimental details are consistent with Office-Home. The comparison results are shown in Table 2.

Table 2. Classification accuracy (%) on the VLCS dataset with backbone of Alexnet architecture.

Method	VOC2007	LabelMe	Caltech-101	SUN	Average
CIDDC [17]	64.38	63.06	88.83	62.10	69.59
DGGS [18]	68.14	58.56	93.23	63.89	70.95
MMD-AAE [9]	67.7	62.6	94.4	64.4	72.3
JiGen [13]	70.62	60.90	96.93	64.30	73.19
MMLD [14]	**71.96**	58.77	96.66	67.13	73.88
DADG [19]	70.77	63.44	96.80	66.81	74.46
EISNet [15]	69.83	63.49	**97.33**	**68.02**	**74.67**
JDAAL(ours)	68.76	**65.30**	96.18	66.75	74.25

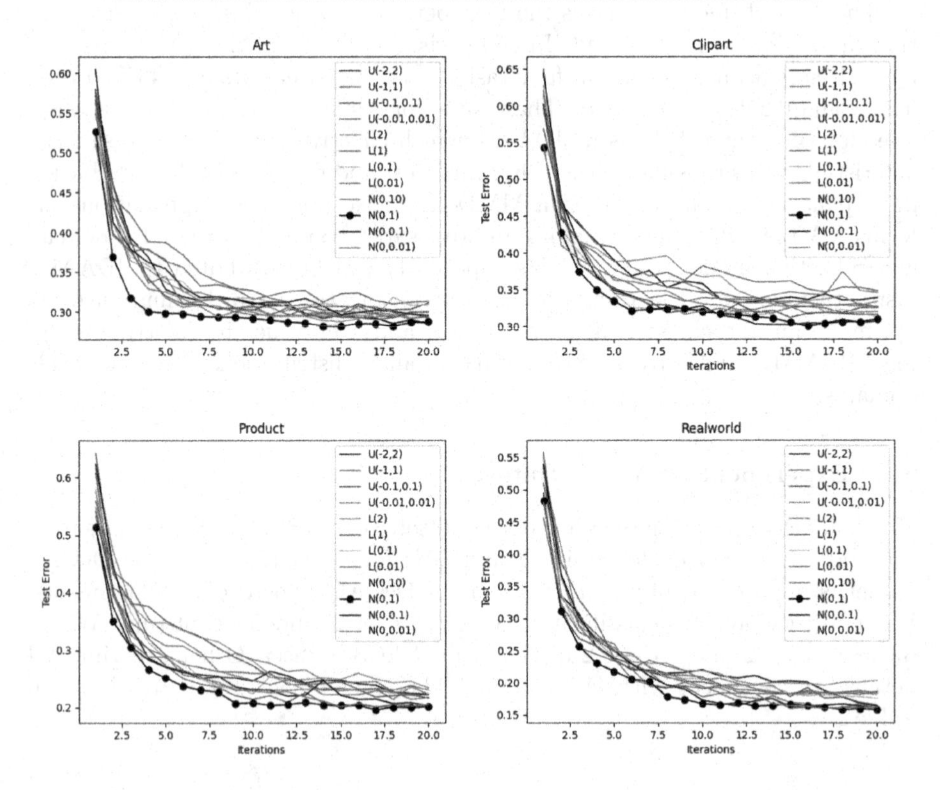

Fig. 3. Test error with different prior distribution in four DG tasks

Based on the results presented in Table 2, it is evident that our JDAAL model has achieved competitive performance across all four domains when compared to several existing methods. Overall, the average accuracy of JDAAL was 74.25%, which is comparable to EISNET (74.67%) and DADG (74.46%), while outperforming CIDDG, MMD-AAE, MMLD and JIGEN by margins ranging from 0.36% to 4.6%. In addition, JDAAL achieves the highest accuracy of 65.30% in Labelme domain, surpassing EISNET by 1.81%. This result further validates that our JDAAL model, which incorporates AAE to constrain multi-kernel maximum mean discrepancy, effectively mitigates overfitting issues associated with source domain data and enhances the adaptability of domain invariant representation.

3.5 Impact on Different Priors

To provide an intuitive understanding of how prior distribution affects the feature space of the domain generalization model, we conducted additional experiments on the Office-Home dataset using different prior distributions. The results are presented in Table 3 and Fig. 3. We compare the average accuracy and test error using Laplace distribution, Normal distribution ($\mathcal{N}$) and Uniform distribution ($\mathcal{U}$) with varying parameters. After analyzing the results presented in Table 3, it is evident that incorporating the Standard Normal distribution as the prior distribution yields superior performance compared to both the Laplace distribution and Uniform distribution. Furthermore, Fig. 3 demonstrates that by examining the test error across four tasks, we can confirm their convergence performance. The Standard Normal distribution exhibits rapid convergence and yields the lowest test error in comparison to other prior distributions, enhancing the generalization capacity of learned features.

Table 3. Classification accuracy (%) with different prior distribution.

Prior Distribution	Art	Clipart	Product	Real world
Laplace (2)	72.51	68.01	85.02	84.99
Laplace (1)	72.06	69.00	84.14	85.45
Laplace (10^{-1})	72.42	69.21	85.29	84.69
Laplace (10^{-2})	72.39	68.19	84.91	84.45
$\mathcal{U}$ (−0.01,0.01)	71.85	69.13	85.35	85.93
$\mathcal{U}$ (−0.1,0.1)	72.34	69.41	85.33	85.54
$\mathcal{U}$ (−1,1)	72.05	69.23	85.17	85.90
$\mathcal{U}$ (−2,2)	72.47	68.54	84.52	84.64
$\mathcal{N}$ (0,0.01)	71.97	68.68	84.90	85.77
$\mathcal{N}$ (0,0.1)	72.55	69.82	84.84	85.46
$\mathcal{N}$ (0,1)	**73.35**	**70.04**	**85.67**	**85.99**
$\mathcal{N}$ (0,10)	72.09	69.04	85.42	83.95

3.6 Impact on Different Kernels

To further analyze the impact of characteristic kernels on domain generalization ability, we conduct several experiments with different numbers of characteristics kernels on the Office-Home dataset. The evaluation metric used was test error for all four DG tasks. Our results, as shown in Fig. 4, indicate that the number of kernel features has a significant effect on performance. Furthermore, the optimal results were obtained when the kernel number was set to seven. It is evident that increasing the number of Gaussian kernels can enhance model generalization to some extent; however, an excessively large quantity will undermine its generalization ability.

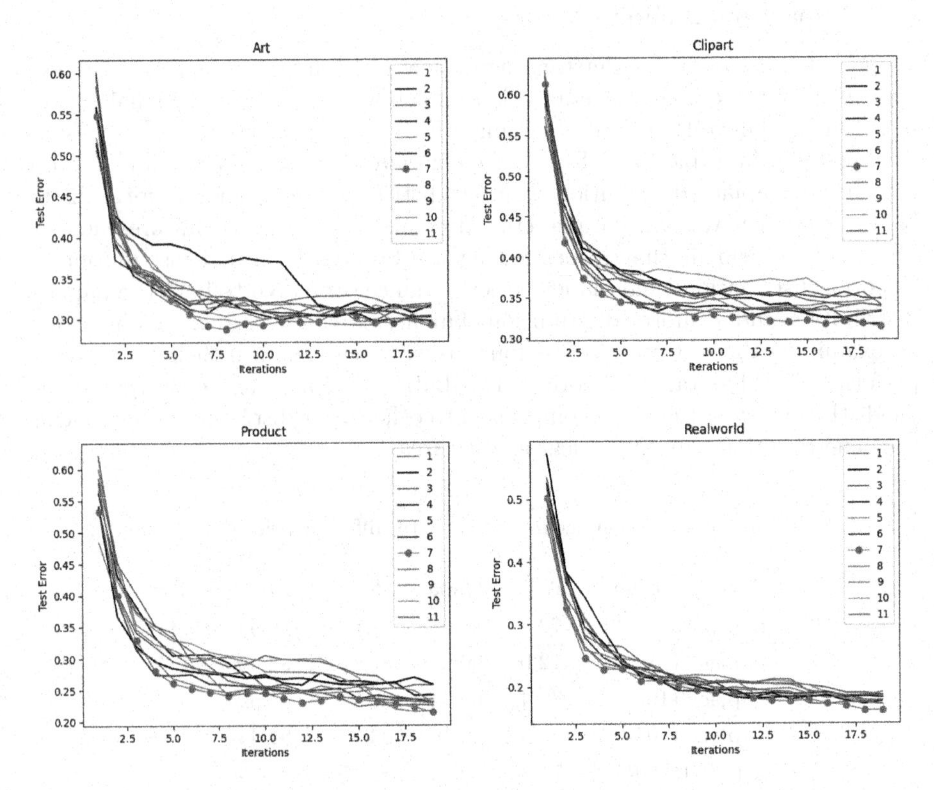

Fig. 4. Test error in different number of kernel in four DG tasks

4 Conclusion

We propose a multi-source domain generalization network, named joint domain alignment and adversarial learning (JDAAL), which leverages the multi-core maximum mean discrepancy measure to align the feature distribution of multiple seen source domains. Moreover, we adopt an optimal multi-core selection strategy to further enhance the embedding matching effect. The adversarial autoencoder is utilized to fuse the multi-kernel maximum mean discrepancy, which makes the domain feature distributions of all seen source domains more indistinguishable. The effectiveness of the JDAAL model is demonstrated through comprehensive theoretical analysis and cross-domain benchmark experiments.

References

1. Doan, K.D., Yang, P., Li, P.: One loss for quantization: deep hashing with discrete Wasserstein distributional matching. In: Proceedings of the IEEE/CVF Conference on Computer Vision and Pattern Recognition, pp. 9447–9457 (2022)
2. Chang, Q., et al.: Data: domain-aware and task-aware self-supervised learning. In: Proceedings of the IEEE/CVF Conference on Computer Vision and Pattern Recognition, pp. 9841–9850 (2022)
3. Zhang, Y., Li, M., Li, R., Jia, K., Zhang, L.: Exact feature distribution matching for arbitrary style transfer and domain generalization. In: Proceedings of the IEEE/CVF Conference on Computer Vision and Pattern Recognition, pp. 8035–8045 (2022)
4. Zhang, X., He, Y., Xu, R., Yu, H., Shen, Z., Cui, P.: Nico++: towards better benchmarking for domain generalization. In: Proceedings of the IEEE/CVF Conference on Computer Vision and Pattern Recognition, pp. 16036–16047 (2023)
5. Rahman, M.M., Fookes, C., Baktashmotlagh, M., Sridharan, S.: Correlation-aware adversarial domain adaptation and generalization. Pattern Recogn. **100**, 107124 (2019)
6. Du, Y., et al.: Learning to learn with variational information bottleneck for domain generalization. In: Vedaldi, A., Bischof, H., Brox, T., Frahm, J.M. (eds.) ECCV 2020. LNCS, vol. 12355, pp. 200–216. Springer, Cham (2020). https://doi.org/10.1007/978-3-030-58607-2_12
7. Harary, S., et al.: Unsupervised domain generalization by learning a bridge across domains. In: Proceedings of the IEEE/CVF Conference on Computer Vision and Pattern Recognition, pp. 5280–5290 (2022)
8. Zhu, W., Lu, L., Xiao, J., Han, M., Luo, J., Harrison, A.P.: Localized adversarial domain generalization. In: Proceedings of the IEEE/CVF Conference on Computer Vision and Pattern Recognition, pp. 7108–7118 (2022)
9. Li, H., Pan, S.J., Wang, S., Kot, A.C.: Domain generalization with adversarial feature learning. In: Proceedings of the IEEE Conference on Computer Vision and Pattern Recognition, pp. 5400–5409 (2018)
10. Gong, B., Shi, Y., Sha, F., Grauman, K.: Geodesic flow kernel for unsupervised domain adaptation. In: 2012 IEEE Conference on Computer Vision and Pattern Recognition, pp. 2066–2073. IEEE (2012)
11. Fang, C., Xu, Y., Rockmore, D.N.: Unbiased metric learning: on the utilization of multiple datasets and web images for softening bias. In: Proceedings of the IEEE International Conference on Computer Vision, pp. 1657–1664 (2013)
12. Li, D., Yang, Y., Song, Y.-Z., Hospedales, T.M.: Deeper, broader and artier domain generalization. In: Proceedings of the IEEE International Conference on Computer Vision, pp. 5542–5550 (2017)
13. Carlucci, F.M., D'Innocente, A., Bucci, S., Caputo, B., Tommasi, T.: Domain generalization by solving jigsaw puzzles. In: 2019 IEEE/CVF Conference on Computer Vision and Pattern Recognition (CVPR) (2020)
14. Matsuura, T., Harada, T.: Domain generalization using a mixture of multiple latent domains. In: Proceedings of the AAAI Conference on Artificial Intelligence, vol. 34, pp. 11749–11756 (2020)
15. Wang, S., Yu, L., Li, C., Fu, C.-W., Heng, P.-A.: Learning from extrinsic and intrinsic supervisions for domain generalization. In: Vedaldi, A., Bischof, H., Brox, T., Frahm, J.M. (eds.) ECCV 2020. LNCS, vol. 12354, pp. 159–176. Springer, Cham (2020). https://doi.org/10.1007/978-3-030-58545-7_10

16. Chen, S.: Decomposed adversarial domain generalization. Knowl.-Based Syst. **263**, 110300 (2023)
17. Li, Y., Tian, X., Gong, M., Liu, Y., Liu, T., Zhang, K., Tao, D.: Deep domain generalization via conditional invariant adversarial networks. In: European Conference on Computer Vision (2018)
18. Mansilla, L., Echeveste, R., Milone, D.H., Ferrante, E.: Domain generalization via gradient surgery. In: Proceedings of the IEEE/CVF International Conference on Computer Vision, pp. 6630–6638 (2021)
19. Chen, K., Zhuang, D., Chang, J.M.: Discriminative adversarial domain generalization with meta-learning based cross-domain validation. Neurocomputing **467**, 418–426 (2022)

Value Creation Model of Social Organization System

Shuming Chen[1], Cuixin Hu[2], and Xiaohui Zou[3](✉)

[1] Tian Yanzhi Technology Industry Co., Ltd., Building 7, No. 30, Shixing Street, Shijingshan District (Jingxi Technology Finance Building 1001-2), Beijing 100041, China
[2] Chongqing Human Resources and Social Security Bureau Vocational Skills Appraisal Guidance Center, Building 3, North District, Human Resources Service Industrial Park, Chongqing, China
[3] Hengqin Searle Technology Co., Ltd., Room 501, Building 1, No. 100, Renshan Road, Hengqin Guangdong-Macao Deep Cooperation Zone, Zhuhai 519000, Guangdong, China
949309225@qq.com

Abstract. It aims to the value creation model of social organization system from the perspective of AI. The method is: first, make a formal expression of the basic overview of social organization system value creation, and then give the functional expression of the social organization system value creativity model, and finally, clarify the Σ factor in the model, and further, try to clearly give the Π factor in the model, and give the $\ddot{E}$ factor in the model. The result is: get a comprehensive summary of the model, and at the same time, give a social application summary of the model. Its significance lies in: through the simple analysis of the three types of factors of the value creativity model of the typical social organization system, not only the three types of value creativity models of individual, family and country and their instance factors are obtained in an unambiguous description.

Keywords: Cognitive Computing · Cognitive sciences and technology · Information presentation · Big data and intelligent information processing

1 Introduction

1.1 A Subsection Sample

It aims to focus on the formal computing research of cognitive system, especially social organization system and its information processing. This article aims to re-understand the value creation model of social organization system from the perspective of AI. Although international evaluation agencies have paid attention to the problem of formal classification of organizations, and there are also micro-cluster analysis models, the overall macro-level situation is still unsatisfactory, especially the lack of formal methods that link macro-micro organizations [1]. The value creation of sustainable development, especially the provision of shareable value creation, the creation of social and financial value, is a focus of attention [2]. The value creation of intangible assets is far better than that of tangible assets [3]. AI autonomous agents and knowledge management attach

F. Sun and J. Li (Eds.): ICCCS 2023, CCIS 2029, pp. 147–155, 2024.
https://doi.org/10.1007/978-981-97-0885-7_13

great importance to formal expression [4]. Organizations can emphasize finance and its social value [5]. The result of increasing value creation with practical implementations in a specific industry is worthy of attention [6]. The value of feature-quantified organizations and their knowledge-sharing communities is very important [7]. In a society that emphasizes the role of enterprises, how to evaluate the value and development potential of a company is a factor that needs to be considered [8]. In addition, from the following two comparative trend graphs, we can also see a spotlight in value creation research (Figs. 1 and 2).

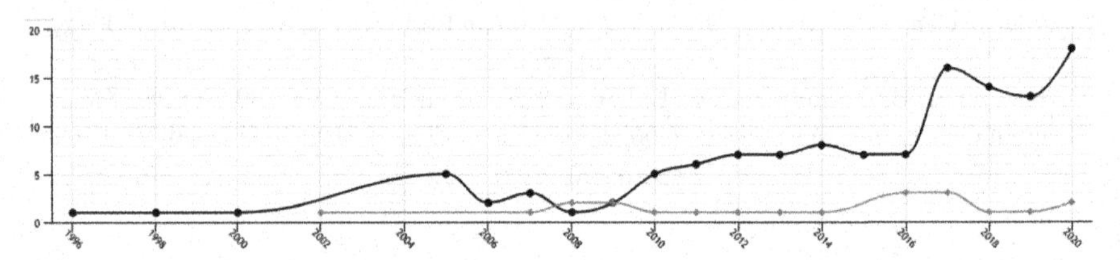

Fig. 1. Compared with value creation research, value creativity research is far from the latter.

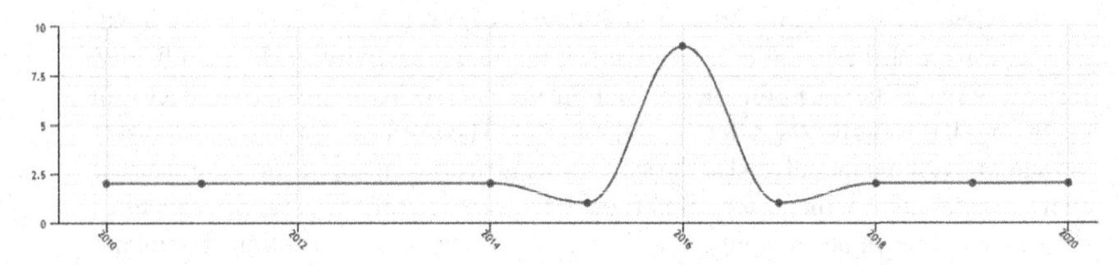

Fig. 2. Social organization research only peaked in 2016

2 Methods

The method is: first, make a formal expression of the basic overview of social organization system value creation, and then give the functional expression of the social organization system value creativity model, and finally, clarify the Σ factor in the social organization system value creativity model, and further, try to clearly give the Π factor in the social organization system value creativity model, and give the $\ddot{E}$ factor in the social organization system value creativity model. Formal Expression: Among the two formal expression methods, direct and indirect, the former is used below. Functional Expression: Given a social organization system ◻, including N agents AI_n divided into three groups of agent (individual or group) positions, the same agent can belong to different positions at the same time: AI_{a1} and AI_{ai} are specific intelligent individuals or intelligent groups or social organization systems in the social organization system. Such agents play the role of "cumulative basic function factors" in the value creation of the social organization system, and are recorded as Σ factors; AI_{b1} and AI_{bj} are specific intelligent individuals or intelligent groups or social organization systems in the social organization system. Such agents play the role of "risk control factors" in the value creation of the social organization system and are recorded as Π factors; AI_{c1} and AI_{ck} are specific intelligent individuals or intelligent groups or social organization systems in the social organization system. Such agents play the role of "strategic amplification factor" in the value creation of the social organization system and are recorded as $\ddot{E}$ factors. The basic form of the value creativity model function corresponding to the social organization system is:

$$\boldsymbol{Vitality}(¤) = \left(b_1(AI_{b1}) \times b_j(AI_{bj}) \times \cdots\right) * \left(a_1(AI_{a1}) + a_{2i}(AI_{ai}) + \ldots\right)^{(1+c_1(AI_{c1})+c_k(AI_{ck})+\cdots)} \tag{1}$$

Including: $a_1(AI_{a1})$ and $a_i(AI_{ai})$ respectively are the Σ factors corresponding to intelligent individuals or intelligent groups AI_{a1} and AI_{ai} in the overall value creation $\boldsymbol{Vitality}(¤)$function of the social organization system; the value range is:

$$a_i \in (-\infty, +\infty) \tag{2}$$

Theoretically, a_i can also be based on the mapping parameters of the value quantity formed $\boldsymbol{Vitality}(¤((AI_{bj}))$in formula (1), namely:

$$a_i = \mathcal{L}\left(Vitality(¤((AI_{ai}))\right) \tag{3}$$

$b_1(AI_{b1})$ and $b_j(AI_{bj})$ respectively are the corresponding Π factors of the intelligent individuals or intelligent groups AI_{b1} and AI_{bj} *in* the overall value creation function of the social organization system; their actual influence is delayed, and the value range is:

$$b_j(t) = [0, 1] \tag{4}$$

Theoretically, b_j can also be based on the mapping parameters of the value quantity formed $\boldsymbol{Vitality}(¤((AI_{bj}))$in formula (1), namely:

$$b_j = \mathcal{L}\left(Vitality(¤\left((AI_{bj})\right)\right) \tag{5}$$

$c_1(AI_{c1})$ and $c_k(AI_{ck})$ respectively are the corresponding $\ddot{E}$ factors of intelligent individuals or intelligent groups AI_{c1} and AI_{ck} in the overall value creation function of the social organization system; their actual influence is delayed, and the value range is:

$$c_k(t) \geq 0 \tag{6}$$

Theoretically, c_k can also be based on the mapping parameters of the value quantity formed $\boldsymbol{Vitality}(¤((AI_{ck}))$in formula (1), namely:

$$c_k = \mathcal{L}\left(Vitality(¤((AI_{ck}))\right) \tag{7}$$

In addition, formula (1) can also be used for comprehensive evaluation of other types of value creation, including: social organization system ¤value creation evaluation of sub-organization systems, social organization system ¤for each participating agent AI_n in a specific single event Evaluation of the value creation of the social organization system, comprehensive evaluation of value creation at all stages, and comprehensive evaluation of the integration of various social organization systems, etc. The follow-up part will analyze in detail the "cumulative basic action factor"-Σ factor, "risk control action factor"-Π factor, and "strategic amplification factor"-$\ddot{E}$ factor. Clarify the Σ Factor: Clarify the Σ factor in the value creativity model of social organization system. The basic concept of Σ factor: The Σ factor refers to: among the N agents AI_n of the social organization system, the sum of the cumulative effects of various daily tasks of a specific intelligent individual or intelligent group or social organization system in the entire life cycle of the social organization system; The value of a single task of a single agent AI_n can be positive or negative, that is, there are losses and gains in specific matters. The Σ factor is the basic content of the social organization system $\mho$. The daily routine work of the social organization system $\mho$involved in life, work, management, production, technology, and operation, basically belongs to the category of the Σ factor. Basic characteristics of Σ factor: The Σ factor illustrates the importance of daily value creation management and control in the social organization system. The total value is accumulated by each Σ factor bit by bit; once the daily work is implemented, the accumulation of achievements will form the value fundamentals of the social organization system $\mho$. From the cumulative logical analysis of a single Σ factor, the success or failure of each node Σ factor will not have a fatal impact on the overall value of the social organization system; the loss of temporary value will not affect the social organization system too much. The fundamental level of value realization; simply put, the Σ factor determines the basic disk. Analysis of Σ Factor Examples: Take an independent technological innovation company as an example. In its development process, the current specific work of each team and each employee belongs to the category of Σ factor, typical types: ① Technical Σ factors, such as: Quaternary Geological Drilling Modeling Module, *3D* BIM Quick Modeling Module, Pipeline Modeling Module, Mountain Torrent Simulation Modeling Module, Technical Framework of Fire Code Knowledge System, Survey BIM Platform Module, etc. ② Production Σ factors, such as: large area geological modeling model project, *3D* BIM intelligent rapid modeling model project, *3D* BIM intelligent rapid modeling training material standard document system, model project video production standard technical system, etc. ③ Marketing type Σ factors, such as: a large central enterprise design institute highway system market development layout, a certain regional land market development layout, etc.④ Enterprise management Σ factors, such as: institutional system construction, talent echelon construction, company brand construction, company standard operation system construction, etc. ⑤ Social resource type Σ factors, such as daily maintenance of relations with a large national association, maintenance of relations with early investors, etc. ⑥ Financial Σ factors, such as: financial scheduling supervision, cash flow supervision, reasonable tax avoidance implementation, capital operation, etc. Clearly Give the Π Factor: Clarify the Π factor in the value creativity model of social organization system. Basic concept of Π factor: The Π factor refers to: among the N agents AI_n of the social organization system, AI_{bj} specific intelligent individuals or intelligent groups or social organization systems are implemented for various difficulties and risks at each key node in the entire life cycle of the social organization system Effective management and control. The Π factor by itself will not bring new value to the social organization system, but it can ensure that at a certain risk point, whether the accumulated value of the social organization system will return to zero (resulting in the sudden death of the social organization system) Or crash) to control the extent of loss. Basic characteristics of Π factor: The Π factor emphasizes that in the social organization system, there must be an agent AI_n aimed at timeliness and decisiveness to manage the overall risk of the system, and to ensure whether the value formed can continue or be suddenly suspended. At a specific time or stage of work, the success or failure of the Π factor can determine the life and death of the entire system; therefore, once the Π factor appears, the social organization system must do everything possible to resolve risks. The formation Π factor of a certain node has a delay in its risk outbreak point, which may be delayed at a special time point. The Π factor does not bring new value creation to the system, but only manages the system to overcome specific difficulties, avoid risks, or reduce risks. The final death or suspension of the social organization system is essentially determined by the specific Π factor; simply put, the Π factor determines life and death. Π factor example analysis: Taking an independent technological innovation company as an example, the cash flow risks and legal risks faced during its development all belong to the category of Π factors, typical types: ① Cash flow risk Π factor: There have been several typical Π factors in the development process of this technology company, including mid-2015, early 2017, early 2018, and early 2019. The key role of cash flow risk Π factors exists; all are due to Only the concerted efforts of relevant key personnel can solve and overcome these obstacles. ② Legal risks/factors: For example, the company is closed due to legal disputes in specific markets, policies, and cooperation. ③ Core health risks/factors: specific backbone health problems lead to sudden collapse of the company's operations, etc. Give the $\ddot{E}$ factor: and give the $\ddot{E}$ factor in the value creativity model of social organization system. The basic concept of Ë factor: The $\ddot{E}$ factor refers to: among the N agents A. In of the social organization system, there are K specific intelligent individuals or intelligent groups or social organization systems. There are strategic layout opportunities at specific key nodes in the entire life cycle of the social organization system. The seemingly trivial strategic layout will bring huge value growth to the social organization system in the future. $\ddot{E}$ Factor implementation requires forward-looking, and the advanced layout and implementation of strategic points determines the maximum possible value realization space of the intelligent system in the future. The basic characteristics of Ë factor: $\ddot{E}$ factor emphasizes opportunity and strategy, forming a magnification of value; $\ddot{E}$ factor is something that can be met but cannot be sought, seemingly simple and irregular to follow; Ë factor is a specific agent A. In in the social organization system, which sets opportunities and strategies, Vision and other integrated products after the integration of wisdom. Only at a specific time or stage of work can $\ddot{E}$ factor achieve a successful strategic layout; if the strategic time point is advanced or delayed, $\ddot{E}$ factor's strategic value creation opportunities may be lost. Once the strategic layout is successful, although it is difficult to see specific benefits in the short term, it will have a profound value impact on the future of the system and form a huge value amplification effect; simply put, the $\ddot{E}$ factor determines the maximum value potential. $\ddot{E}$ factor example analysis: Take an independent technological innovation company as an example. The various strategic technologies, talents, corporate structure, social resources, capital, etc. possessed during its development all belong to the category of $\ddot{E}$ factors, typical types: ① Strategic technology $\ddot{E}$ factor, typical subcategories: Algorithm theory system category original by independent technology company; High-threshold intelligent technology modules such as geological body intelligent modeling, point cloud intelligent classification modeling, *3D* BIM rapid intelligent modeling; *I3D* intelligent development framework, distributed object database, cross-tool platform bottom interface seamless intermodulation container technology and other integrated underlying technologies; National invention patent technologies that integrate core technologies and business models. ② Strategic talents $\ddot{E}$ factor: It mainly refers to how independent technology companies can focus on various complementary world heroes that can form resonance with independent technology companies through the work of strategic talents and factors on a global scale; The core entrepreneurial personnel of independent technology companies are all strategic talents. ③ Strategic company structure $\ddot{E}$ factor: The corporate constitution, platform strategy and framework, and industrialization framework of independent technology companies will form a valuable and meaningful strategic company structure $\ddot{E}$ factor in the future. ④ Strategic social resources $\ddot{E}$ factor: The close strategic cooperation relationship cultivated between independent technology companies and well-known domestic associations will form a valuable and meaningful strategic social resource factor in the future. ⑤ Strategic capital $\ddot{E}$ factor: Independent technology companies' own industrialized capital system, etc., will form valuable and meaningful strategic capital factors in the future.

3 Results and Discussion

The result is: get a comprehensive summary of the social organization system value creativity model, and at the same time, give a social application summary of the social organization system value creativity model. As a result, a comprehensive summary of the social organization system value creativity model: Comparative analysis of the three types of factors of Σ, Π and $\ddot{E}$: The Σ factor determines the fundamentals of the development of the social organization system, the Π factor determines the risk, life and death of the social organization system, and the $\ddot{E}$ factor determines the maximum value creation of the social organization system. Starting from the essential logic of the three types of factors of Σ, Π, and $\ddot{E}$, the evaluation and cognition of the value creation of job roles corresponding to different agents in the social organization system can be systematically optimized and straightened out. Some seem to be reasonable. The typical misunderstanding: ① According to the risk control factor (Π factor), due to the particularity of risky work, once you have gone through it, you will pass it, leaving little valuable traces at the company level; it will often affect the importance and value of previous risk work Lack of sufficient knowledge. ② For the strategic magnification factor ($\ddot{E}$ factor), due to the advanced nature of the strategic layout, it is not necessary to do the work; and the realization of value is delayed, often lacking sufficient recognition of the importance and value of the previous strategic factor work know. ③ Regarding the cumulative basic factor (Σ factor), they often only recognize the importance of their own work, fail to see the overall needs of the overall situation, strategy, and long-term work, and lack sufficient awareness of current risks and future trends. Based on the value creativity model of the social organization system, it lays a theoretical foundation for solving the relatively fair and reasonable distribution logic within the organization system under the transparent logic framework: ① Cumulative basic action factor (Σ factor): The short-term value is the main factor and the medium-term value is supplemented to reflect the realization of related values. ② Risk control factor (Π factor): The medium-term value is the main factor and the short-term value is supplemented to reflect the realization of the relevant value. ③ Strategic amplification factor ($\ddot{E}$ factor): The medium and long-term value is the main factor, and the short-term value is the supplement to reflect the realization of the relevant value. In addition, the Σ factor of the social organization system ¤value creativity model helps the system ¤recognize the value of the screw and provides a quantitative analysis method; only these Σ factor basic work has accumulated to a certain degree, and the collaboration between each other to a certain level, The system can form a fundamental system with self-survivability (market competitiveness), otherwise risk control, strategic layout, etc. cannot be discussed. At the same time give a summary of the social application of the social organization system value creativity model: The value creativity model of social organization system, oriented to the fields of sociology, economics, management, human resources, etc., provides quantifiable evaluation, improvement, optimization, and implementation of the value creation of each agent in the organization for the social organization system. The model method has far-reaching industry application prospects. It can be expected that the value creativity model of social organization system: It can provide quantitative methods for talent training at all levels of the country; Provide quantitative methods for the selection of talents in various fields for the whole society; It can establish a systematic value distribution system for the organization system and provide a quantitative method; It can establish a value contribution evaluation system for each position and provide a quantitative method for the organizational system; It can build a scientific human resource management system for the social organization system and provide quantitative methods (Table 1).

Table 1. Three kinds of factor, formal expression with functional expression.

Table Head	Table Column Head: Formal Expression with Functional Expression		
	Three kinds of factor	Formal Expression	Functional Expression
A as Σ	clarify the Σ factor	AI_{a1} and AI_{ai}	Accumulation function
B as Π	clearly the Π factor	AI_{b1} and AI_{bi}	Risk control function
C as Ë	clearly the Ë factor	AI_{c1} and AI_{ci}	Magnification function

4 Conclusions

In the family system, the typical work content that related individuals or organizations can form the cumulative Σ factor are: daily family life, getting along, etc. In the family system, the typical work content that related individuals or organizations can form risk/factors are: betrayal of marriage, domestic violence, etc. In the family system, the typical work content that can be formed by related individuals or organizations as strategic factors include: giving birth to children, supporting specific careers, etc. Instance factors of national value creativity model: In the national organization system, the typical work contents that related individuals or organizations can form the cumulative Σ factor include: daily operation of the country, economic development, social governance, border order protection, etc. In the national organization system, the typical work contents that can be formed by related individuals or organizations are: territorial integrity, political legitimacy, military controllability, and social stability. In the national organization system, the typical work content that related individuals or organizations can form strategic factors include: specific political figures, specific strategic technology industries, specific social system governance systems, etc. Instance factors of individual value creativity model: In an individual agent, the typical work content that can be formed by related individuals or organizations that can accumulate Σ factors are: daily healthy life, study, work, etc. In an individual agent, the typical work content that can be formed by related individuals or organizations are: drug use, gambling, game addiction, suicide, etc. In the individual agent, the typical work content that can be formed by related individuals or organizations are: family background, meeting nobles, choosing opportunities in the times, and creating specific inspirations. Some Conclusions on the Model Function of Value Creativity of Social Organization System: 1. The social organization system value creativity model function is an original formula born in China. It provides a powerful theoretical analysis tool for the evaluation of social organization system value creativity, as well as management, decision-making, and benefit distribution based on this. 2. Social organization system value creativity model function can be widely used in various social organization systems, including but not limited to family organizations (family, family), political organizations (political parties, government, departments, etc.), economic organizations (enterprises, Interest groups, etc.), cultural organizations (associations, theater troupes, etc.), military organizations (troops, etc.), religious organizations, etc.; and used for the evaluation of value creation of sub-organization systems in the social organization system ¤, the social organization system ¤targets a specific single event The value creation evaluation of each participating agent in AI_n. 3. The value creativity model function of social organization system, although it is only a basic mathematical function at present; but based on this, for the calculation, analysis and evaluation of the value creativity of various social organization systems, a series of subsequent system studies can be constructed The formula system and theoretical system are further original and developed into a new discipline-"Quantitative Analysis of Value Creation in Social Organization System".

Acknowledgments. Thanks to the experts of ICCCS 2023 Program Committee for their understanding and support!

References

1. Espadoto, M., et al.: Deep learning multidimensional projections. Inf. Vis. **19**, 247–269 (2019)
2. Grieco, C., et al.: Measuring value creation in social enterprises. Nonprofit Voluntary Sect. Quart. **44**, 1173–1193 (2015)

3. Pirson, M.A.: Social entrepreneurship - a model for sustainable value creation (2010)
4. Córdova, F., et al.: Cognitive strategic model applied to a port system. ISC Int. J. Inf. Secur. **11**, 73–78 (2019)
5. Jessica, Gu., Wang, H., Fanjiang, Xu., Chen, Yu.: Simulation of an organization as a complex system: agent-based modeling and a gaming experiment for evolutionary knowledge management. In: Kaneda, T., Kanegae, H., Toyoda, Y., Rizzi, P. (eds.) Simulation and Gaming in the Network Society. TSS, vol. 9, pp. 443–461. Springer, Singapore (2016). https://doi.org/10.1007/978-981-10-0575-6_30
6. Liu, S., et al.: Using a social entrepreneurial approach to enhance the financial and social value of health care organizations. J. Health Care Finance **40**(3), 31–46 (2014)
7. Guidi, M., et al.: An empirical approach to a community based integrated waste management system: a case study for an implementation and a start of community engagement model for municipal waste in Bolivia (2016)
8. Dankulov, M.M., et al.: The dynamics of meaningful social interactions and the emergence of collective knowledge. Sci. Rep. **5**, 12197 (2015)

Decision Making and Systems

Mixed Orientation ProMPs and Their Application in Attitude Trajectory Planning

Jian Fu, Zhu Yang(✉), and Xiaolong Li

School of Automation, Wuhan University of Technology, Wuhan 430070, China
fujian@whut.edu.cn, 1075227479@qq.com

Abstract. The application of motion primitives to encode robot motion has garnered considerable attention in the field of academic research. Existing models predominantly focus on reproducing task trajectory in relation to position, often neglecting the significance of orientation. Orientation Probabilistic Movement Primitives (ProMPs) indirectly encode motion primitives for attitude by utilizing their trajectory probabilities on Riemannian manifolds, specifically the 3-sphere $\mathcal{S}^3$. However, assuming a Gaussian distribution imposes constraints on its abilities. We propose Mixed Orientation ProMPs to enhance trajectory planning and minimize the occurrence of singular configurations. This model consists of multiple separate Gaussian distributions in the tangent space, enabling the approximation of any distribution. Furthermore, optimization objective functions of the Lagrangian type can incorporate constraints, such as singularity avoidance, and others. Finally, the effectiveness and reliability of the algorithm were validated through trajectory planning experiments conducted on the UR5 robotic arm.

Keywords: Orientation Probabilistic Movement Primitives · GMM Algorithm · Trajectory Planning

1 Introduction

Currently, Euclidean geometry is widely used to describe the position, velocity, and force of robot movement characteristics, as well as non-Euclidean space movement characteristics such as position and orientation, stiffness, and operability. Most of the current skill imitation learning uses a regression model based on a Gaussian distribution to model the time-position and position-velocity of the robot in Euclidean space. However, accurately modeling the pose data in Euclidean space is challenging. Pastor et al. [1] applied DMP to generate poses that did not meet the requirement of unit norm, and Silvério [2] and Kim [3] adopted GMM to model the pose, which also had the same problem. When studying the position and orientation of the robot end-effector, the constraints of the orientation expression must be considered. For example, the quaternion $q \in \mathrm{S}^3$ needs to satisfy $q^\top q = 1$, and the rotation matrix needs to be an orthogonal matrix, which means $R^\top R = I$. In the problem of learning quaternion poses, traditional

This work was supported by the National Natural Science Foundation of China under Grant 61773299.

F. Sun and J. Li (Eds.): ICCCS 2023, CCIS 2029, pp. 159–173, 2024.
https://doi.org/10.1007/978-981-97-0885-7_14

imitation learning methods have additional constraints on learning poses in Euclidean space. Therefore, Ude [4] pointed out that dynamic movement primitives (DMPs) are an effective approach for learning and controlling complex robot behaviors, specifically for representing point-to-point periodic movements. However, one issue that arises when using DMPs to define control strategies in Cartesian space is the lack of a minimum singularity-free directional representation. The use of the orientation quaternion effectively eliminates these issues and avoids the direct integration of orientation parameters, ensuring that the resulting parameters are located in SO(3). It has been experimentally verified that the DMPs on modulation and coupling can be utilized to expand the orientation DMPs. Based on the mapping relationship between quaternions and Euclidean space, the position variables in the original DMP model are replaced [5]. The distance between the current position and orientation and the target position and orientation is converted into Euclidean space, replacing the location distance $g - \xi$ of the traditional DMP. However, the main disadvantage of this model is that it does not encode the variability of the demonstration, resulting in an inability to adapt to new orientations. Huang [6] and Abu-Dakka [7] proposed the Kernelized Movement Primitives (KMP) method and extended it to the learning of orientation trajectories. This method can learn different orientation trajectory tasks and generalize the learned methods to new situations. Leonel Rozo et al. [8] mapped quaternion trajectory data onto each other on the manifold and tangent space using logarithmic mapping and exponential mapping methods on Riemannian manifolds. This allowed for the learning of orientation trajectory data from ProMPs, enabling the reproduction of robot posture trajectories and passing via point tasks. ProMPs is a motion primitive that utilizes probability distributions to describe trajectories, and it provides better recurrence and generalization of robot trajectories. Du Jinyu et al. [9] integrated the expected maximum algorithm and Gaussian mixture model algorithm into ProMPs, and achieved the fusion of multiple ProMPs, which can effectively switch between point crossing tasks and human-machine collaboration tasks. However, when dealing with data in non-Euclidean space, such as attitude and position data with surface constraints, accurately modeling in Euclidean space becomes challenging. Riemannian manifolds, on the other hand, can easily calculate concepts such as curvature and divergence, making them convenient for learning in metrics and optimization. Therefore, after incorporating the pose, the advantages of ProMPs and the Riemannian manifold can be integrated.

The orientation of the robot end-effector is typically described using a unit quaternion, denoted as $q \in S^3$, which is an element of a 3-sphere (a three-dimensional manifold). Quaternions are considered a minimal representation compared to Euler angles and rotation matrices, offering robust stability in closed-loop directional control. Quaternions can represent a three-dimensional sphere in a four-dimensional quaternion space. Riemannian manifolds allow for the computation of concepts such as length, angle, area, curvature, or divergence, which are useful in various applications, including statistics, optimization, and metric learning [10,11]. Therefore, in this paper, the quaternion is indirectly fitted to a Gaussian distribution on the Riemannian manifold. This extends the classic ProMPs framework to the Riemannian manifold for research purposes, enabling the motion planning of the orientation trajectory [12].

In this paper, we build a Mixed Orientation ProMPs model, and the main contributions are as follows:

(1) The quaternion is indirectly fitted to the Gaussian distribution on the Riemannian manifold, and the mean and variance are updated iteratively using the maximum likelihood estimation and Gauss-Newton optimization algorithm. The minimization evaluation of the multivariate geodesic regression model is then implemented using the Riemannian manifold parallel transport operation. The weight vector for each set of demonstrations is obtained from the minimized calculation results of the loss function, and the Gaussian distribution of weight parameters for different teaching trajectories is obtained. The Orientation ProMPs model is constructed in this way.
(2) We propose Mixed Orientation ProMPs to enhance trajectory planning and minimize the occurrence of singular configurations. The Mixed Orientation ProMPs utilize the expected maximum algorithm and Gaussian mixture model algorithm. It can be seen from the experiments that the trajectory planned by the Mixed Orientation ProMPs is obviously smoother and has less fluctuation compared to that of the single Orientation ProMPs.
(3) To achieve smoother passing via point, an optimization goal is set for the distance between the desired via point and the mean value of each ProMP. Map the passing via point task to each group of small ProMPs models, and then realize the passing via point task through weighted combination.

2 The ProMP Algorithm

2.1 Probabilistic Movement Primitives

ProMPs are frameworks for probabilistically learning and synthesizing motor skills in robotics. ProMPs can represent spatially and temporally related trajectories as distributions. For a single trajectory, a weight vector $\boldsymbol{\omega}$ is used to compactly represent it. At time step t, each degree of freedom (DoF) is represented by its position q_t and velocity $\dot{q}_t$. We use $\mathbf{y}_t$ to represent a d-dimensional vector of joint configuration or Cartesian space position at time t. When the underlying weight vector $\boldsymbol{\omega}$ is given, the probability of observing a trajectory $\boldsymbol{\tau}$ is modeled as a linear basis function.

$$\mathbf{y}_t = \begin{bmatrix} q_t \\ \dot{q}_t \end{bmatrix} = \boldsymbol{\Psi}_t \boldsymbol{\omega} + \epsilon_y \Rightarrow \mathcal{P}(\mathbf{y}_t \mid \boldsymbol{\omega}) = \mathcal{N}(\mathbf{y}_t \mid \boldsymbol{\Psi}_t \boldsymbol{\omega}, \boldsymbol{\Sigma}_{\mathbf{y}}) \tag{1}$$

where $\boldsymbol{\Psi}_t = \left[\psi_t, \dot{\psi}_t\right]^T$ defines the $2 \times n$ dimensional time-dependent basis matrix.

2.2 Orientation ProMPs

For the motion trajectory in Cartesian space, multiple teaching trajectories can be collected, and their trajectory data can be divided into position data and orientation data of the end effector. Each teaching trajectory $\tau = \{\mathbf{y}_t\}_{t=1}^{T}$ is composed of the end pose point $\mathbf{y}_t \in \mathbb{R}^3 \times \mathrm{S}^3$ of the robot at each time step from 1 to T. For robot orientation trajectory data, it is not suitable for learning classical ProMPs models, so it needs to be extended to manifolds. In order to concentrate on the core of the issue, we will solely discuss the trajectory of attitudes hereafter.

The classical ProMPs parameterized trajectory model, which uses linear basis functions, is shown in Eq. 1. For the multivariate Gaussian regression on the Riemannian manifold, an estimated value $\hat{\mathbf{y}}_i$ in pose ProMPs is a d-dimensional variable, expressed as a combination of the M-dimensional geodesic basis W, based on the tangent plane of the point p. This can be further expressed using a pattern similar to ProMPs, where each component of $\hat{\mathbf{y}}_i$ is constructed using Md basis functions. The equivalent expression of $\hat{\mathbf{y}}_i$ can be written as:

$$\hat{\mathbf{y}}_i = \mathrm{Exp}_{\mathrm{p}}\left(\mathbf{W}\phi_t\right) \equiv \mathrm{Exp}_{\mathrm{p}}\left(\boldsymbol{\Psi}_t\omega\right) \tag{2}$$

where $\boldsymbol{\Psi}_t = \mathrm{blockdiag}\left(\boldsymbol{\phi}_1^{\mathrm{T}}, \boldsymbol{\phi}_2^{\mathrm{T}}, \ldots, \boldsymbol{\phi}_M^{\mathrm{T}}\right) \in \mathbb{R}^{d\times Md}, \quad \boldsymbol{\omega} = \left[\omega_1^{\mathrm{T}}, \omega_2^{\mathrm{T}}, \ldots, \omega_M^{\mathrm{T}}\right]^{\mathrm{T}} \in \mathbb{R}^{Md}, \quad \mathbf{W} \in \mathbb{R}^{d\times M}$ and $\boldsymbol{\phi}_t \in \mathbb{R}^M$.

For each teaching trajectory, this paper estimates its weight vector using a multivariate general linear model. Firstly, it should be noted that Eq. 2 is based on the minimum expression of the manifold dimension M. Considering the conventional approach in machine learning, the dimension in the Orientation ProMPs formula is typically increased from M to N_ϕ. According to Eq. 2, it can be observed that $\mathrm{Exp}_{\mathrm{p}}\left(\mathbf{W}\phi_t\right) \equiv \mathrm{Exp}_{\mathrm{p}}\left(\boldsymbol{\Psi}_t\omega\right)$, where $\mathbf{W} = \left[\boldsymbol{\omega}_1 \ldots \boldsymbol{\omega}_{N_\phi}\right], N_\phi$ is the number of Gaussian basis functions. Then, we extract the quaternion orientation trajectory $\tau_n = \{\mathbf{y}_t\}_{t=1}^T$ from the trajectory, where $\mathbf{y}_t \in S^3$. Next, we can utilize the weight parameter solving method of the multivariate general linear regression model to minimize the loss function and estimate the weight parameter $\widehat{\omega}_n$. The specific expression is as follows:

$$(\hat{\mathbf{p}}, \hat{\boldsymbol{\omega}}_m) = \underset{(\mathbf{p},\omega_m)\in\mathbf{TM}\forall m}{\mathrm{argmin}} \underbrace{\frac{1}{2}\sum_{t=1}^{T} d_{\mathrm{M}}\left(\mathrm{Exp}_{\mathbf{p}}\left(\mathbf{W}\phi_t\right), \mathbf{y}_t\right)^2}_{E(\mathbf{p},\omega_m)} \tag{3}$$

where, ϕ_t is the Gauss function vector at time t, and $\mathbf{W}$ is the N_ϕ weight tangent vectors $\widehat{\omega}_m \in \mathrm{T}_{\hat{\mathbf{p}}}\mathrm{M}$ to be estimated in the tangent space of point $\mathbf{p} \in \mathrm{M}$, that is, $\mathbf{W} = \mathrm{blockdiag}\left(\hat{\omega}_1^1, \ldots, \hat{\omega}_{N_\phi}^1\right), E\left(\mathbf{p}, \boldsymbol{\omega}_m\right)$ represents the loss function of each teaching trajectory. The weight parameter vector of this trajectory can be obtained by minimizing the loss function of each teaching trajectory.

In order to solve the minimum value problem of the formula, a method similar to the parallel transmission operation on Riemannian manifolds can be adopted. Based on this operation, the gradient on the Riemannian manifold is computed. Specifically, we need to solve for the gradients of $E\left(\mathbf{p}, \boldsymbol{\omega}_m\right)$ with respect to $\mathbf{p}$ and the gradient of each $\boldsymbol{\omega}_m$. These gradients are computed by transferring each error term $\log_{\hat{y}_t}(y_t)$ from the tangent space $\mathcal{T}_{\hat{\mathbf{y}}_t}\mathcal{M}$ of the estimated value $\hat{y}_t = \mathrm{Exp}_{\mathbf{p}}(\mathbf{W}\phi)$ to the tangent space $\mathcal{T}_{\mathbf{p}}\mathcal{M}$ of the reference point $\mathbf{p}$. Therefore, the expression of the loss function is redefined as follows:

$$E\left(\mathbf{p}, \boldsymbol{w}_m\right) = \frac{1}{2}\sum_{t=1}^{T} \left\|\Gamma_{\hat{\mathbf{y}}_t\to\mathbf{p}}\left(\log_{\hat{\mathbf{y}}_t}\left(\mathbf{y}_t\right)\right)\right\|^2 \tag{4}$$

The approximate gradient of the loss function $E(\mathbf{p}, \boldsymbol{w}_m)$ is expressed as follows:

$$\begin{aligned} \nabla_{\mathbf{p}} E(\mathbf{p}, \omega_m) &\approx -\sum_{t=1}^{T} \Gamma_{\hat{\mathbf{y}}_t \to \mathbf{p}} \left(\log_{\hat{\mathbf{y}}_t}(\mathbf{y}_t)\right) \\ \nabla_{\omega_m} E(\mathbf{p}, \omega_m) &\approx -\sum_{t=1}^{T} \phi_{t,m} \Gamma_{\hat{\mathbf{y}}_t \to \mathbf{p}} \left(\log_{\hat{\mathbf{y}}_t}(\mathbf{y}_t)\right) \end{aligned} \tag{5}$$

By iteratively calculating Eq. 4 and Eq. 5, the vector $\boldsymbol{p}_n$ and weight matrix W_n of each teaching trajectory n can be obtained.

Note that the estimated weight vector, w, lies on the tangent space of point $\mathbf{p}$. Therefore, for different teaching trajectories, the estimated point $\mathbf{p}$ will also vary. Therefore, the tangent space of each teaching trajectory is also different. In other words, the weight vector of each teaching trajectory is in a different tangent space. This can be inconvenient for actual calculations. However, the multiple teaching trajectories collected are usually very similar. Therefore, it can be assumed that all the teaching trajectories share the tangent space of the same point $\mathbf{p}$. In summary, the training of the Orientation ProMPs model only requires estimating the point $\mathbf{p}$ of a single demonstration. This estimated point is then used to estimate the tangent-space weight vectors for all taught trajectories.

3 GMM Algorithm

The Gaussian distribution can be used to fit most data models. However, for complex models, a single Gaussian distribution may not provide a good approximation. At this time, multiple Gaussian distributions can be combined, resulting in a mixed Gaussian distribution that approximates this complex model. Therefore, most data models can construct their Gaussian Mixture Models (GMMs) [13].

Assuming that there is an observed variable X, GMM is now used to establish its probability distribution model. There are K clusters defining GMM. The kth cluster obeys a Gaussian distribution with mean μ_k and variance Σ_k. The probability that the observed variable X belongs to each cluster is defined as π_k. Then its log-likelihood function is expressed as:

$$\begin{aligned} \log p(X \mid \theta) &= \log \prod_{n=1}^{N} p(x_n \mid \theta) \\ &= \sum_{n=1}^{N} \log \sum_{k=1}^{K} \pi_k p(x_n \mid \mu_k, \Sigma_k) \end{aligned} \tag{6}$$

The logarithmic function in Eq. 6 includes a summation operation, making it difficult to directly use the partial derivative method for optimization. However, following the idea of the EM algorithm, Eq. 6 can be transformed into:

$$\log p(X \mid \theta) \geq \sum_{n=1}^{N} \sum_{k=1}^{K} \log \pi_k p(x_n \mid \mu_k, \Sigma_k) \tag{7}$$

After incorporating hidden variables, the lower bound Q of the logarithmic likelihood function can be expressed as:

$$Q\left(\theta, \theta^{i}\right)=\sum_{n=1}^{N} \sum_{k=1}^{K} p\left(\gamma_{nk} \mid x_{n}, \theta^{i}\right) \log \pi_{k} \mathrm{N}\left(x_{n} \mid \theta_{k}\right) \tag{8}$$

Among them, $\mathrm{N}\left(x_{n} \mid \theta_{k}\right)$ represents the Gaussian distribution that follows the parameter θ_{k}, and $p\left(\gamma_{nk} \mid x_{n}, \theta^{i}\right)$ represents the probability that x_{n} is observed to belong to the kth Gaussian distribution by fixing the parameter θ. According to the EM algorithm, the value of $p(\gamma_{nk} \mid x_{n}, \theta^{i})$ needs to be obtained in step E. According to the Bayesian formula:

$$\begin{aligned} \gamma_{nk}^{*}=p\left(\gamma_{nk} \mid x_{n}, \theta^{i}\right) &=\frac{p\left(\gamma_{nk} \mid x_{n}, \theta^{i}\right)}{p\left(x_{n} \mid \theta^{i}\right)} \\ &=\frac{\pi_{k} \mathrm{N}\left(x_{n} \mid \mu_{k}, \Sigma_{k}\right)}{\sum_{k=1}^{K} \pi_{k} \mathrm{N}\left(x_{n} \mid \mu_{k}, \Sigma_{k}\right)} \end{aligned} \tag{9}$$

Substituting Eq. 9 into the lower bound Q of the logarithmic likelihood function of Eq. 8:

$$Q\left(\theta, \theta^{i}\right)=\sum_{n=1}^{N} \sum_{k=1}^{K} \gamma_{nk}^{*} \log \pi_{k} \mathrm{N}\left(x_{n} \mid \theta_{k}\right) \tag{10}$$

Therefore, the maximization of the log-likelihood function is transformed into maximizing Eq. 10. Next, the partial derivative of Eq. 10 is calculated for each Gaussian distribution, and the update formula for the M-step parameters is obtained.

$$\left\{\begin{array}{ll} \pi_{k}^{new} & =\frac{1}{N} \sum_{n=1}^{N} \gamma_{nk}^{*} \\ \mu_{k}^{new} & =\frac{1}{N \pi_{k}} \sum_{n=1}^{N} \gamma_{nk}^{*} x_{n} \\ \Sigma_{k}^{new} & =\frac{1}{N \pi_{k}} \sum_{n=1}^{N} \gamma_{nk}^{*}\left(x_{n}-\mu_{k}^{new}\right)\left(x_{n}-\mu_{k}^{new}\right)^{T} \end{array}\right. \tag{11}$$

To summarize, the algorithm repeatedly iterates the E-step and M-step, calculating K groups of Gaussian distributions with varying weights and parameters. These distributions are then combined to form a Gaussian Mixture Model (GMM), which represents an approximate probability distribution model of an observed variable.

4 Mixed Orientation ProMPs

When the desired via point of the ProMPs on a Riemannian manifold deviates significantly from the reference trajectory, the effectiveness of trajectory planning may be compromised. In particular, it is possible to plan the trajectory that allows the robot to pass through the singular point. However, simply adding the teaching trajectory to include the expected via point may also lead to fluctuations in the planned trajectory.

Therefore, a Mixed Orientation ProMPs model is proposed. This model utilizes an alternate estimation method for both parameter space and weight space. It enables the planned trajectory to not only meet the desired via point requirements, but also ensures the smoothness of the trajectory. The main process involves using the Orientation ProMPs algorithm to obtain the parameters of each teaching trajectory during the training phase. These parameters are then divided into multiple small parameter sets using a Gaussian mixture model. Each small parameter set corresponds to a Gaussian distribution in Orientation ProMPs. Then, by utilizing the concept of the EM algorithm, we can update the parameters of the corresponding Orientation ProMPs when passing through the via point in the parameter space. Additionally, we can update the probability of the desired via point in each Orientation ProMPs in the weight space. This allows the model to smoothly pass through the via point.

4.1 Mixed Orientation ProMPs

In order to construct Mixed Orientation ProMPs, a demonstration trajectory with multiple training sets is trained using Orientation ProMPs to obtain the parameters $\mathbf{p}$ and the corresponding $\boldsymbol{\omega}$ for each demonstration trajectory. Then, a Gaussian mixture model of parameter $\boldsymbol{\omega}$ is established in the Orientation ProMPs. This model transforms the Gaussian distribution of the original Orientation ProMPs parameters into a mixed Gaussian distribution. This model contains K sets of different Gaussian distributions, with each small training set corresponding to a Gaussian distribution. The probability distribution of the kth Gaussian distribution is denoted as $p\left(\boldsymbol{\omega}; a^k, \boldsymbol{\theta}^k\right)$. Here, $\boldsymbol{\theta}^k = \left\{\boldsymbol{\mu}^k, \boldsymbol{\Sigma}^k\right\}$ represents the mean and variance of the kth Gaussian distribution. The value of k belongs to the set $\{1, 2, \ldots, K\}$, and $a^k = p(k)$ represents the probability of selecting the kth Gaussian distribution. Therefore, the weight parameter ω can be expressed as:

$$p(\boldsymbol{\omega}) = \sum_{k=1}^{K} p(k) p(\boldsymbol{\omega} \mid k) = \sum_{k=1}^{K} a^k \, \mathrm{N}\left(\boldsymbol{\omega} \mid \boldsymbol{\mu}^k, \boldsymbol{\Sigma}^k\right) \tag{12}$$

In the training process of using a Gaussian mixture model for parameter estimation, the EM algorithm is used to train the parameter ω. At step E, the algorithm calculates the probability γ_k that the parameter $\boldsymbol{\omega}$ belongs to class k.

$$\gamma_{ik} = p\left(k \mid \boldsymbol{\omega}_i\right) = \frac{a^k \, \mathrm{N}\left(\boldsymbol{\omega}_i \mid \boldsymbol{\mu}^k, \boldsymbol{\Sigma}^k\right)}{\sum_{j=1}^{K} a^j \, \mathrm{N}\left(\boldsymbol{\omega}_i \mid \boldsymbol{\mu}^j, \boldsymbol{\Sigma}^j\right)} \tag{13}$$

Then obtain the parameters a^k, $\boldsymbol{\mu}_k$and$\boldsymbol{\Sigma}_k$ of each Gaussian distribution in step M:

$$\begin{cases} n^k = \sum_{i=1}^{N} \gamma_{ik}, a^k = \dfrac{n^k}{N} \\ \boldsymbol{\mu}^k = \dfrac{\sum_{i=1}^{N} \gamma_{ik} \omega_i}{n^k} \\ \boldsymbol{\Sigma}^k = \dfrac{1}{n^k} \left(\sum_{i=1}^{N} \gamma_{ik} \left(\boldsymbol{\omega}_i - \boldsymbol{\mu}^k\right) \left(\boldsymbol{\omega}_i - \boldsymbol{\mu}^k\right)^{\mathrm{T}} \right) \end{cases} \tag{14}$$

The parameters $a^{1:K}$, $\mu^{1:K}$, and $\Sigma^{1:K}$ can be obtained by iteratively performing the E and M steps of the EM algorithm until convergence. After establishing a GMM for the parameters $\boldsymbol{\omega}$ of Orientation ProMPs, the original training set was successfully divided into K groups of small training sets. Each group of small training sets is represented by different sub Orientation ProMPs, and the parameters of the kth sub Orientation ProMPs are $\omega^k \sim \mathrm{N}\left(\omega^k \mid \mu^k, \Sigma^k\right)$. The model of Mixed Orientation ProMPs can be expressed as a weighted combination of K sub Orientation ProMPs. Since each $\Psi \mu^k$ is in the tangent space of $\mathbf{p}$, the result of the weighted combination is still in the tangent space of $\mathbf{p}$. The Mixed Orientation ProMPs model is essentially equivalent to disassembling the original Orientation ProMPs with large variances into multiple sub Orientation ProMPs with small variances. These sub orientation ProMPs are then weighted and combined to obtain a mixed probability model that can smoothly achieve the trajectory planning of passing via point.

4.2 Trajectory Planning of Mixed Orientation ProMPs

For the newly proposed Mixed Orientation ProMPs, it is also necessary to incorporate the functionality of passing the expected via point. Assume that the point $\mathbf{y}_t^*$ needs to be passed, that is, ω in the conditional probability $p\left(\boldsymbol{\omega} \mid \mathbf{y}_t^*\right)$ of the corresponding parameter when passing this via point is required. For the desired via point on the manifold, it needs to be converted to the tangent space based on the reference point $\mathbf{p}$. This is done by finding $p\left(\boldsymbol{\omega} \mid \tilde{\mathbf{y}}_t^*\right)$, where $\tilde{\mathbf{y}}_t^* = \log_{\mathrm{p}}\left(\mathbf{y}_t^*\right)$. For Mixed Orientation ProMPs, it is necessary to find the parameters $a^{1:K}$ and $\boldsymbol{\omega}^{1:K}$, so that the probability of $\tilde{y}_t^*$ occurring is maximized. The goal now is to maximize the log-likelihood function of the observed variable $\tilde{\mathbf{y}}_t^*$ with respect to the parameters $a^{1:K}$ and $\boldsymbol{\omega}^{1:K}$.

$$L(\boldsymbol{\theta}) = \log p\left(\tilde{\mathbf{y}}_t^* \mid a^{1:\mathrm{K}}, \boldsymbol{\omega}^{1:\mathrm{K}}\right) = \log \sum_{k=1}^{\mathrm{K}} a^k p\left(\tilde{\mathbf{y}}_t^* \mid \boldsymbol{\omega}^k\right) \tag{15}$$

For the logarithmic likelihood function of Eq. 15, it is impossible to estimate the parameters directly using partial derivatives. Then we need to use the idea of the EM algorithm. First, we define a latent variable, z, and then we define $q_k(z)$ to represent the probability of selecting the k-th Orientation ProMPs.

Applying Jenson's inequality to Eq. 15 gives:

$$\begin{aligned} L(\boldsymbol{\theta}) &= \log \sum_{k=1}^{K} a^k p\left(\tilde{\mathbf{y}}_t^* \mid \boldsymbol{\omega}^k\right) \\ &= \log \sum_{k=1}^{\mathrm{K}} q_k(z) \frac{a^k p\left(\tilde{\mathbf{y}}_t^* \mid \boldsymbol{\omega}^k\right)}{q_k(z)} \\ &\geq \sum_{k=1}^{\mathrm{K}} q_k(z) \log \frac{a^k p\left(\tilde{\mathbf{y}}_t^* \mid \boldsymbol{\omega}^k\right)}{q_k(z)} \end{aligned} \tag{16}$$

where:

$$\begin{cases} Q(\boldsymbol{\theta}) = \sum_{k=1}^{\mathrm{K}} q_k(z) \log \frac{a^k p\left(\tilde{\mathbf{y}}_t^* \mid \boldsymbol{\omega}^k\right)}{q_k(z)} \\ \sum_{k=1}^{\mathrm{K}} q_k(z) = 1 \end{cases} \tag{17}$$

In Eq. 17, $Q(\boldsymbol{\theta})$ can be considered as the lower bound of the logarithmic likelihood function $L(\boldsymbol{\theta})$. In order to maximize the logarithmic likelihood function, it is necessary to make the lower bound of the logarithmic likelihood function $L(\boldsymbol{\theta})$ as large as possible. From Eq. 16, it is easy to find that the equal sign is obtained only when $a^k p\left(\tilde{\mathbf{y}}_t^* \mid \boldsymbol{\omega}^k\right) = c q_k(z)$ where c is a constant.

According to the first and second steps of the Eq. 16, we can get:

$$c = \sum_{k=1}^{\mathrm{K}} a^k p\left(\tilde{\mathbf{y}}_t^* \mid \omega^k\right) \tag{18}$$

Therefore, in the E step of the EM algorithm, to determine the specific value of $q_k(z)$, it is necessary to optimize the lower bound to approach the target logarithmic likelihood function. From the given conditions, the following can be derived:

$$q_k(z) = \frac{a^k p\left(\tilde{\mathbf{y}}_t^* \mid \boldsymbol{\omega}^k\right)}{\sum_{k=1}^{K} a^k p\left(\tilde{\mathbf{y}}_t^* \mid \boldsymbol{\omega}^k\right)} \tag{19}$$

According to the $q_k(z)$ obtained from Eq. 19, in the M step of the EM algorithm, it is only necessary to maximize the lower bound. The formula for the lower bound of maximization is:

$$\underset{a^{1:K}, \omega^{1:K}}{\arg\max} Q(\boldsymbol{\theta}) = \underset{a^{1:K}, \omega^{1:K}}{\arg\max} \sum_{k=1}^{\mathrm{K}} q_k(z) \log a^k p\left(\tilde{\mathbf{y}}_t^* \mid \boldsymbol{\omega}^k\right) \tag{20}$$

It can be inferred from Eq. 20 that maximizing the lower bound involves finding the appropriate values for two parameters, a^k and ω^k, for each Orientation ProMP. This ensures that the probability of successfully passing the expected via point is maximized.

Therefore, in the E step of the EM algorithm, the distribution of the latent variable $q_k(z)$ is determined. This distribution helps approximate the lower bound $Q(\boldsymbol{\theta})$ approximate to the logarithmic likelihood function $L(\boldsymbol{\theta})$ as a whole. Then, in the M step, the optimal parameter a_{new}^k can be found by taking the partial derivative of a^k.

$$a_{\text{new}}^k = q_k(z) = \frac{a^k p\left(\tilde{\mathbf{y}}_t^* \mid \boldsymbol{\omega}^k\right)}{\sum_{k=1}^{\mathrm{K}} a^k p\left(\tilde{\mathbf{y}}_t^* \mid \boldsymbol{\omega}^k\right)} \tag{21}$$

For the update of $\boldsymbol{\omega}^k$, the original passing via point method of Orientation ProMPs can be directly used to process the expected via point and each group of Orientation ProMPs, and then combined according to the weight a^k. However, this method does not take into account the distance between the desired via point and each set of training data, and the impact of the planned trajectory is still generic. Based on this, multiple Orientation ProMPs can be considered as bases. For each expected via point, the projection point of this point on different bases is obtained to find the closest base. The following objective function is then defined:

$$\mathbf{J} = \sum_{k=1}^{K} \left\| \mathbf{y}_t^k - \left(\mathbf{y}_t^{(\text{mean})}\right)^k \right\|^2 \quad \text{s.t.} \quad \tilde{\mathbf{y}}_t^* = \sum_{k=1}^{K} a_t^k \mathbf{y}_t^k, \tag{22}$$

Among them, $\mathbf{y}_t^k$ represents the optimal mapping point of the expected via point $\tilde{\mathbf{y}}_t^*$ on the k-th type of Orientation ProMPs. Additionally, $\left(\mathbf{y}_t^{(\text{mean})}\right)^k$ denotes the average orientation of the k-th type of Orientation ProMPs at time t. The objective of this function is to minimize the Euclidean norm while considering the mean value of the corresponding Orientation ProMPs. This is done under the condition that the sum of the weights of the mapping points of each Orientation ProMPs equals the expected via point.

For each parameter, it can be written in the following matrix form:

$$\boldsymbol{a}_t=\begin{bmatrix} a_t^1 \\ a_t^2 \\ \vdots \\ a_t^K \end{bmatrix}, \mathbf{Y}_t=\begin{bmatrix} \mathbf{y}_t^1 \\ \mathbf{y}_t^2 \\ \vdots \\ \mathbf{y}_t^K \end{bmatrix}, \mathbf{Y}_t^{(\text{mean})}=\begin{bmatrix} \left(\mathbf{y}_t^{(\text{mean})}\right)^1 \\ \left(\mathbf{y}_t^{\text{mean})}\right)^2 \\ \vdots \\ \left(\mathbf{y}_t^{(\text{mean})}\right)^K \end{bmatrix} \tag{23}$$

Then combining Eq. 21 and Eq. 23, the objective function can be transformed into:

$$\mathrm{J}=\left\|\mathbf{Y}_t-\mathbf{Y}_t^{(\text{mean}))}\right\|^2 \quad \text{s.t.} \quad \tilde{\mathbf{y}}_t^*=\boldsymbol{a}_t^{\mathrm{T}}\mathbf{Y}_t. \tag{24}$$

For Eq. 24, it is actually a Lagrangian problem with equality constraints. The Lagrangian multiplier λ_t can be introduced, and the following Lagrangian equation can be formulated:

$$\mathrm{L}\left(\mathbf{Y}_t\right)=\left\|\mathbf{Y}_t-\mathbf{Y}_t^{(\text{mean})}\right\|^2+\lambda_t\left\|\boldsymbol{a}_t^{\mathrm{T}}\mathbf{Y}_t-\tilde{\mathbf{y}}_t^*\right\|^2 \tag{25}$$

For the Lagrangian equation of Eq. 25, the goal of optimization is to find a value $\mathbf{Y}_t^*=\arg\min_{\mathbf{Y}_t}\mathrm{L}\left(\mathbf{Y}_t\right)$, that minimizes the objective function, then the optimal mapping point can be obtained by taking partial derivatives of the Lagrange equation:

$$\mathbf{Y}_t^*=\left(\mathbf{I}+\lambda_t a_t \boldsymbol{a}_t^{\mathrm{T}}\right)^{-1}\left(\mathbf{Y}_t^{(\text{mean})}+\lambda_t \boldsymbol{a}_t \tilde{\mathbf{y}}_t^*\right) \tag{26}$$

For the optimal mapping point in Eq. 26, they are the K mapping points where the desired via point lies on the tangent space of point $\mathbf{p}$. Then, remap these K mapping points back to the manifold using exponential mapping to obtain the point $\mathrm{Exp}_{\mathrm{p}}\,\mathbf{Y}_t^*$ that needs to be projected onto the manifold. At this time, the new $\boldsymbol{\mu}_\omega^k$ for each type of Orientation ProMPs can be obtained by using the passing via point trajectory formula of the Orientation ProMPs. Finally the final passing via point trajectory is obtained by combining weight parameters.

The difference between this algorithm and the ordinary EM algorithm is that it does not require multiple iterations because the optimal parameters can be directly calculated by the update algorithm of Orientation ProMPs. When there are multiple expected via points, only one iteration is required for each via point, and the previous result can be used as the basis for the next iteration. Under the condition of ensuring rapid algorithm convergence, it can also smoothly realize passing via point.

5 Experiments

This section will verify the correctness of the Mixed Orientation ProMPs proposed in this paper for learning the reproduction of the teaching trajectory and the generalization of the passing via point trajectory. It also introduces the specific process of the experiment, including the acquisition phase, learning phase, and planning phase of the trajectory. Additionally, it compares and analyzes the passing via point trajectory of single Orientation ProMPs and Mixed Orientation ProMPs.

5.1 Training Set Collection Phase

Ten teaching trajectories of the UR5 robot were collected. These teaching trajectories represent the orientation trajectory data of the robot's end effector. The collected orientation trajectory is shown in Fig. 1, including 10 pieces of orientation data. The unit quaternion at each moment is represented by the gray curve. The time points of each piece of teaching data are fixed at 185.

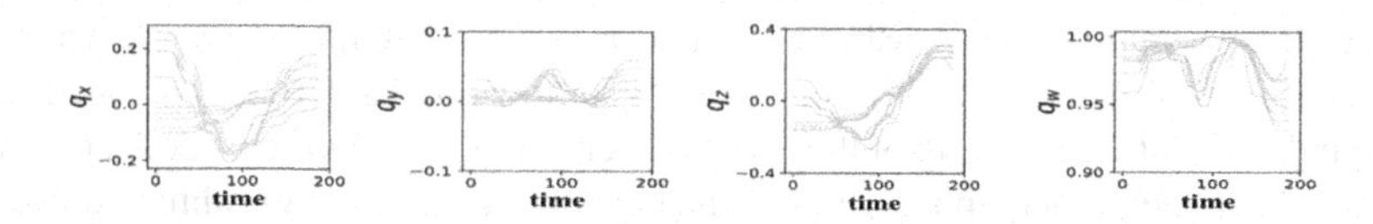

Fig. 1. Data for 10 orientation trajectories of the robot end-effector

5.2 Model Learning Phase

At this stage, the model parameters are obtained by training the training set collected in the previous section using the Mixed Orientation ProMPs described in this paper. Subsequently, the trajectory planned by the model can be obtained based on the acquired model parameters.

In the Mixed Orientation ProMPs learning stage, the algorithm also learns the parameters of each teaching trajectory through the Orientation ProMPs algorithm. It then establishes a Gaussian mixture model to reproduce the task trajectory based on these parameters. Set the number of basis functions as $N_\phi = 30$. The initial learning rate η of this model is 1, and its maximum learning rate $\eta_{\max}$ is also 1. Then, 10 sets of demonstration weight parameter sets $\left\{\omega^{(n)}\right\}_{n=1}^{10}$ can be obtained through the learning of the Orientation ProMPs model. Each large weight parameter vector contains 30 small weight data, that is, $\omega^{(n)} = \left[\omega_1^{(n)\top} \ldots \omega_{30}^{(n)}\right]^{\mathrm{T}}$. The GMM algorithm is used to divide the learned weight parameters into two categories for modeling. Two Orientation ProMPs models are obtained, with weights a^1 and a^2, and parameters ω^1 and ω^2 respectively. Then, the parameters are weighted and combined according to their respective weights to obtain the task trajectory of the Mixed Orientation PROMPs. As shown in Fig. 2, the trajectory distribution learned by Mixed Orientation ProMPs on the collected training set is given.

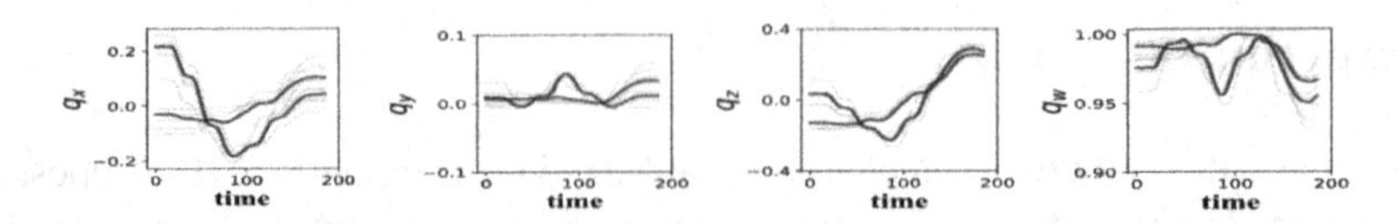

Fig. 2. The distribution of task trajectories learned by Mixed Orientation PROMPs

In Fig. 2, the two black curves represent the average trajectory distribution of the two types of sub-ProMPs learned by Mixed Orientation ProMPs from the teaching trajectory data, while the gray curve represents the training set data. It can be observed from the figure that the trajectory distribution shape learned by the Mixed Orientation ProMPs model is similar to the shape of the trajectory data in the training set as a whole.

5.3 Trajectory Planning Stage

In the stage of trajectory planning, it is necessary for the model to utilize the parameters learned from the teaching trajectory in order to generate a new trajectory that aligns with the characteristics of the new task. This ensures that the trajectory fulfills the requirements of the passing via point task. Select a desired via point at time t = 40, where the unit quaternion of the orientation coordinates is set to $(-0.102, 0.016, -0.158, 0.982)$. Through the trajectory planning of the Mixed Orientation ProMPs on the collected training set is given mentioned above, refer to the concept of the EM algorithm and use formula 19 to update the weight of each Orientation ProMPs in step E. Then, use Eq. 24 to calculate the mapping of the desired via point on each orientation ProMPs. Finally, in the M step, the mean and covariance of the new weight parameters are obtained by using the passing via point trajectory planning method proposed above to map the points. By combining this new parameter with the new weight, a smooth passing via point trajectory can be achieved.

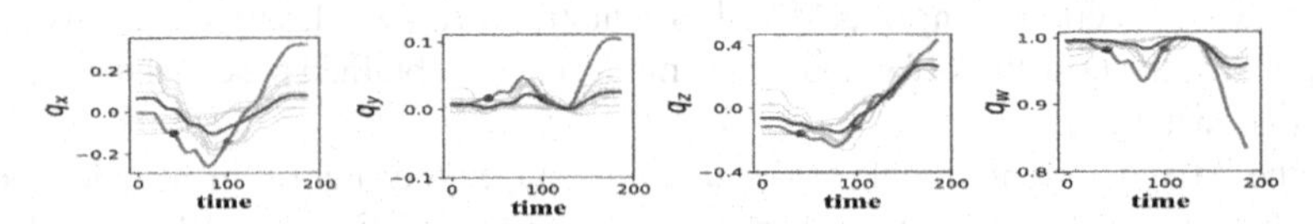

Fig. 3. The effect of passing via point trajectory planning of single Orientation ProMPs (Color figure online)

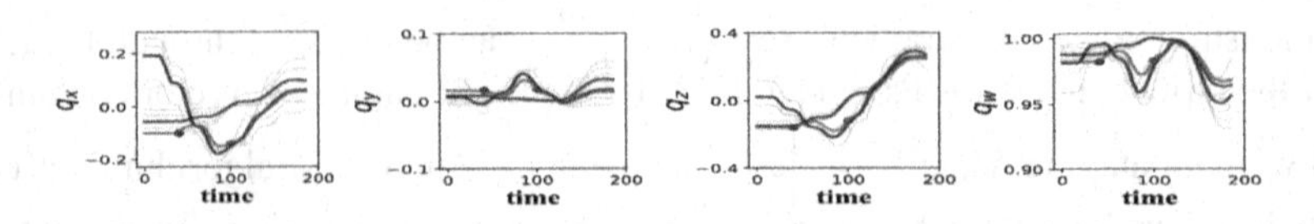

Fig. 4. The effect of passing via point trajectory planning of Mixed Orientation ProMPs (Color figure online)

Figure 3 and Fig. 4 respectively show the effect of passing via point trajectory planning using single Orientation ProMPs and Mixed Orientation ProMPs. The gray curve represents the teaching trajectory data, while the black curve represents the mean trajectory distribution learned from the training set. The dark blue dots and light blue dots in the figure represent the expected via points that need to be passed at 40 and 99, respectively. The red curve represents the new passing via point trajectory planning obtained after model re-learning. It can be seen from the figure that the trajectory planned by the Mixed Orientation ProMPs is obviously smoother and has less fluctuation compared to that of the single Orientation ProMPs. When the planned trajectory fluctuates significantly, it becomes easy for the robot to deviate from the intended teaching trajectory, increasing the likelihood of encountering singular points.

5.4 Physical Trajectory Planning

Similar to the three stages of simulation, the first step is to collect the physical teaching trajectory. Next, we train the 10 trajectories using single Orientation ProMPs and Mixed Orientation ProMPs separately. Finally, we set two desired via points. At time t = 40, we select a desired via point and set the unit quaternion of its orientation coordinates to $(-0.889, 0.424, -0.149, 0.088)$. Select a desired via point at time t = 99, and set the unit quaternion of its orientation coordinates to $(-0.756, 0.605, -0.023, 0.247)$. The results are shown in Figs. 5 and 6.

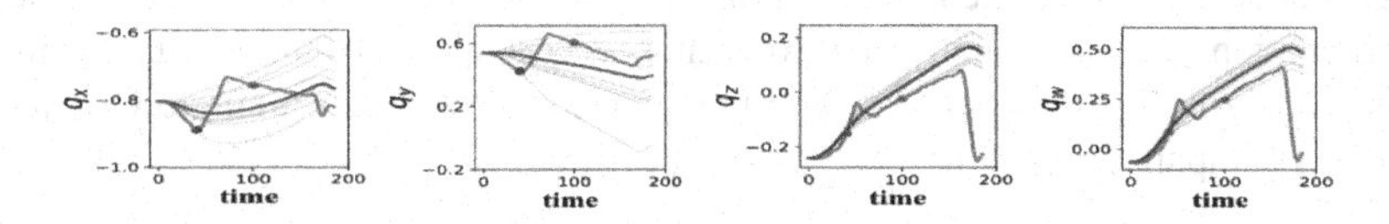

Fig. 5. The effect of passing via point trajectory planning of single Orientation ProMPs of real objects

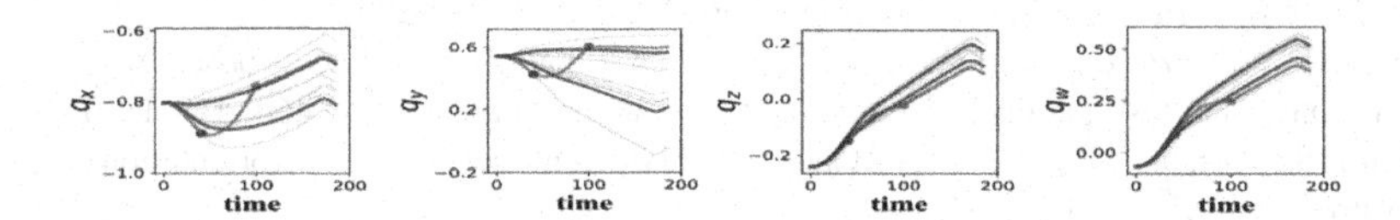

Fig. 6. The effect of passing via point trajectory planning of Mixed Orientation ProMPs of real objects

From Fig. 5 and Fig. 6, it can be seen that Mixed Orientation ProMPs are better than single Orientation ProMPs. The advantage of Mixed Orientation ProMPs is that they can utilize the most probable primitives to achieve passing via points, thereby reducing trajectory fluctuations. This allows the robot to minimize the risk of passing through singular points. The actual path taken by UR5 is shown in Fig. 7.

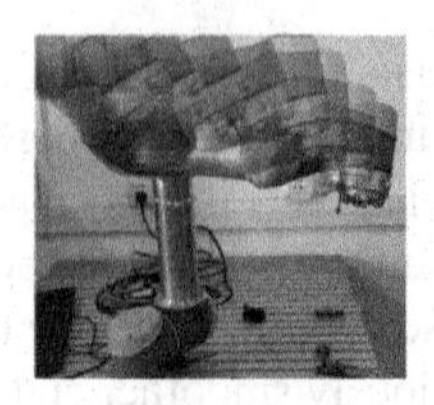

Fig. 7. The effect of passing via point trajectory planning of Mixed Orientation ProMPs of real objects

6 Conclusion

Firstly, the Gaussian mixture model algorithm is introduced, along with the process of deriving the expected maximum algorithm using Jensen's inequality method and variation method. Based on this, an algorithm for combining multiple Orientation ProMPs is proposed. In terms of model prediction, the double-space alternate estimation method considers both local optimization and global optimization when updating the trajectory of the passing via point. This method effectively fulfills the requirements of the passing via point task. In order to achieve smoother via point passing, an optimization goal is formulated to map the via point passing task to each group of small Orientation ProMPs models. This is done in terms of the average distance of the expected point and the Orientation ProMPs. Finally, the experimental verification of the mixture of multi-Orientation ProMPs is carried out, and the superiority of it and a single Orientation ProMPs is compared in detail.

References

1. Pastor, P., Hoffmann, H., Asfour, T., et al.: Learning and generalization of motor skills by learning from demonstration, Anchorage, pp. 763–768. IEEE (2009)
2. Silvério, J., Rozo, L., Calinon, S., et al.: Learning bimanual end-effector poses from demonstrations using task-parameterized dynamical systems, Hamburg, pp. 464–470. IEEE (2015)
3. Kim, S., Haschke, R., Ritter, H.: Gaussian mixture model for 3-DoF orientations. Robot. Auton. Syst. **87**, 28–37 (2017)
4. Ude, A., Nemec, B., Petrič, T., et al.: Orientation in cartesian space dynamic movement primitives, Hong Kong, pp. 2997–3004. IEEE (2014)
5. Koutras, L., Doulgeri, Z.: A correct formulation for the orientation dynamic movement primitives for robot control in the cartesian space, Cambridge, pp. 293–302. PMLR (2020)
6. Huang, Y., Abu-Dakka, F.J., Silvério, J., et al.: Generalized orientation learning in robot task space, Montreal, pp. 2531–2537. IEEE (2019)
7. Huang, Y., Abu-Dakka, F.J., Silvério, J., et al.: Toward orientation learning and adaptation in cartesian space. IEEE Trans. Robot. **37**(1), 82–98 (2020)
8. Rozo, L., Dave, V.: Orientation probabilistic movement primitives on riemannian manifolds. CoRL (2021)
9. Fu, J., Du, J., Teng, X., et al.: Adaptive multi-task human-robot interaction based on human behavioral intention. IEEE Access **9**, 133762–133773 (2021)
10. Ratliff, N., Toussaint, M., Schaal, S.: Understanding the geometry of workspace obstacles in movement optimization, Washington, pp. 4202–4209. IEEE (2015)

11. Hauberg, S., Freifeld, O., Black, M.: A geometric take on metric learning. In: Advances in Neural Information Processing Systems, vol. 25 (2012)
12. Suomalainen, M., Abu-Dakka, F.J., Kyrki, V.: Imitation learning-based framework for learning 6-D linear compliant movements. Auton. Robot. **45**(3), 389–405 (2021)
13. Legeleux, A., Buche, C., Duhaut, D.: Gaussian mixture model with weighted data for learning by demonstration. In: The International FLAIRS Conference Proceedings (2022)

Multi-population Fruit Fly Optimization Algorithm with Genetic Operators for Multi-target Path Planning

Ke Cheng[1], Qingjie Zhao[1(✉)], Lei Wang[2(✉)], Wangwang Liu[2], Shichao Hu[1], and Kairen Fang[1]

[1] School of Computer Science and Technology, Beijing Institute of Technology, Beijing, China
Zhaoqj@bit.edu.cn
[2] Beijing Institute of Control Engineering, Beijing, China
15413869@qq.com

Abstract. Automatic path planning is very important for many applications such as robots exploring unknown environments and logistics delivery. In this paper, we propose a discrete multi-population fruit fly optimization algorithm with genetic operators, where a greedy strategy is used to obtain good initial population, 3-opt heuristic search simulating olfactory to make the algorithm achieve higher convergence accuracy, multiple population collaborative strategy simulating vision to avoid the algorithm falling into local optima, and the genetic mechanism of selection-crossover-mutation prompts the population easily getting the optimal solution. The proposed planning algorithm has fewer parameters to be adjusted and has the advantages of high accuracy and fast convergence. The experimental results prove that the proposed planning algorithm performs best compared with other several algorithms.

Keywords: Path planning · fruit fly optimization algorithm · robot exploring

1 Introduction

Due to limited prior knowledge about remote or complex environments such as space asteroids, intelligent mobile robots are usually the first ones to execute exploration tasks. When it is necessary to explore multiple target locations on the sphere, a mobile robot needs to be able to plan the optimal path on its own, which is very important for it to quickly complete the tasks of exploring multiple target positions. This multi-target path planning can be regarded as a traveling salesman problem (TSP) [1], which is to find an optimal path to make a traveling salesman traverse all cities along the shortest path length.

Nowadays, swarm intelligence algorithms are usually used to solve those combination optimization problems similar to the traveling salesman one, which simulate group

Pre-research Project on Civil Aerospace Technologies of China National Space Administration (Grant No. D010301).

F. Sun and J. Li (Eds.): ICCCS 2023, CCIS 2029, pp. 174–191, 2024.
https://doi.org/10.1007/978-981-97-0885-7_15

behaviors and use heuristic algorithms to search the optimal solution. The core of a swarm intelligence algorithm is evolutionary computation and information interaction in the group. Searching is carried out continuously until the optimal solution is obtained by individual competition, information exchange and evolutionary iteration. For their robustness and adaptability, swarm intelligence algorithms have attracted great attentions from the scholars in various fields. For example, Genetic algorithm (GA) [2, 3] models the natural phenomena that new individuals will be obtained when chromosome selection, crossover and mutation occurs, and has been widely used to deal with various practical problems. Particle swarm optimization (PSO) algorithm [4–6] simulates the behavior of a bird swarm. Ant colony optimization (ACO) algorithm [7–9] simulates the ant foraging process. [10] combined simulated annealing (SA) algorithm and differential evolution algorithm for the optimization problem of heat exchange networks. [11] studied meta-strategy simulated annealing (SA) and tabu search (TS) algorithms for the vehicle routing problem. In 2012, Pan [12] proposed a fruit fly optimization algorithm (FOA) simulating the foraging behavior of fruit fly to model the financial distress.

As an emerging population intelligence optimization algorithm with few parameters and fast convergence, FOA has received attentions from a wide range of scholars for optimization problems. [13] proposed a multi-objective FOA to deal with the resource scheduling problem. [14] applied the FOA to feature selection in machine learning to improve the performance of the model. Some other applications of FOA include path planning [15], joint replenishment [16], neural network parameter tuning [17] et al.

In this paper, addressing the need for mobile robots exploring multiple target locations on the surface of asteroids, we propose a discrete multi-population fruit fly optimization algorithm with genetic operators (GA-DMFOA) to enable mobile robots to automatically plan their exploration paths. The algorithm not only takes into account the olfactory sensitivity of individual fruit flies and their vision based group collaboration ability, but also optimizes the initialization method, heuristic strategy, collaborative search method, individual improvement and other aspects to ensure that the algorithm has a high convergence success rate and convergence speed. (a) Greedy strategy is used in the initialization which makes the population can acquire higher adaptation and can fasten the algorithm convergence. (b) In the olfactory search stage, the 3-opt heuristic search strategy is used to improve the convergence accuracy. (c) In the visual search stage, the multiple population collaborative strategy promotes the algorithm avoid falling into a local optimum. (d) The genetic mechanism of selection-crossover-mutation is used to enhance the algorithm's ability of finding the best solution. Experimental results on the publicly available standard data-set TSPLIB demonstrate that our algorithm can provide an excellent solution for the path planning of exploring multiple target locations.

2 Multi-population Fruit Fly Optimization Algorithm with Genetic Operators

2.1 Path Planning Problem Formal Description

The goal of multi-target path planning for an asteroid exploring robot is to find the best location sequence so that the robot can traverse all positions and return to the start point at the least cost. The robot can be looked as a traveling salesman and the locations can be considered as cities that the traveling salesman has to pass through. Thus, the mathematical definition of this combination optimization problem is as follows.

Given a set of n locations, finding an optimal location sequence $x = \{x_1, x_2, \ldots, x_n\}$ for the mobile robot, so that the total travel length $F(x)$ of the robot traversing all locations and back to the start point is the shortest, where the distance between the locations x_i and x_j are defined as d_{x_i,x_j}. The definition of $F(x)$ is shown in Eq. (1).

$$F(x) = \sum_{i=1}^{n-1} d(x_i, x_{i+1}) + d(x_n, x_1) \tag{1}$$

2.2 Basic Fruit Fly Optimization Algorithm

The olfactory and visual organs of fruit flies are very sensitive compared to other insects. When searching for food, fruit fly populations will fly towards food by their olfactory organs firstly. They also visually observe the other fruit flies in the colony and, by comparison, learn where the highest odor concentration in the colony is located and then fly towards it. This is how fruit fly populations continue to search for the location of food.

In the fruit fly optimization algorithm, the olfactory search determines the search ability of fruit fly population, and the visual search determines that the individuals can observe each other and fly towards where with the highest odor concentration or the food location. Both of the searches are the key steps of the algorithm for population search, and each step is a continuous exploration in the fruit fly population. The specific steps of the fruit fly optimization algorithm are as follows.

Step 1: Initialize the size of the fruit fly population *PopSize*, the maximum number of iterations of search *MaxIter*, and the starting search position of the fruit fly population X_{axis}, Y_{axis}.

Step 2: The olfactory search phase: given the random search operator *Random*(), each individual performs a random search near the population search location.

$$X_i = X_{axis} + Random() \tag{2}$$

$$Y_i = Y_{axis} + Random() \tag{3}$$

Step 3: Calculate the position $Dist_i$ of the individual from the origin, and then calculate the concentration $Smell_i$ of the individual fruit fly according to the concentration determination value function f.

$$Dist_i = \sqrt{X_i^2 + Y_i^2} \tag{4}$$

$$Smell_i = f(X_i, Y_i) \tag{5}$$

Step 4: Find the most concentrated individual in the fruit fly population and mark its location *BestIndex* and odor concentration *BestSmell*.

$$[BestSmell, BestIndex] = \max(Smell_i) \tag{6}$$

Step 5: The visual search phase: each fruit fly moves toward the individual with the best concentration *BestIndex*, and the position of this individual is updated as the search position of the population; if *BestSmell* is the currently known individual with the highest concentration, it is recorded as the optimal solution *Best*.

$$Best = BestSmell \tag{7}$$

$$X_{axis} = X\,(BestIndex) \tag{8}$$

$$Y_{axis} = Y(BestIndex) \tag{9}$$

Step 6: Repeat the above operations from Step 2 to Step 5 until the number of iterations reaches *MaxIter* and then end and output the optimal solution *Best*.

From the above computational steps of the fruit fly optimization algorithm (FOA), it can be seen that the core idea is the random search of individual fruit flies in the population. Each individual not only has its own olfactory search ability, but also can interact with the information of the current population optimal solution and perform visual search in the local area according to the location of the current optimal solution, and move toward the location of food by iteration. FOA based on olfactory and visual search mechanism has the advantages of less parameters, fast convergence and easy implementation. However, it also has some shortcomings such as negative problem for complex function optimization, tending to fall into local optimal solution if the distribution of fruit fly individuals is sparse, with small flight distance at each iteration, and relying on the heuristic operator in discrete problems.

2.3 Fruit Fly Optimization Algorithm with Genetic Operators

1) *Algorithm Description.* When applying the fruit fly optimization algorithm, we need to develop discrete strategies and improvement schemes related. Real number coding and binary coding are two common coding methods for this problem.

Binary encoding is not intuitive and requires decoding operations on the results. In this paper we use a real number encoding approach. Let the number of locations be n and the ith location is represented by a real number label x_i. The sequence $x = (x_1, x_2,\ldots,$ $x_{\mathrm{n}})$ is the location sequence for a robot to visit and also the value of a fruit fly individual. The task is to acquire the optimal sequence x so that the robot can travel the shortest path length.

2) *Greedy Strategy Initialization.* When generating the initial fruit fly population, a random generation method is usually used to obtain the path sequence. However, the initial individuals obtained by this method tend to have lower concentrations and represent longer path lengths. Therefore, we adopt an initialization method based on greedy strategy, which can make the starting adaptation of the population higher and can improve the convergence speed of the algorithm.

The idea of greedy strategy is selecting the next operation at the least cost. As for the path planning problem, i.e., when generating a segment of the planning sequence, each time the point that is closest to the current point and has not been visited is put into the path sequence as the next planning point. Specifically, assume that the number of path points is *N*, the set of visited path points is *V*, the initial set of path points is *S*, and each individual in the population is a segment of the path sequence. Since the greedy strategy initialization yields different path sequences when different starting points are chosen, the calculated path lengths will different. In order to obtain the optimal initial position of the fruit fly population, a more complete initialization scheme is used in this paper. Each path point is used as the starting point to initialize the path using the greedy strategy, and the individual with the highest concentration (with shortest path length) is used as the initialized position of the fruit fly population in the *N* path sequence.

Compared with random initialization, greedy strategy can obtain a stable initial population with better fitness. From the point of view of the number of iterations, using greedy strategy initialization can greatly reduce the number of iterations needed to find the best path, and improve the search efficiency of the algorithm in the subsequent heuristic search. The specific operation process is shown in the Algorithm 1.

Algorithm 1: Greedy strategy initialization

1: Initialize all control parameters;
2: Initialize the population starting position (sequence) *init* and starting fitness f_{init};
3: Calculate the distance between each path point and obtain the distance matrix *Dist*;
4: **for** i =1 to N **do**
5: Select x_i as the current path point current;
6: Initialize the set of unvisited path points $V = \{x_1, x_2, \dots, x_{i-1}, x_{i+1}, \dots x_n\}$;
7: Initialize the current individual sequence Cur = $\{x_i\}$;
8: **while** *V* not empty **do**
9: Select the element in the set *V* with the shortest distance to the point current, labeld c_i, according to the distance matrix *Dist*;
10: Update the set $V = V - c_i$;
11: Update the set $Cur = Cur + c_i$;
12: Update the current path point current = c_i;
13: **end while**
14: Calculate the concentration F(*Cur*) of the sequence *Cur*;
15: **if** F(*Cur*) > f_{init} **then**
16: f_{init} = F(*Cur*);
17: *init* = *Cur*;
18: **end if**
19: **end for**
20: **return** the optimal sequence *init* and its fitness f_{init}

3) *3-opt Heuristic Search.* After the greedy initialization, we have obtained a better initial position for fruit fly population. Then each individual performs olfactory searching based on this initial position. In this stage, each fruit fly will perform a random search based on the current position, which means generating a different sequence of paths.

k-opt (k ≥ 2) is a heuristic algorithm based on element swapping. When the sequence is broken into *k* pieces, these pieces can be combined into many different sequences, and there are $(k-1)!2^{k-1}$ kinds of combinations for the new sequence. As shown in Fig. 1, 2-opt algorithm randomly selects two points *i* and *j* to divide the sequence into two segments, and if changing the connection between segments i and j in the sequence, the segments will be inverted. 3-opt algorithm divides the sequence into three segments, which can generate eight different combinations, as shown in Fig. 1. The larger k is, the more the combinations there will be. In order to verify the influence of k on the algorithm, we have tested the methods with 2-opt and 3-opt on the data sets with different sizes. The experimental results show that the method with 3-opt has better convergence accuracy. As for $k > 3$, although the search range becomes larger, the search time will get longer. Therefore, we adopt 3-opt algorithm as a heuristic search strategy. The process of generating new individuals using 3-opt search is shown in Algorithm 2.

Algorithm 2: 3-opt algorithm steps

1: Perform 3-opt operation on a fruit fly individual x, producing 8 new individuals, denoted as x_i;
2: Count the concentration $F(x_i)$ of these 8 new individuals, higher concentration means shorter path length;
3: Return the highest concentration as a new fruit fly individual

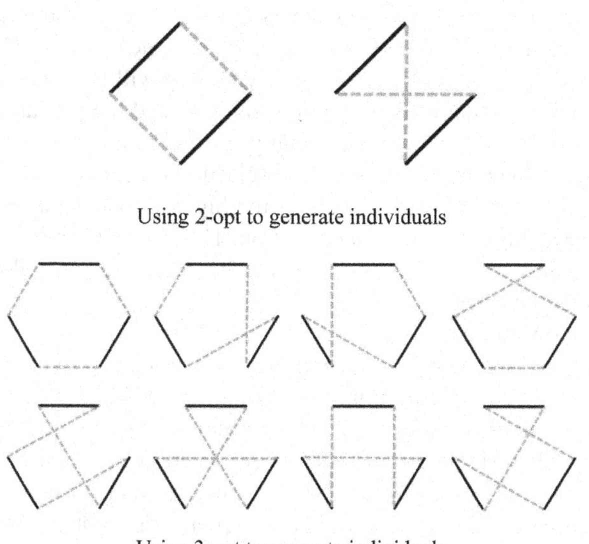

Using 2-opt to generate individuals
Using 3-opt to generate individuals

Fig. 1. K-opt operation diagram

4) *Multiple Population Collaborative Search.* In the traditional fruit fly algorithm, its visual search phase is around the optimal position of the current fruit fly population. Such an operation converges quickly, but has a higher probability to fall into a local optimum solution. The multiple population search scheme has been proven to be an effective improvement [18], so we use it on the basis of the 3-opt algorithm. Assuming that there are *Group* populations, the top *Group* individuals with the highest concentrations are selected when calculating the starting positions of the populations. Each of the

individuals is used as the starting search position of its population, thus expanding the exploration area and enhancing the search ability of the algorithm.

Since there is a possible problem that the total population size *PopSize* may not divisible by *Group*, the population sizes generated by the top *Group* individuals with the highest concentrations are not exactly the same. Let $PopSize = Group \times P + Q$, the population size including the individual with the highest concentration x_{top1} as the starting position is $P + Q$, and the population sizes generated by the other $Group - 1$ individuals with higher concentrations are P.

5) *Genetic Operators.* It is difficult for the population to jump out of the local optimal solution if no information interaction between individuals. In this paper, we improve fruit fly optimization algorithm by incorporating genetic operators to information interaction between individuals to raise the probability of jumping out of a local optimal solution.

(1) Select operation

Genetic algorithm needs to select two parent individuals before performing crossover operations, and a common method is to use a roulette wheel approach for selection where each individual has a certain probability of being selected and the sum of the probabilities is 1, but the individual with higher concentration has a higher probability being selected. Therefore, the roulette selection method needs to design a better probability calculation function to get reasonable selected probabilities for the individuals with different concentrations. Otherwise, If the probabilities are close, the selection effect will not be good. However, if they are very different, the sampling will be imbalanced seriously. This operation will need more computation resources, and the probability calculation function also needs to be modified accordingly for different types of data which will introduces more parameters. We improve the selection operation that when reforming each individual X_i, the individual *Best* with the highest concentration in the current population is selected for crossover and mutation. This method can not only reduce the computational load with no extra parameters to be considered, but also has a higher probability to obtain the best individual.

(2) Crossover operation

The crossover of segments of chromosomes in genes to produce new offspring is an important way of information interaction among individuals in a population. In genetic algorithms, there are also various improvement schemes for crossover operations, mainly including single-point crossover and multi-point crossover. The more the crossover points, the more the samples will be produced, but the more the computing time will be needed. In this paper, we use two-point fragment crossover method to produce relatively abundant progeny individuals and avoid excessive time cost. In the selection operation, we select a pair of parent individuals, one is the known optimal solution *Best* of the current population, and the other is the current fruit fly individual X_i, and the crossover operation inserts a randomly selected fragment from *Best* into the corresponding position in X_i.

Since we have constrained that each path point must be passed through once and only once, and real number encoding is used for the expression of solution, we need to deal with the sequence conflict problem after the crossover. The rule to handle the conflict is to find the mapping element of a conflicting element, and the conflicting mapping is recursively searched in the intersection fragments until the element of the conflicting mapping is not in the intersection fragments. Taking the number of path points 10 as an example, the operation procedure for the crossover of fruit fly individuals is shown in Fig. 2. For the problem of conflicting path points generated by individuals after the crossover, we establish the conflict mappings as follows: $2 \leftrightarrow 10$, $4 \leftrightarrow 9$, $5 \leftrightarrow 8$, and we can get a new individual X_i after dealing with the conflicts on the mapping relations of the elements.

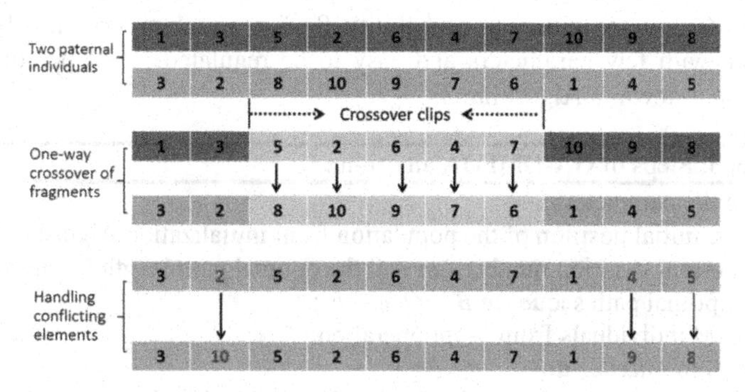

Fig. 2. Schematic diagram of crossover operation

(3) Mutation operation

In human reproduction, mutation is an important means to increase the diversity of the population and is an important process for human reproduction. The mutation operation in genetic algorithm is the same, in which the individual after crossover mutates at a certain probability will become a new individual.

Mutation includes shift mutation, insertion mutation, flip mutation and reciprocal mutation. Among them, reversal mutation, shift mutation and insertion mutation are fragment changes of individuals. Since both of 3-opt operation and crossover operation help to generate new individuals by changing the fragments of individuals, therefore we use reciprocal mutation as a way of individual mutation. Reciprocal mutation randomly selects two positions i and j in an individual and swaps the element values of these two positions. Again taking the number of 10 path points as an example, the flow of the mutation operation is shown in Fig. 3.

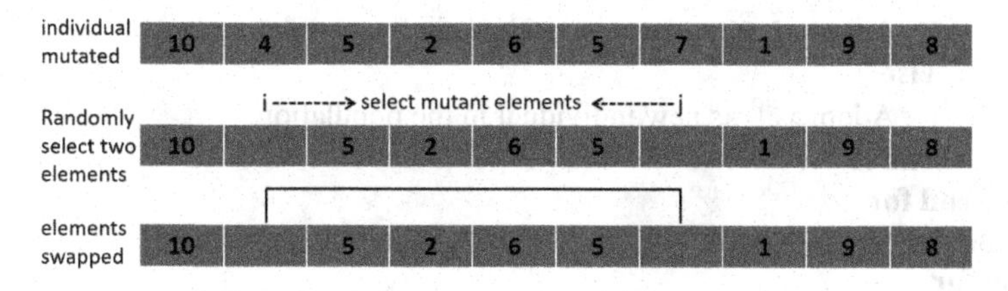

Fig. 3. Schematic diagram of mutation operation

The crossover and mutation can greatly improve the search ability of the genetic algorithm. This process continuously uses the known optimal solution *Best* in the current population to reform each individual, and this reformation can improve the algorithm accuracy.

6) *Algorithm Steps.* We propose the discrete multi-population fruit fly optimization algorithm with genetic operators (GA-DMFOA), containing greedy strategy for initialization, 3-opt combined with genetic operators for heuristic search, and multiple population collaborative method. There are only four parameters in GA-DMOFA: population size M, the maximum number of iterations $MaxIter$, group number of multi-population collaboration $Group$ and mutation probability P_{mutate}, which shows the advantage of the algorithm with few parameters and easy to be regulated. The operation flow of GA-DMOFA is shown in Algorithm 3.

Algorithm 3: Steps of GA-DMFOA algorithm

1: Initialize the parameters;
2: Obtain the initial position of the population from initialization Algorithm1;
3: Let the current iteration number $iter = 0$,the optimal path length be $BestLen = \infty$, and the optimal path sequence $BestQue = \{\}$;
4: Generate M individuals from 3-opt operation;
5: **for** $iter = 0$ to $MaxIter$ **do**
6: Calculate the concentration $F(x_i)$ of each individual in the current population, the individual with the highest concentration $F(x_{top})$ being x_{top};
7: **if** $1./F(x_{top}) < BestLen$ **then**
8: $BestLen = 1./F(x_{top}), BestQue = x_{top}$;
9: **end if**
10: Take the top $Group$ individuals with the highest concentrations and calculate the size of each population Num_i;
11: **for** $i = 1$ to $Group$ **do**
12: Take the fruit fly individual x_i with the ith highest concentration as the start position of the search for this population;
13: **for** $cnt = 1$ to Num_i **do**
14: Obtain the new individual x_i^{3opt} by applying the 3-opt operation described in Algorithm 2 to x_i ;
15: Do crossover and mutation of genetic operators, the kth individual x_k in the population is improved by $BestQue$ to obtain the individual x_k^{ga} ,where $k = cnt + Group * (i-1)$;
16: **if** $F(x_k^{ga}) > F(x_i^{3opt})$ **then**
17: Adopt x_i^{3opt} as new individuals in the population;
18: **else**
19: Adopt x_k^{ga} as new individual in the population;
20: **end if**
21: **end for**
22: **end for**
23: **end for**
24: Output the optimal path length $BestLen$ and the optimal path sequence $BestQue$

3 Experimental Results and Analysis

3.1 Test Environment

The operating system for this experiment is MacOS 10.14.6, the CPU is Intel Core i5 with quad-core and 2.4 GHz main frequency, running memory is 8G, the programming language is Python 3.7, and the programming software is Pycharm 2019.3.

The standard dataset used is TSPLIB [19], which is an international standard database for the Traveling Merchant Problem (TSP) and other related problems, with numerous datasets of different sizes, and each dataset gives a standard optimal solution for researchers to verify the effectiveness of their algorithms. The city sizes and standard optimal solutions (Please see Table 1) can be taken from TSPLIB. We can choose any dataset to test different algorithms in our experiments. Since population intelligence algorithms are somewhat stochastic, we run the algorithm 20 times repeatedly in order to test the performance more accurately. The criterions used to evaluate the effectiveness of the algorithm include the optimal value, the average value, the worst value, the average error value and the average number of iterations to reach the optimal solution. The error value is mainly used to compare the performance on different datasets, and is calculated using the Eq. (10).

$$Error(\%) = \frac{Average - StandardOptimumSolution}{StandardOptimumSolution} * 100\% \tag{10}$$

Table 1. Test Datasets

TSPLIB dataset	City Size	Standard Optimum Solution
Berlin52	52	7542
Pr76	76	108159
Rat99	99	1211
kroA100	100	21282
Ch130	130	6110
Ch150	150	6528
kroA200	200	29368

3.2 Effect of Greedy Strategy Initialization

Besides the greedy strategy used in this paper, random greedy initialization and only once greedy strategy are commonly used. In this paper, the algorithm with random initialization scheme is named Rand-GA-DMFOA, that with only once greedy strategy initialization is named GA1- DMFOA, and that with complete greedy strategy proposed in this paper is GA-DMFOA. The population size $M = 40$, the maximum number of iterations $MaxIter = 800$, the number of groups with multi-population collaboration

Group = 3, and the mutation probability is *Pmutate* = 0.05. Experiments are conducted on the datasets of berlin52, pr76, rat99, ch130, and the results are shown in Table 2.

The bolded values in Table 2 are the optimal solutions. From the results, we can see that the initial, optimal, average, and worst values obtained by GA-DMFOA are better than those of Rand-GA-DMFOA and GA1-DMFOA. And the smaller the size of the dataset is, the smaller the difference between different initialization schemes will be. As for the number of iterations to reach the optimal solution, we can see that GA-DMFOA needs much smaller iteration times, which means it has a fastest convergence speed.

Table 2. Effects of Different Initializing Methods

Dataset	Algorithm	Initial value	Optimal value	Average value	Worst value	Average number of iterations
Berlin52	Rand-GA-DMFOA	25144.62	7578.19	7891.77	8167.45	227
	GA1-DMFOA	8318.76	7566.23	7745.90	8011.42	128
	GA-DMFOA	**8182.19**	**7544.36**	**7724.84**	**7914.03**	**126**
Pr7	Rand-GA-DMFOA	510567.09	110922.83	112978.51	129552.99	253
	GA1-DMFOA	13107.31	110991.78	111799.13	129632.74	149
	GA-DMFOA	**130921.00**	**110893.31**	**111590.25**	**129393.13**	**135**
Rat99	Rand-GA-DMFOA	4929.13	1277.59	1357.51	1438.43	441
	GA1-DMFOA	1459.41	1298.62	1310.46	1409.77	152
	GA-DMFOA	**1369.53**	**1251.62**	**1290.75**	**1399.27**	**143**
Chl30	Rand-GA-DMFOA	42333.53	6538.4	6780.25	6940.88	752
	GA1-DMFOA	7294.19	6470.27	6601.44	6789.29	504
	GA-DMFOA	**7216.17**	**6419.18**	**6527.48**	**6776.16**	**492**

3.3 k-opt Parameter Analysis

We will compare the performance of 2-opt and 3-opt by experiments. We set the population size $M = 40$, the maximum number of iterations $MaxIter = 800$, the number of populations of multiple population collaboration $Group = 3$, and the mutation probability $P_{\text{mutate}} = 0.05$. The experiments are performed on the dataset rat99, and the comparison results are shown in Fig. 4. In the figure, we can see that the algorithm with 3-opt converges faster and the final accuracy is higher than that with 2-opt; the number of individuals generated by the algorithm with 3-opt is 8 which needs reasonable computing resources, so we select 3-opt as the heuristic search strategy in our later experiments.

3.4 Effect of Population Number

In order to determine the influence of the population number *Group* to the algorithm, experiments were carried out on TSPLIB and we can obtain a more reasonable number value of the populations *Group* through analysis. First, we set the probability of mutation in the genetic operator $Pmutate = 0.05$, the population size $M = 60$, the maximum number of iterations $MaxIter = 800$, the set of population numbers as $Group = \{1, 2, 3, 4, 5, 6\}$, and the evaluation criterion is the error value. The data sets used for the test are berlin52, kroA100 and ch150, and the related test results are shown in Fig. 5.

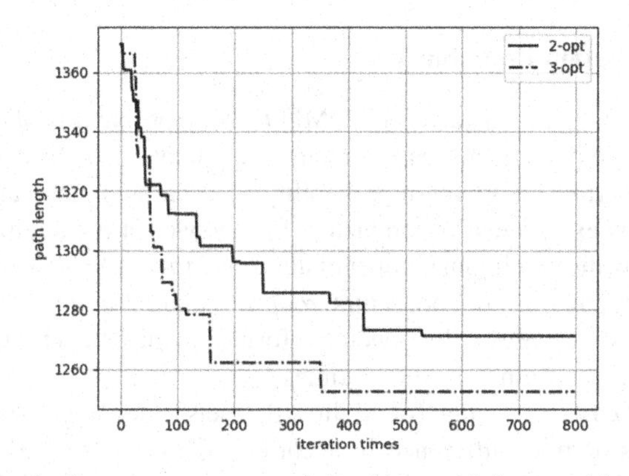

Fig. 4. Convergence curves of algorithms with 2-opt and 3-opt

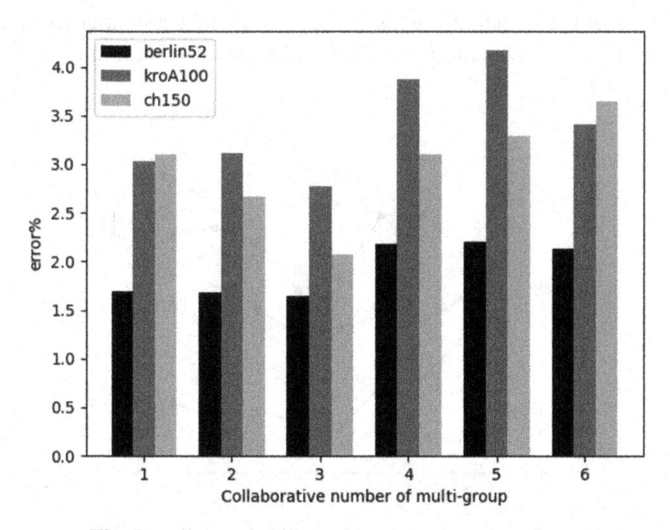

Fig. 5. Effects of different population number Group

From the results in Fig. 5, we can conclude that $Group = 3$ is a more reasonable value. On small data sets, multiple groups bring less improvement, which is because under the same parameter conditions, the smaller the data size is, the easier it is to get the optimal solution. We can also see that the error value does not decrease gradually with the increase of Group, which is because when multiple groups evolve, we take the first Group individuals with the highest concentrations as the initial positions for evolution. We also find that the concentrations of fruit fly populations differ greatly, and especially at the beginning of the iteration, the highest concentration is much higher than the other individuals, and the difference decreased slowly during the iteration. This shows that if the Group is very high, some individuals with very low concentrations will be used as the initial position for searching, and the search ability will be naturally poor, and the final accuracy will be low.

3.5 Effect of Mutation Probability

One more parameter to be adjusted in GA-DMFOA is the probability of mutation *Pmuate* in the genetic operators. In this paper, reciprocal mutation is used to generate new individuals. To determine a mutation probability *Pmutate* with generalization ability, a parametric analysis experiment is conducted. Firstly, we set the number of population individuals $M = 60$, the maximum number of iteration $MaxIter = 800$, and the population number of multi-population collaboration $Group = 3$ and $Pmutate = \{1\%, 2\%, 3\%, \ldots, 10\%\}$ for experimental analysis. Tests were performed in the data sets berlin52, kroA100 and ch150, and the algorithm was run 20 times.

From Fig. 6, it can be seen that for all three datasets tested, the variation in mutation probability causes a small difference in accuracy. We can also find that $Pmutate = 5\%–10\%$ has less impact on the performance of the algorithm, and the solution obtained around $Pmutate = 5\%$ is the optimal. Considering that the larger the mutation probability is, the larger the computational resources are needed, we take a smaller value as possible, so we take $Pmutate = 5\%$ as the benchmark value of the mutation probability in this paper.

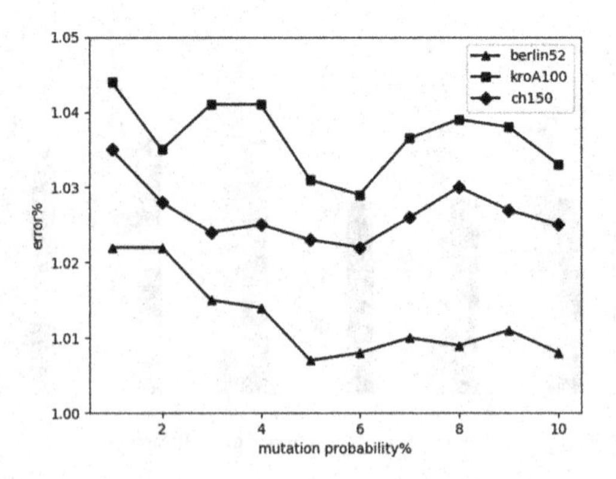

Fig. 6. Effects of mutation probability P_{mutate}

3.6 Ablation Experiments

In GA-MFOA, we adopt greedy strategy for initialization. The performance of the algorithm is improved by using 3-opt heuristic method for population searching, genetic operators for population improving, and multiple population collaboration to enhance the search capability. The final operational accuracy of the algorithm is the result of the combination of the several modules. In this subsection we conduct ablation experiments to verify the influence of each module on the final performance of the algorithm. The test dataset is a moderately sized rat99 dataset. The criteria used for the test are the average of the path lengths obtained from 20 runs and the average number of iterations to reach the optimal solution. The experimental parameters were set as the population size M = 60, the maximum number of iteration $MaxIter = 500$, the number of multiple populations Group = 3, and the mutation probability $P_{mutate} = 5\%$.

All ablation experiments use the complete greedy strategy to initialize the starting position of the population, ensuring that the starting position state is consistent across the experiments. We conducted four sets of ablation experiments for comparison, and the experimental results are shown in Table 3, and it can be seen from the average values that both multi-population collaborative search and genetic operators can improve the accuracy of the algorithm. However, the multi-population collaborative search is more conducive to reducing mean value and average number of iterations. The last row shows the experimental results of the GA-DMFOA proposed in this paper, which is the optimal. It shows that the multiple population search and genetic improvement can play a complementary role when used together to produce better optimization paths and converge faster.

Table 3. Ablation Experiment Results

Method	3-opt	group	GA	Average value	Average number of iterations
3-opt	√			1281.13	235
3-opt + group	√	√		1269.29	187
3-opt + GA	√		√	1271.41	193
3-opt + group + GA	√	√	√	1250.77	172

3.7 Performance of GA-DMFOA

To verify the performance of GA-DMFOA proposed in this paper, we compare it with other population intelligence algorithms such as genetic algorithm (GA), particle swarm algorithm (PSO), ant colony algorithm (ACO), simulated annealing algorithm (SA), and tabu search algorithm (TS). The GA population size is 60, the number of iterations is 500, the proportion of parent retention is 20%, and the probability of mutation is 5%; the PSO population size is 150, the maximum number of iterations is 500; the number of ACO ants is 40, the weight of pheromone volatility is 0.25, the weight of pheromone importance is

2, the total amount of pheromone release is 1, and the importance of heuristic function factor is 2; the initial temperature of SA is 2000, the ending temperature is 0.001, the cooling rate is 0.97, and the total number of individuals is 150; the forbidden table capacity of TS is 20, and the number of iterations is 500; the population size of GA-DMFOA is 60, the number of multiple populations is 3, the probability of mutation is 5%, and the maximum number of iterations is 500. In order to test their performances, the algorithms are executed 20 times and the running results are recorded.

The test results on the pr76 dataset are shown in Table 4, where GA-DMFOA has best performance in all four kinds of criterion including optimal values, average value, worst value, and the average number of iterations to reach the optimal solution. The convergence speed and accuracy of GA-DMFOA are much higher than those of the other algorithms.

Table 4. Performance of Different Algorithms on PR76

Method	Optimum value	Average value	Worst value	Average number of iterations
PSO	114259.88	120246.72	135256.39	453
GA	110281.14	115789.76	123986.66	488
SA	110291.47	112590.37	122478.50	204
ACO	112466.39	117536.79	123839.37	353
TS	115776.38	124380.70	126429.69	222
GA-DMFOA	**101736.47**	**111506.74**	**122456.29**	**134**

In order to verify the performance of the algorithm on a larger dataset, we conducted an experimental comparison on dataset ch150 with the same parameter settings. The experimental results are shown in Table 5, and the accuracy of GA-DMFOA on this dataset is still the best. Figure 7 shows the convergence curves of the algorithms, and it shows that TS and SA fall into the local optimal solutions; ACO, on the other hand, obtains a poor solution; the convergence curves of PSO and SA show lower efficiency in the early search process; GA's search ability decreases in the late stage; only GA-DMFOA is the best which converges fast and can arrive the optimal solution very close to what given by this dataset. Figure 8 is the visualization result of the path planned by GA-DMFOA.

The above experimental results demonstrate that GA-DMFOA is a population intelligence optimization algorithm with excellent performance and achieves good experimental results.

Table 5. Performance of Different Algorithms on CH150

Method	Optimum value	Average value	Worst value	Average number of iterations
PSO	6991.85	7042.05	7187.67	411
GA	6743.26	6813.42	7014.59	212
SA	7078.44	7078.44	7078.44	272
ACO	7103.28	7473.89	7530.89	372
TS	6698.32	6923.69	7102.47	387
GA-DMFOA	**6592.34**	**6697.50**	**7098.76**	**129**

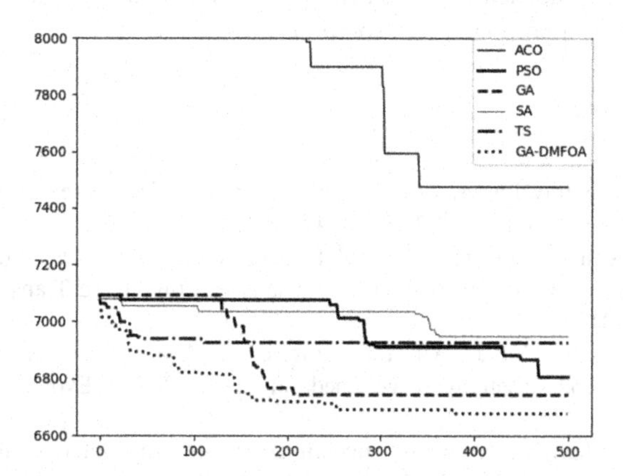

Fig. 7. Convergence curves of different algorithms on ch150

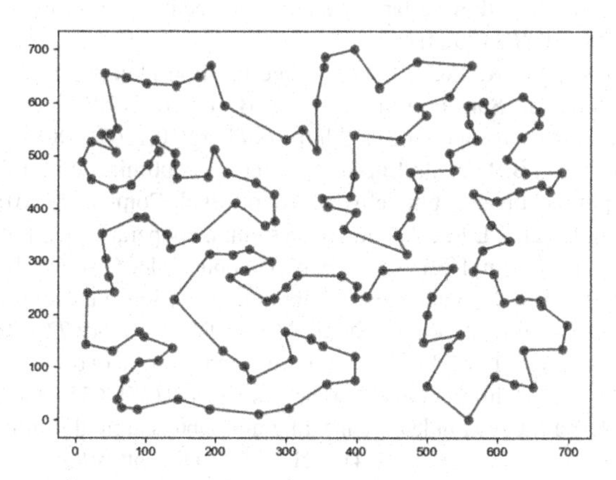

Fig. 8. Planned path by GA-DMFOA on ch150

4 Conclusion

We propose a discrete multi-population fruit fly optimization algorithm with genetic operators which can be adopted for many applications such as a mobile robot automatically planning its path for traversing multi-locations when exploring unknown environments or a manipulator planning its welding path. We have improved the fruit fly optimization algorithm in several ways. Greedy strategy is used for the initialization to ensure the population can acquire higher adaptation and make the algorithm converge faster; in the olfactory search stage, the 3-opt heuristic search strategy is used to impel the algorithm achieve higher convergence accuracy; in the visual search stage, the multiple population collaborative strategy is used to avoid the algorithm falling into a local optimum; the genetic mechanism of selection-crossover-mutation is used to enhance the algorithm's ability of finding the best solution.

References

1. Laporte, G.: The traveling salesman problem: an overview of exact and approximate algorithms. Eur. J. Oper. Res. **59**(2), 231–247 (1992)
2. Kalakova, A., KumarNunna, H.S., Jamwal, P.K., Doolla, S.: A novel genetic algorithm based dynamic economic dispatch with short-term load forecasting. IEEE Trans. Ind. Appl. **57**(3), 2972–2982 (2021)
3. Song, Y.J., Ou, J.W., Wu, J., Wu, Y.T., Xing, L.N., Chen, Y.W.: A cluster-based genetic optimization method for satellite range scheduling system. Swarm Evol. Comput. **79**, 101316 (2023)
4. Kennedy, J., Eberhart, R.: Particle swarm optimization. In: IEEE International Conference on Neural Networks, pp. 1942–1948 (2002)
5. Amirali, M., Andries, E., Beatrice, O.B.: Cooperative coevolutionary multi-guide particle swarm optimization algorithm for largescale multi-objective optimization problems. Swarm Evol. Comput. **78**, 101262 (2023)
6. Chen, Y., Zhao, Q., Xu, R.: AG-DPSO: landing position planning method for multi-node deep space explorer. In: Sun, F., Liu, H., Fang, B. (eds.) ICCSIP 2020. CCIS, vol. 1397, pp. 206–218. Springer, Singapore (2021). https://doi.org/10.1007/978-981-16-2336-3_19
7. Ren, T., Luo, T.Y., Jia, B.B., et al.: Improved ant colony optimization for the vehicle routing problem with split pickup and split delivery. Swarm Evol. Comput. **77**, 101228 (2023)
8. Dorigo, M., Gambardella, L.M.: Ant colony system: a cooperative learning approach to the traveling salesman problem. IEEE Trans. Evol. Comput. **1**(1), 53–66 (1997)
9. Yang, X., Dong, H., Yao, X.: Passenger distribution modelling at the subway platform based on ant colony optimization algorithm. Simul. Model. Pract. Theory **77**, 228–244 (2017)
10. Aguitoni, M.C., Pavao, L.V., Da, M.A.: Heat exchanger network synthesis combining simulated annealing and differential evolution. Energy **181**(15), 654–664 (2019)
11. Osman, I.H.: Metastrategy simulated annealing and tabu search algorithms for the vehicle routing problem. Ann. Oper. Res. **41**(4), 421–451 (1993). https://doi.org/10.1007/BF02023004
12. Pan, W.T.: A new fruit fly optimization algorithm: taking the financial distress model as an example. Knowl.-Based Syst. **26**(26), 69–74 (2012)
13. Wang, L., Zheng, X.L.: A knowledge-guided multi-objective fruit fly optimization algorithm for the multi-skill resource constrained project scheduling problem. Swarm Evol. Comput. **38**, 54–63 (2018)

14. Hou, Y., Li, J., Yu, H., Lia, Z.: BIFFOA: a novel binary improved fruit fly algorithm for feature selection. IEEE Access **7**(99), 177–181 (2019)
15. Tao, J., Wang, J.Z.: Study on path planning method for mobile robot based on fruit fly optimization algorithm. Appl. Mech. Mater. **536**, 970–973 (2014)
16. Wang, L., Shi, Y., Liu, S.: An improved fruit fly optimization algorithm and its application to joint replenishment problems. Expert Syst. Appl. **42**(9), 4310–4323 (2015)
17. Hu, R., Wen, S., Zeng, Z., Huang, T.: A short-term power load forecasting model based on the generalized regression neural network with decreasing step fruit fly optimization algorithm. Neurocomputing **221**(19), 24–31 (2017)
18. Yuan, X., Dai, X., Zhao, J., Qian, H.: On a novel multi-swarm fruit fly optimization algorithm and its application. Appl. Math. Comput. **233**(3), 260–271 (2014)
19. Reinelt, G.: TSPLIB — a traveling salesman problem library. ORSA J. Comput. **3**(3), 376–384 (1991)

Task Assignment of Heterogeneous Robots Based on Large Model Prompt Learning

Mingfang Deng[1(✉)], Ying Wang[2], Lingyun Lu[3], Huailin Zhao[1], and Xiaohu Yuan[4]

[1] School of Electrical and Electronic Engineering, Shanghai Institute of Technology, Shanghai, China
dengmingfang2021@163.com

[2] School of Physics and Electronic Information, Yantai University, Yantai, Shandong, China

[3] Nanjing Research Institute of Electronic Engineering, Nanjing, China

[4] Department of Computer Science and Technology, Tsinghua University, Beijing, China

Abstract. In order to ensure that the heterogeneous robot clusters performing the task can complete the target search at the specified location of the target object that the user needs to search according to their own field of view capabilities, we propose a task assignment algorithm for heterogeneous indoor robot clusters based on the robot's own field of view constraints. In particular, the heterogeneous robot clusters need to be parsed by linguistic commands to obtain the assignment results. Therefore, we solve the task assignment of heterogeneous robot clusters by performing cue learning on a large model to achieve the maximum utilization of heterogeneous robots while satisfying the field-of-view constraint; the simulation verifies the effectiveness of the task assignment of this method.

Keywords: Heterogeneous robots · prompt learning · task assignment · large models

1 Introduction

Indoor search robots have movement modes such as moving, rotating, and rolling wheels, which can help or replace humans in accomplishing a number of indoor search tasks [1].

Heterogeneous robot swarms in indoor scenarios provide an effective way to accomplish complex operations, as individual robots are constrained by their own high level of physical constraints, leading to challenges in performing Complex challenges in performing complex and diverse tasks [2–4]. In the system, the robots with different forms interact with each other through information and cooperate with each other to accomplish the search task of indoor scenes.

At present, the algorithms used for Multi-Robot Task Allocation (MRTA) problem are mainly intelligent optimization algorithms, neural network algorithms and so on. Among them, the population intelligence algorithms include particle swarm algorithm, ant colony algorithm, genetic algorithm, etc. Among them, population intelligence algorithms include particle swarm algorithm, ant colony algorithm, genetic algorithm, etc.

F. Sun and J. Li (Eds.): ICCCS 2023, CCIS 2029, pp. 192–200, 2024.
https://doi.org/10.1007/978-981-97-0885-7_16

For example, literature [8] et al. solved the heterogeneous UAV task allocation configuration problem using ant colony hybrid algorithm. Literature [9] et al. proposed a self-organizing mapping neural network based dual competitive strategy algorithm for task allocation of heterogeneous robots.

However, the above method requires direct information about the target object that the user is searching for and does not incorporate a reasoning process about the user's natural language. In addition to that, the above approach requires specific mathematical modeling for each heterogeneous robot group. Meanwhile, since 2012, natural language processing has begun to transition from statistical learning to end-to-end neural network deep learning approaches, which have greatly improved the performance of processing systems [10]. Insufficient linguistic data resources are a serious problem in the research of deep learning and neural networks. If the language data is insufficient, then the accuracy of any model extraction will be greatly reduced. In order to solve this problem, scholars began to try to maximize the model performance on small-scale datasets. Pre-trained language models were proposed by Feng et al. [10]

Pre-trained models have shown excellent performance on almost all downstream tasks of natural language processing.2022 Generative Pre-trained Transformer (GPT), a Transformer-based generative pre-training model developed by OpenAI, Inc. Has become the current natural language processing research GPT [12] utilizes the encoder and decoder of the Transformer model to obtain rich linguistic knowledge from linguistic data, and reaches a fairly high level in natural language generation tasks. In the same year, Tsinghua University proposed the chatglm pre-training model with performance comparable to that of GPT 3.5.

Due to the rapid development of large models in various fields, a large number of methods on how to better train large models have emerged, among which the typical one is prompt-learning [13]. This method is suitable for low-resource scenarios, i.e., high accuracy outputs can be obtained given a small number of prompt datasets, and there is no need to re-assign modeling for specific heterogeneous groups given the good generalization ability of the large models and the fact that good outputs can be obtained directly by designing appropriate prompts.

Therefore, in this paper, we investigate the task assignment of heterogeneous robots in the presence of constraints on the field of view range by parsing the user's natural language with a large model.

The main innovations of this paper are as follows:

1) the first allocation method that proposes cue learning on large models to obtain allocation results;
2) five natural language allocation optimization datasets of varying difficulty are created.
3) a link between heterogeneous robotics groups and users is constructed through the WeChat platform.

2 Mathematical Model of Task Assignment

2.1 Description of the Problem

The heterogeneous robot task assignment problem is defined as the execution of m different tasks by n robots with different morphology and size, and the minimization of the objective function is achieved by deciding a specific sequence of tasks to be accomplished by each robot. Heterogeneous robot task assignment can be modeled as a quaternion {R, T, VC, F}. Where R = {R1, R2,..., Rn} denotes the set of heterogeneous robot groups.

Each robot contains basic information about its height and field of view, etc. T = {T1, T2,..., Tm} denotes the set of tasks. Each task should contain basic parameters such as the orientation information to be searched, the type of task, etc. VC is the set of field of view constraints. F denotes the objective function of the task assignment system. The objective function considered in this paper is to minimize the error rate of the orientation accuracy navigated by the heterogeneous robots while completing the task.

2.2 Target Function

The task assignment scheme needs to be evaluated by metrics, and the task assignment model established in this paper takes the minimization of the error rate of the navigation orientation accuracy of heterogeneous robots in indoor scenarios as the objective function. The objective function established is as follows:

$$minF = \sum_{i}^{N}[1 - D(HR_i)] \tag{1}$$

where the $D(HR_i)$ stands for the accurate of the final task assignment.

2.3 Capacity Constraints to Implement the Mandate

A robot can only perform one task at a time, i.e.:

$$\sum_{j=1}^{T} x(i,j) \le 1, i = 1, 2, \ldots, N \tag{2}$$

where x(i, j) denotes that robot i performs the j-th task, 1 denotes going to the current task, and 0 denotes not going to the current task.

2.4 Capacity Constraints

Due to the limitation of each robot's own height, there is a dead zone in the field of vision, so the robots are required to cooperate with each other to complete the target search according to their own size advantages, such as tall robots need to search for the position above the task, and short robots need to search for the position below the task. TK = [TK1, TK2] denotes the type of the task. tk1 denotes the task above the search space, tk2 denotes the task below the search space. RK = [RK1, RK2] denotes the type of robot. The task below, e.g., TK = [1 0] indicates that the task is a search space. RK = [RK1, RK2] indicates the robot type. RK = [1 0] represents tall and large robots, and RK = [0 1] represents short and small robots. Search robots for heterogeneous indoor scenarios need to satisfy the following constraints:

$$TK_{T_i}(l) \le RK_{HR_i}(l)l = 1, 2 \tag{3}$$

2.5 Evaluation Metric

For task assignment, the heterogeneous robot obtains the final task assignment result by parsing the natural language instructions proposed by the user. As shown in Fig. 1, we learn the hints for the large model chatglm by using the four datasets designed in Fig. 1.

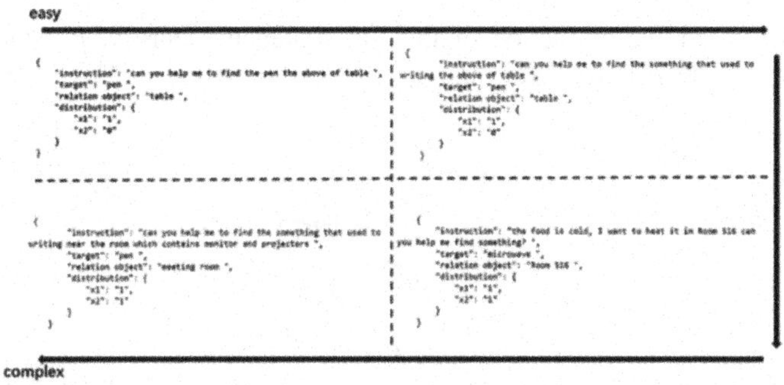

Fig. 1. The four difficulty gradient assignment datasets designed in this paper.

Meanwhile, in order to better improve the accuracy of the large model for flexible natural language processing, we designed the target object search dataset containing certain inferences, as shown in Fig. 2.

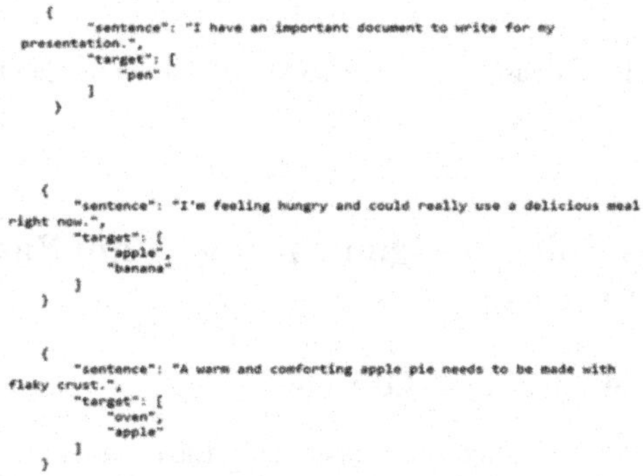

Fig. 2. Targeted object search dataset with reasoning.

We take the three results of target object parsing accuracy, pending execution search target scene accuracy and task assignment result parsing accuracy in the allocation result of chatglm as the final evaluation metrics respectively. The constructed evaluation metrics are shown below:

$$Ac_{target} = \frac{\sum_{i=1}^{M} P_c^i}{\sum_{i=1}^{M} P_c^i + P_f^i} \quad (4)$$

$$Ac_{ro} = \frac{\sum_{i=1}^{M} P_c^i}{\sum_{i=1}^{M} P_c^i + P_f^i} \tag{5}$$

$$Ac_{allocation} = \frac{\sum_{i=1}^{M} P_c^i}{\sum_{i=1}^{M} P_c^i + P_f^i} \tag{6}$$

where P_c^i stands for the extraction of the i-th sentence accuracy. P_f^i stands for the extraction of the i-th sentence error rate.

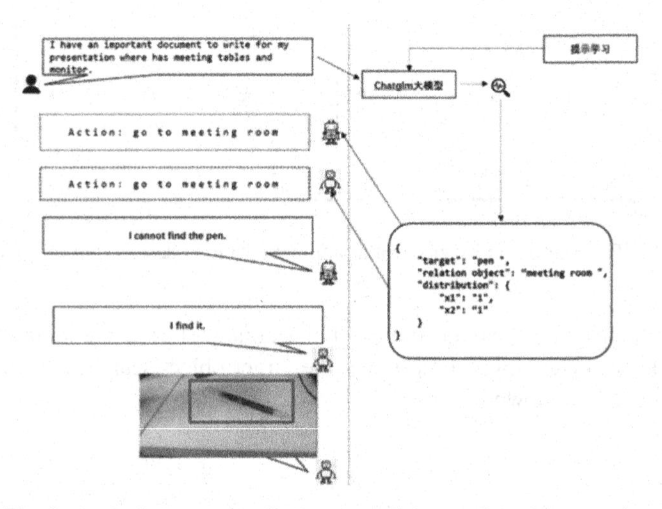

Fig. 3. Intrinsic Reasoning for Large Models Based on Cue Learning.

3 Heterogeneous Robot Assignment Algorithm Based on Large Model Prompt Learning

3.1 The Fundamentals of Prompt Learning

Since GPT, EMLO, and Bert's successive proposals, the model of pre-trained model plus fine-tuning has been widely used in many natural language tasks, which starts with pre-training a language model on a large-scale unsupervised corpus in the pre-training stage, and then fine-tuning again the model based on the trained language model on specific downstream tasks to obtain a model adapted to the downstream task. Then, the trained language model is fine-tuned on specific downstream tasks to obtain a model adapted to the downstream tasks. However, in most of the downstream task fine-tuning, the gap between the target of the downstream task and the pre-trained target is too large, which leads to insignificant improvement. Up to this point, a new fine-tuning paradigm based on training language models, cue learning, has been proposed, led by GPT-3, which aims to avoid introducing additional parameters by adding templates, so that language models can achieve the desired results in small or zero sample scenarios.

Prompt learning is to train a model P(y|x), where x is the input and y is the predicted output, and the input x is tuned to x′ in the perfect fill-in-the-blank format by introducing a suitable template template, and the tuned input x′ contains certain empty slots, which are filled in by using the language model P to fill in the empty slots and infer the predicted result y. This is shown in Fig. 4. The result y. As shown in Fig. 4, it is the prompt template we set up, in which the prompt case is put in the prompt template template, and the words to be processed are the sentences to be predicted.

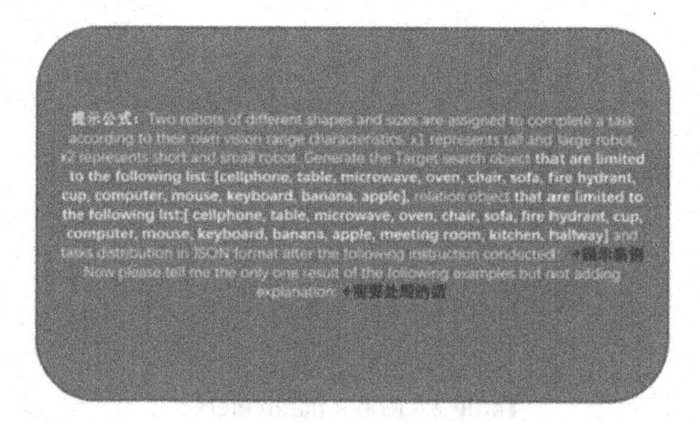

Fig. 4. Template for assigning task prompts for learning design.

3.2 Algorithmic Implementation

In this paper, we use the method based on large model hint learning to complete the heterogeneous robot group task assignment, and its general flow is shown in Fig. 3. The specific flow is as follows.

Step1: Initialize {R1, R2,..., Rn},{T1, T2,..., Tm} location information and related parameters of the chatglm model. The user's output statement S in the WeChat interface;
Step2: The target object name T of S is extracted by the chatglm model, the associated object RO, and the assignment result A;
Step3: Based on the results of A until the robot travels to RO for searching;
Step4: IF any bot searches for T;
Step5: return The image recognized by yolov5;
Step6: ELSE Eliminate the ROs that have already been searched, and return to Step2;
Step7: END
Step8: The entire search is over;

4 Simulation Experiment and Analysis

For the task allocation problem in the heterogeneous multi-robot indoor scenario, the number of robots is set to N = 2 and the number of tasks is set to M = 10. The specific parameters are shown in Table 1 and Table 2. Using pycharm software and chatglm interface, as well as NVIDIA3090ti graphics card one respectively for the five datasets created above for simulation validation and analysis, and finally, in order to highlight the groundedness of the algorithms in this paper, we integrated the physical experiments demonstrated in Fig. 7.

Table 1. Task parameters

task	position	Type TK
Task1	(10.69, 11, 25)	[1 0]
Task2	(50.30, 89.43)	[1 0]
Task3	(70.52, 65.41)	[0 1]
Task4	(140.62, 89.38)	[1 0]
Task5	(250.96, 161.86)	[0 1]
Task6	(75.21, 56.48)	[0 1]
Task7	(−32.65, 36.74)	[1 0]
Task8	(23.66, 128.70)	[1 0]
Task9	(77.25, 98.32)	[0 1]
Task10	(41.66, −198.76)	[0 1]

Table 2. Robot parameters

R	position	Type TK
R1	(168.30, 25.96)	160
R2	(−5.88, 9.37)	50

4.1 Simulation Experiment

In order to verify the effectiveness of the heterogeneous robot task assignment algorithm based on large model prompt learning proposed in this paper. In this paper, by parsing the natural language instructions on the four created datasets, we obtain the effect of chatglm on the linguistic reasoning ability as well as the task assignment accuracy under the effect of different number of cueing models, as shown in Fig. 5.

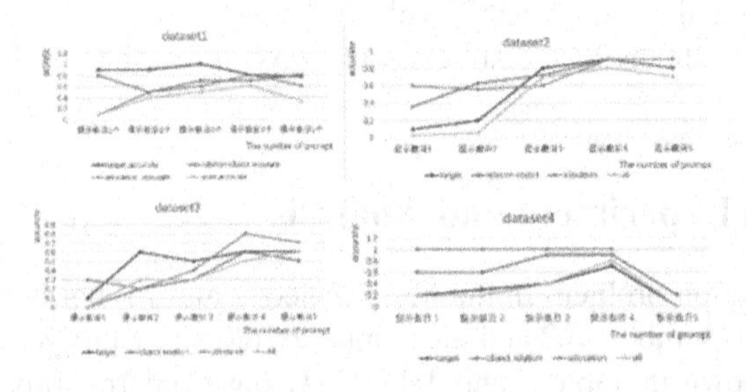

Fig. 5. Prompts based on the first four datasets Learning evaluation results

From the allocation results in Fig. 5, it can be seen that the heterogeneous robot allocation results are affected by the number of hints. The highest allocation accuracy is achieved when the number of hints is 3 or 4.

Figure 6 shows the graph of the accuracy of target object inference on the last most difficult dataset with the number of hints. The highest accuracy is achieved when the number of hints is 3, and the accuracy is around 60%.

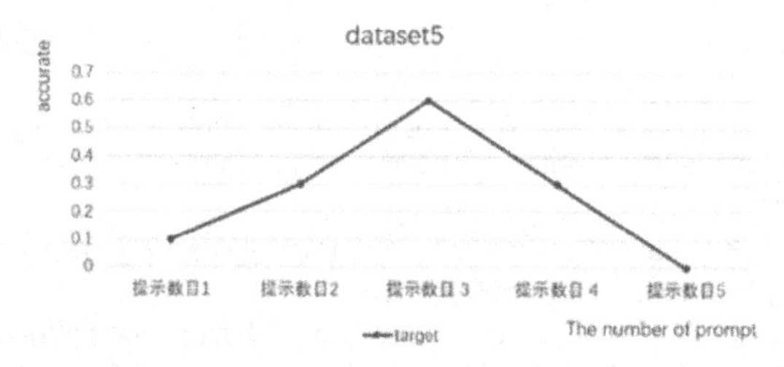

Fig. 6. Demonstration of target object search accuracy based on the fifth dataset.

In addition to this, we explored on the physics experiment, as shown in Fig. 7. Where the left one in Fig. 7 indicates the robot needed to build the physics experiment in this paper, and the process of going to find the apple according to the user's needs, this physics experiment is built on the WeChat platform.

Fig. 7. Physical Platform Demonstration of Heterogeneous Robot Task Assignment (Looking for Apple Example)

5 Conclusion

In this paper, we study the problem of multi-robot task allocation under field of view constraints for the limitations of robot search affected by field of view. Simulation experimental results show the effectiveness of this paper's large model-based task assignment. The next work will analyze and study the problem of reducing energy consumption during task allocation for heterogeneous robot groups based on large model prompt learning.

Acknowledgments. This paper is funded by the Key Laboratory of Information System Requirement, NO: LHZZ202X-MOX.

References

1. 刘雅丽， et al.： 通信距离约束下异构水下爬游机器人任务分配. 计算机测量与控制 **29**(09), 204–209 (2021). https://doi.org/10.16526/j.cnki.11-4762/tp.2021.09.038
2. Liu, Y., Song, R., Bucknall, R.W., Zhang, X.: Intelligent multi-task allocation and planning for multiple unmanned surface vehicles (USVs) using self-organising maps and fast marching method. Inf. Sci. **496**, 180–197 (2019)
3. 李文，万晓冬，周文文.：基于自适应粒子群算法的多无人机混合编队技术. 计算机测量与控制 **29**(02)，132–136+154 (2021). https://doi.org/10.16526/j.cnki.11-4762/tp.2021.02.027
4. Sun, J., Zhang, G., Lu, J., Zhang, W.: A hybrid many-objective evolutionary algorithm for flexible job-shop scheduling problem with transportation and setup times. Comput. Oper. Res. **132**, 105263 (2021)
5. Li, M., Liu, C., Li, K., et al.: Multi-task allocation with an optimized quantum particle swarm method. Appl. Soft Comput. **96**, 106603 (2020)
6. Lerman, K., Jones, C., Galstyan, A., et al.: Analysis of dynamic task allocation in multi-robot systems. Int. J. Robot. Res. **25**(3), 225–241 (2006)
7. 翟羽佳, 杨惠珍, 王强.: 基于资源均衡函数的异构多AUV任务分配方法研究. In: 中国自动化学会控制理论专业委员会 (Technical Committee on Control Theory, Chinese Association of Automation) , 中国自动化学会 (Chinese Association of Automation) ,中国系统工程学会 (Systems Engineering Society of China). 第三十八届中国控制会议论文集 (7)[出版者不详], 123–127 (2019)
8. 曹宗华, 吴斌, 黄玉清, 等.: 基于改进蚁群算法的多机器人任务分配. 组合机床与自动化加工技术 (2), 34–37 (2013)
9. Chen, M., Zhu, D.: A workload balanced algorithm for task assignment and path planning of inhomogeneous autonomous underwater vehicle system. IEEE Trans. Cogn. Dev. Syst. **11**(4), 483–493 (2018)
10. 冯志伟, 张灯柯, 饶高琦.: 从图灵测试到 ChatGPT—人机对话的里程碑及启示. 语言战略研究 **8**(2), 20–24
11. Lund, B.D., Wang, T.: Chatting about ChatGPT: how may AI and GPT impact academia and libraries? Libr. Hi Tech News **40**(3), 26–29 (2023)
12. Khattak, M.U., Rasheed, H., Maaz, M., et al.: MaPLe: multi-modal prompt learning. In: Proceedings of the IEEE/CVF Conference on Computer Vision and Pattern Recognition, pp. 19113–19122 (2023)

Quickly Adaptive Automated Vehicle's Highway Merging Policy Synthesized by Meta Reinforcement Learning with Latent Context Imagination

Songan Zhang[1(✉)], Lu Wen[2], Hanyang Zhuang[3], and H. Eric Tseng[4]

[1] Global Institute of Future Technology, Shanghai Jiao Tong University, 800 Dongchuan RD. Minhang, Shanghai 200240, China
songanz@umich.edu

[2] University of Michigan, 500 S State St, Ann Arbor, MI 48109, USA
lulwen@umich.edu

[3] University of Michigan-Shanghai Jiao Tong University Joint Institute, Shanghai Jiao Tong University, 800 Dongchuan RD. Minhang, Shanghai 200240, China
zhuanghany11@sjtu.edu.cn

[4] Ford Motor Company, 18900 Michigan Ave, Dearborn, MI 48126, USA

Abstract. Under a wide range of traffic cultures and driving conditions, it is essential that an automated vehicle performs highway merging with appropriate driving styles - driving safely and efficiently without annoying or endangering other road users. Despite the extensive exploration of Meta Reinforcement Learning (Meta RL) for quick adaptation to different environments and its application to automated vehicle driving policies, most state-of-the-art algorithms require a dense coverage of the task distribution and extensive data for each of the meta-training tasks, which is extremely expensive for the automotive industry. Our paper proposes IAMRL, a context-based Meta RL algorithm in which meta-imagination reduces real-world training tasks and data requirements. By interpolating the learned latent context space with disentangled properties, we perform meta-imagination. As a result of our autonomous highway merging experiments, IAMRL outperforms existing approaches in terms of generalization and data efficiency.

Keywords: Meta reinforcement learning · Quick adaptation · Automated vehicle

1 Introduction

Through Meta Reinforcement Learning (Meta RL), it is possible to learn "how to learn" across a set of meta-training tasks to enable quick adaptation to new tasks. Adaptation efficiency and generalization to new tasks have been greatly improved through state-of-the-art meta-learning algorithms. In the real world,

F. Sun and J. Li (Eds.): ICCCS 2023, CCIS 2029, pp. 201–211, 2024.
https://doi.org/10.1007/978-981-97-0885-7_17

however, these algorithms may not always have access to abundant data from a variety of meta-training tasks and dense coverage of task distributions. In the absence of sufficient meta-training tasks and training data, the meta-learner is prone to overfitting, and as a result, it is unable to generalize.

Among the most popular Meta Reinforcement Learning (Meta RL) methods is the context-based method [1] that learns latent representations of the distribution of tasks and optimizes policies based on those representations. This type of method allows for off-policy learning and usually provides rapid adaptation within several test steps. The challenge of generalization for these methods is twofold. It is necessary that the encoder infers unseen tasks correctly, and the policy should then interpret unseen task representations and output optimal behaviors based on latent task contexts.

Amongst all decision-making policy designs in autonomous driving, highway merging is an important feature for automated vehicles to maintain mobility. A lane change that is urgent (e.g. highway merging) is considered a mandatory lane change, as opposed to a discretionary lane change (e.g. overtaking a slower vehicle). In this work, we formulate the highway merging problem as a quasi-static Markov Decision Process (MDP), and the automated vehicle is supposed to learn to adapt to a wide range of different MDPs. Meanwhile, we present IAMRL, a context-based MetaRL algorithm that improves policies' generalization ability without additional test-time adaptation through imagination.

In the remainder of this paper, we first introduce the related work in Sect. 2 and the preliminaries in Sect. 3. And in Sect. 4, first, we design to learn an encoder that can represent tasks in a disentangled latent context space (Sect. 4.1), which enables efficient and reasonable interpolation. Then, we introduce latent-context-imagination (meta-imagination) by sampling on the learned disentangled latent context space and conditionally generating imaginary rollouts to augment training data (Sect. 4.2). The simulation environments of the highway merging scenario and the experiment results will be introduced in Sect. 5. At last, we conclude the paper in Sect. 6.

2 Related Work

Meta Reinforcement Learning. The current Meta RL algorithms fall into three categories: 1) context-based Meta RL methods [1–5], as outlined in Sect. 1; 2) gradient-based Meta RL methods [6,7], which aims to learn a policy initialization that adapts quickly to new tasks after a few policy gradient steps, but these methods are usually data-intensive during meta-training and online adaptation; 3) RNN-based Meta RL methods [8,9], which preserve previous experience in a Recurrent Neural Network (RNN), but the meta learner in this type lacks mathematical convergence proof and is prone to meta-overfitting [6].

Data Augmentation and Imagination in Machine Learning. Data augmentation has been widely used to improve image-based tasks within machine learning [10–12] and meta-learning [13–15]. The techniques inspire imagination and are then used in RL [16,17] and Meta RL [4,18–20] to improve generalization. However, imagination in RL simply augments data for a single task. In Meta RL, previous work generated imaginary tasks that were either limited in diversity [20] or uncontrollable or non-interpretable [4,18,19]. Our approach is similar to LDM [18], but it produces imaginary tasks much less efficiently, lacks control, focuses only on reward-variant MDPs, and requires much more complex training.

3 Preliminaries

Meta RL. The notion of meta-learning in Meta RL is firstly described in [6], which states that a meta-learner $\mathcal{M}(*)$ aims to learn how to do fast adaptation by learning from meta-training tasks. $\mathcal{T}_i$ represents each task with $\mathcal{T}_i = (S, A, P_i, R_i, \gamma)$, where S represents the state space, A represents the action space, P_i represents the transition probability function, R_i represents the reward function and γ is the discount factor. In the following equation, a meta learner $\mathcal{M}$ can be optimized with parameters θ:

$$\begin{aligned} \mathcal{M}_\theta := \arg\max_\theta \sum_{i=1}^{N} \mathcal{J}_{\mathcal{T}_i}(\phi_i) \\ \phi_i = f_\theta(\mathcal{T}_i). \end{aligned} \tag{1}$$

where f is the adaptation operation, and $\mathcal{J}$ denotes the objective function (usually in the expected reward form). Confronted with a new task $\mathcal{T}_i$, the meta-learner adapts from θ to ϕ_i by performing f_θ operation, which varies among different types of Meta RL methods.

A Context-Based Meta RL Algorithm. This study leveraged **PEARL** [1], a prominent context-based, off-policy Meta RL algorithm, which captures the features of the current task in a latent probabilistic context variable $\mathbf{z}$ and does policy adaptation based on the posterior context $\mathbf{z}$. The PEARL is composed of two key components: a probabilistic encoder $q(\mathbf{z}|\mathbf{c})$ for task inference, where $\mathbf{c}$ refers to context (at timestep t, context $\mathbf{c}_t = (\mathbf{s}_t, \mathbf{a}_t, r_t, \mathbf{s}_{t+1})$), and a task-conditioned policy $\pi(\mathbf{a}|\mathbf{s}, \mathbf{z})$. During the meta-training process, PEARL trains the meta policy on top of Soft Actor Critic (SAC), and trains the encoder with two losses: one is from the critic of SAC, and another is from a KL-divergence term which can be interpreted as the result a variational approximation to an information bottleneck that constrains the mutual information between $\mathbf{z}$ and $\mathbf{c}$. We built our IAMRL on top of PEARL because of its data efficiency and rapid adaptation.

4 Imagination Augmented Meta Reinforcement Learning

In this work, we build Imagination Augmented Meta Reinforcement Learning (IAMRL) as an off-policy context-based MetaRL algorithm, preserving good

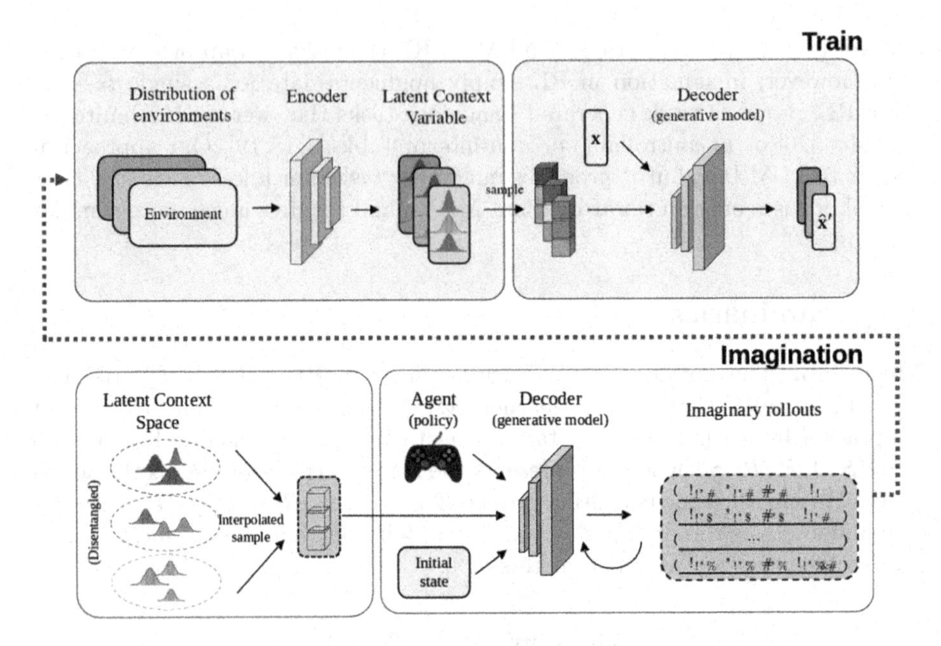

Fig. 1. IAMRL schematic. We introduce meta-imagination in the context-based Meta RL framework to improve policy generalizability.

data efficiency and quick adaptation. The framework consists of an encoder for task inference, a decoder that works as a generative model, and a meta policy to be trained with different task variants. The schematic of IAMRL is shown in Fig. 1.

4.1 Task Inference

Encoder Structure. An encoder is supposed to extract task-related information from high-dimensional input data, context $\mathbf{c}$, and represent the task with a latent context vector $\mathbf{z}$ for later use in adaptation. According to [1] and [21], researchers normally model the representation as a product of independent Gaussian factors based on the permutation invariant property of the fully observed MDP. It is inefficient in information gathering and can be problematic with sparse information tuples since this inference architecture treats all context tuples equally. Our solution to this problem is to utilize a GRU to analyze the chain of sequential data and determine which information to keep and which information to discard. The context tuples are recurrently fed into the GRU and our posterior inference of the task per time step is stored in its hidden state.

Disentangled Latent Context Space. A low-dimensional latent representation (latent context) of different tasks can be obtained from the encoder. Our following meta-imagination will sample from these latent representations, which constitute the latent context space. However, previous context-based Meta RL methods do not provide special regularization on the latent context, resulting in a convoluted and unintelligible latent context space. Conducting interpolation on such unorganized latent context space has major drawbacks, including that there is no control over how generated tasks are distributed, that the interpolated latent context space does not have a reasonable map to the task domain, or is too concentrated on a small range of the task domain.

Our structure enforces disentanglement on the latent context to produce efficient and maneuverable imaginative tasks by following the beta-VAE [22] technique. According to [23,24], a disentangled representation defines a latent context in which this dimension is sensitive to variations in a single generative factor while remaining relatively invariant to variations in other generative factors [25]. Here we want to further clarify that the *generative factor* means a hidden factor that affects the environment's transition function, reward function, etc., while the *latent context* means a vector in the learned encoder latent space that represents the meta policy's belief over what environment the agent is playing in. Although a lot of works [22,26–28] have explored learning disentangled factors, we choose to leverage the β-VAE method [22] for disentanglement performance, training stability, and implementation complexity. Similarly to β-VAE, our disentangled representation learning assumes that tasks can be described with a number of generative factors that are conditionally independent and interpretable.

The key idea behind β-VAE is to augment the original Variational Autoencoder (VAE) objective with a single hyperparameter β (usually $beta > 1$) that modulates the learning constraints on the capacity of the latent information bottleneck. And to encourage disentanglement, [29] suggests stronger constraints on $\mathbf{z}$'s capacity.

By leveraging the β-VAE technique in training our task encoder, we can learn a latent context space with the following properties: 1) The latent context is disentangled; 2) One active dimension is used for one generative factor in the latent context space, while other dimensions remain invariant to any generative factor and close to normal distributions; 3) The latent context extracts minimal sufficient information, capturing all and only task-variants-relevant information.

4.2 Imagination Using Generative Models

After being well-trained with the encoder in βVAE style, the decoder continues to play the role of a generative model to enable imagination. Imagination is done through sampling a set of imaginary (denoted with $\mathcal{I}$) latent context $\mathbf{z}^{\mathcal{I}} = \{\mathbf{z}^{\mathcal{I}(0)}, ..., \mathbf{z}^{\mathcal{I}(n)}\}$ and then providing them to the conditioned generative model. The derived conditioned generative model $p_\theta(\mathbf{x}|\mathbf{z}^{\mathcal{I}(n)})$ is an approximated MDP simulator for the imaginary task $\mathcal{T}^{\mathcal{I}(n)}$ with latent context $\mathbf{z}^{\mathcal{I}(n)}$, where (n) denotes the index of imaginary task.

With disentangled latent context space, we can take advantage by evaluating tasks with their latent context inferred by the encoder. Concretely, let's say that we have a set of meta-training tasks $\mathcal{T} = \{\mathcal{T}^{(0)}, ..., \mathcal{T}^{(N-1)}\}$ with dataset $\mathcal{D}_{\mathcal{T}}$ collected by interacting with corresponding environments. After doing task inference with the encoder, we can get a latent variable set $\mathbf{Z}^{\mathcal{T}} = \{\mathbf{z}^{\mathcal{T}^{(0)}}, ..., \mathbf{z}^{\mathcal{T}^{(N-1)}}\} = (Z_0 \times Z_1 ..... \times Z_{M-1})$, where M is the dimension number of the latent context variable vector and N is the total number of meta-training tasks. We then applied the β-VAE [22] technique to disentangle the latent context. Since not necessarily all components/dimensions of the latent context vector can be disentangled or has an interpretable generative factor representation, we define a linear injective mapping f between the generative factor index and the latent context vector index, $f : \{0, 1, ..., K-1\} \rightarrow \{0, 1, 2, ..., M-1\}$, where K is the number of generative factors and $K \leq M$.

Given the latent representation set $\mathbf{Z}^{\mathcal{T}}$, IAMRL applies the task interpolation separately on the disentangled latent variable space following:

$$\begin{aligned} Z_k^{\mathcal{I}} = \{ & \left(\lambda_k z_{f(k),i-1} + (1-\lambda_k)\, z_{f(k),i}\right) : \\ & i \in \{1, ..., I_k\}\, ; \lambda_k = \frac{j}{d_k} + \epsilon, j \in \{0, ..., d_k\}\} \end{aligned} \tag{2}$$

where k is the index of generative factors, ϵ is a small noise, d_k is integer which representing the interpolation density, and I_k is the number of possible values for the kth generative factor. Here the $Z_k^{\mathcal{I}}$ contains all interpolated elements of the kth dimension of $Z^{\mathcal{I}}$.

With the above latent context interpolation, we can generate interpolated tasks by combining samples on each generative factor, with notation $\mathbf{Z}^{\mathcal{I}} = (Z_0^{\mathcal{I}} \times Z_1^{\mathcal{I}} ... \times Z_{K-1}^{\mathcal{I}})$. From the perspective of generative factors, we can divide interpolated tasks into three types: 1) all generative factors are sampled from the latent space of real tasks $\mathbf{Z}^{\mathcal{T}}$, which is a special case where $\lambda_k = 0$ or $1, \forall k$, but with different combinations from those in $\mathbf{Z}^{\mathcal{T}}$ so as to generate new tasks; 2) generative factors are hybrid combinations of interpolated and existing value of factors, where $\exists k$ s.t. $\lambda_k = 0,$ or 1, and $\exists k'$ s.t. $\lambda_{k'} \in (0, 1)$; 3)generative factors are all with interpolated values, where $\lambda_k \in (0, 1), \forall k$. This task interpolation process by manipulating the disentangled latent context space gives us a very high level of interpretation and freedom to look into generalization domains that interest us most.

5 Simulation Environments and Results

In this section, we conduct experiments to test and evaluate the performance of IAMRL. We first describe the simulation environment and task distributions of the highway merging environments in Section **Simulation environments and tasks description**. And later in Section **Adaptation evaluation results**, we compare the policy's adaptation and generalization of IAMRL to state-of-the-art Meta RL algorithms.

Simulation Environments and Tasks Description. It is necessary to perform mandatory lane changes in order for an automated vehicle to follow the navigation. Additionally, we require the automated vehicle to be driven in a manner that respects its surrounding traffic culture, e.g., merging efficiently without irritating or endangering other drivers, and responding to crash threats swiftly and appropriately without creating hazards to other road users. In addition to being safe to drive cautiously, being overcautious is not recommended. By considering variants that can reflect complex real-life driving environments, we modify a classical highway merging environment [30] into a simulator suitable for meta-RL testing.

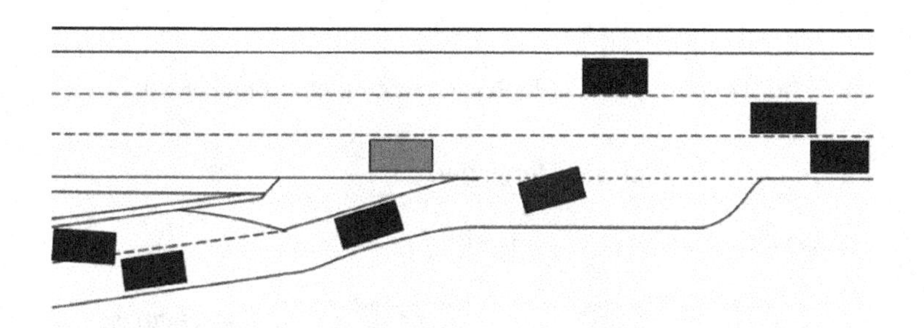

Fig. 2. *Highway-merging.* The ego vehicle (in green) is expected to learn a merging policy that can adapt quickly to different driving environments, considering task-variant generative factors: traffic speed, and longitudinally interactive model parameters.

As shown in Fig. 2, the ego vehicle (in green) must make a mandatory lane change in order to merge into the main lane before the merging area ends. Surrounding vehicles on the main lane follow the Intelligent Driver Model (IDM) [31] longitudinally and ignore lateral dynamics in our setup. Before the ego vehicle successfully merges, the vehicle right behind it (in yellow) follows a longitudinally interactive model. Different longitudinally interactive models are designed for different complexity levels of highway-merging tasks:

Highway-exp-v0 (simple version): the yellow vehicle uses a proportional interactive model that follows: $a_p = p \cdot a_{ego}$, where a_{ego} is ego vehicle's acceleration. To mimic different interactive styles, we consider p as a variant with the value range $p \in [-1, 1]$. Generally speaking, vehicles with higher positive p tend to have more opponents while those with smaller negative p behave more cooperatively.

Highway-exp-v1 (hard version): use the Hidas' interactive model [32], among the parameters of which we consider two as task variants with MDP configurations similar to [30]:

- Maximum speed decrease $Dv \in [5, 15]$ (mph);
- Acceptable deceleration rage $b_f \in [1.0, 3.5]$ (m/s^2)

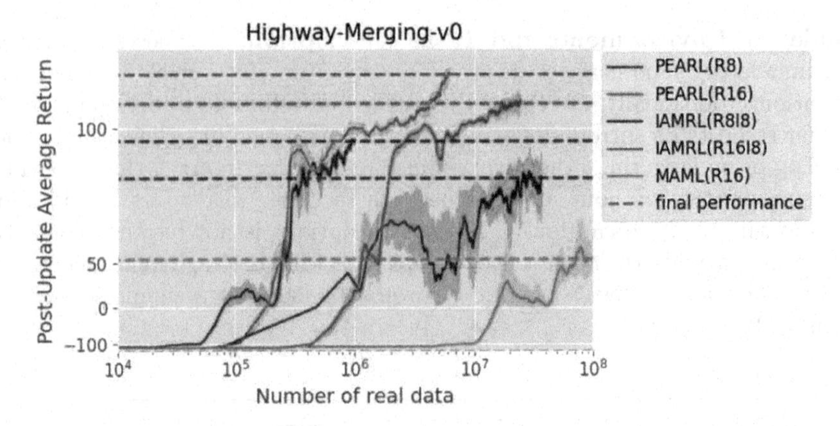

(a) Training curve in highway merging environment v0.

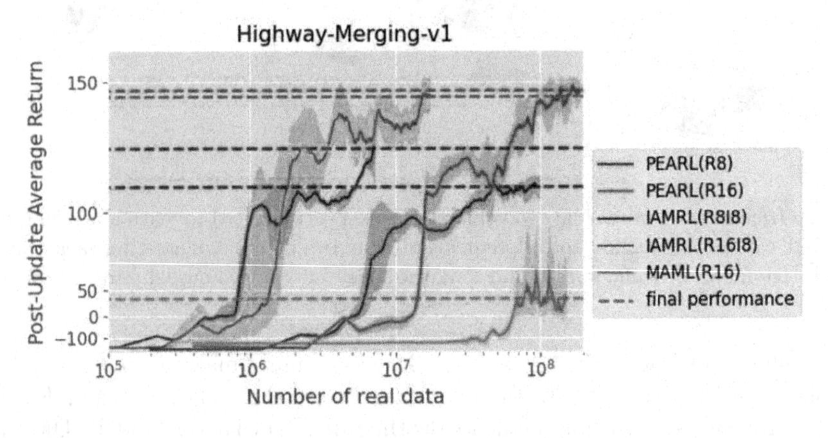

(b) Training curve in highway merging environment v1.

Fig. 3. Comparison of meta policy's adaptation performance w.r.t the number of real data used. Post-adapted policy's episodic return on meta-testing tasks during the meta-training phase is illustrated. Each learning curve shows the mean (solid lines) and variance (shades) of three independent runs.

Adaptation Evaluation Results. With a well-trained encoder-generative model, we can evaluate the effectiveness of imagination in improving policy learning. By measuring the performance of IAMRL on meta-testing tasks as a function of the number of samples used during meta-training, we first evaluate its data efficiency. We compare policies trained with different sizes and sources of data with two state-of-the-art Meta RL baselines: PEARL [1], an off-policy algorithm, and MAML [6], a gradient-based model-agnostic meta-learning algorithm.

In Fig. 3, we present the after-adaptation performance of each agent with respect to the number of real data. We denote the number of real tasks with the letter R and a number following, and the number of imaginary tasks with the letter I and a number following. For example, $R8I8$ means the policy is trained with 8 real tasks and 8 imaginary tasks. From the meta-learning curves in Fig. 3, we can observe all IAMRL training (R8I8 and R16I8) using 10–100x less real data to reach the same level of post-adaptation performance (for a simple and fair comparison, we use the number of real data required when post-update average return reach the blue dash line as the comparison metric) in comparison to PEARL training (R8 and R16).

We then analyze IAMRL's generalization capability. We use PEARL(R8) and PEARL(R16) as two important baselines to evaluate how imagination contributes. We can observe that IAMRL(R8I8) outperforms PEARL(R8), as well as IAMRL(R16I8) outperforms PEARL(R16), despite the less obvious superiority of IAMRL(R16I8) vs. PEARL(R16). This is because the benefits of meta-imagination highly depend on the property of a task and the potential improvements based on available real data. It is demonstrated that the agent can more efficiently learn a meta policy by leveraging imagination.

6 Conclusion

This paper presents a novel off-policy context-based meta-reinforcement learning algorithm we named as IAMRL that improves the generalization capability and data efficiency by using meta-imagination. By combining training tasks with imaginary tasks through efficient sampling on the learned disentangled latent context space, the policy's generalization ability is improved. Through the use of an autonomous driving problem in which we evaluate and illustrate task inference performance, we validate the encoder's task inference performance, and the meta policy's higher generalization and data efficiency.

References

1. Rakelly, K., Zhou, A., Finn, C., Levine, S., Quillen, D.: Efficient off-policy meta-reinforcement learning via probabilistic context variables. In: International Conference on Machine Learning, pp. 5331–5340. PMLR (2019)
2. Zhang, S., Wen, L., Peng, H., Tseng, H.E.: Quick learner automated vehicle adapting its roadmanship to varying traffic cultures with meta reinforcement learning. In: 2021 IEEE International Intelligent Transportation Systems Conference (ITSC), pp. 1745–1752. IEEE (2021)
3. Mendonca, R., Gupta, A., Kralev, R., Abbeel, P., Levine, S., Finn, C.: Guided meta-policy search. In: Advances in Neural Information Processing Systems, vol. 32 (2019)
4. Kirsch, L., van Steenkiste, S., Schmidhuber, J.: Improving generalization in meta reinforcement learning using learned objectives. arXiv preprint arXiv:1910.04098 (2019)

5. Gupta, A., Mendonca, R., Liu, Y., Abbeel, P., Levine, S.: Meta-reinforcement learning of structured exploration strategies. In: Advances in Neural Information Processing Systems, vol. 31 (2018)
6. Finn, C., Abbeel, P., Levine, S.: Model-agnostic meta-learning for fast adaptation of deep networks. In: International Conference on Machine Learning, pp. 1126–1135. PMLR (2017)
7. Nichol, A., Achiam, J., Schulman, J.: On first-order meta-learning algorithms. arXiv preprint arXiv:1803.02999 (2018)
8. Wang, J.X., et al.: Learning to reinforcement learn. arXiv preprint arXiv:1611.05763 (2016)
9. Duan, Y., Schulman, J., Chen, X., Bartlett, P.L., Sutskever, I., Abbeel, P.: Rl^2: fast reinforcement learning via slow reinforcement learning. arXiv preprint arXiv:1611.02779 (2016)
10. Wong, S.C., Gatt, A., Stamatescu, V., McDonnell, M.D.: Understanding data augmentation for classification: when to warp? In: 2016 International Conference on Digital Image Computing: Techniques and Applications (DICTA), pp. 1–6. IEEE (2016)
11. Zhu, J.-Y., Park, T., Isola, P., Efros, A.A.: Unpaired image-to-image translation using cycle-consistent adversarial networks. In: Proceedings of the IEEE International Conference on Computer Vision, pp. 2223–2232 (2017)
12. Zhang, H., Goodfellow, I., Metaxas, D., Odena, A.: Self-attention generative adversarial networks. In: International Conference on Machine Learning, pp. 7354–7363. PMLR (2019)
13. Ni, R., Goldblum, M., Sharaf, A., Kong, K., Goldstein, T.: Data augmentation for meta-learning. In: International Conference on Machine Learning, pp. 8152–8161. PMLR (2021)
14. Yao, H., et al.: Improving generalization in meta-learning via task augmentation. In: International Conference on Machine Learning, pp. 11 887–11 897. PMLR (2021)
15. Khodadadeh, S., Zehtabian, S., Vahidian, S., Wang, W., Lin, B., Bölöni, L.: Unsupervised meta-learning through latent-space interpolation in generative models. arXiv preprint arXiv:2006.10236 (2020)
16. Hafner, D., Lillicrap, T., Ba, J., Norouzi, M.: Dream to control: learning behaviors by latent imagination. arXiv preprint arXiv:1912.01603 (2019)
17. Hafner, D., Lillicrap, T., Norouzi, M., Ba, J.: Mastering Atari with discrete world models. arXiv preprint arXiv:2010.02193 (2020)
18. Lee, S., Chung, S.-Y.: Improving generalization in meta-RL with imaginary tasks from latent dynamics mixture. In: Advances in Neural Information Processing Systems, vol. 34, pp. 27222–27235 (2021)
19. Lin, Z., Thomas, G., Yang, G., Ma, T.: Model-based adversarial meta-reinforcement learning. In: Advances in Neural Information Processing Systems, vol. 33, pp. 10161–10173 (2020)
20. Mendonca, R., Geng, X., Finn, C., Levine, S.: Meta-reinforcement learning robust to distributional shift via model identification and experience relabeling. arXiv preprint arXiv:2006.07178 (2020)
21. Wen, L., Zhang, S., Tseng, H.E., Singh, B., Filev, D., Peng, H.: Prior is all you need to improve the robustness and safety for the first time deployment of meta RL. arXiv preprint arXiv:2108.08448 (2021)
22. Higgins, I., et al.: beta-VAE: learning basic visual concepts with a constrained variational framework. In: International Conference on Learning Representations (2016)

23. Chen, R.T., Li, X., Grosse, R., Duvenaud, D.: Isolating sources of disentanglement in VAEs. In: Proceedings of the 32nd International Conference on Neural Information Processing Systems, vol. 2615, p. 2625 (2019)
24. Tokui, S., Sato, I.: Disentanglement analysis with partial information decomposition. arXiv preprint arXiv:2108.13753 (2021)
25. Burgess, C.P., et al.: Understanding disentangling in *beta*-VAE. arXiv preprint arXiv:1804.03599 (2018)
26. Chen, R.T., Li, X., Grosse, R.B., Duvenaud, D.K.: Isolating sources of disentanglement in variational autoencoders. In: Advances in Neural Information Processing Systems, vol. 31 (2018)
27. Chen, X., Duan, Y., Houthooft, R., Schulman, J., Sutskever, I., Abbeel, P.: InfoGAN: Interpretable representation learning by information maximizing generative adversarial nets. In: Advances in Neural Information Processing Systems, vol. 29 (2016)
28. Tran, L., Yin, X., Liu, X.: Disentangled representation learning GAN for pose-invariant face recognition. In: Proceedings of the IEEE Conference on Computer Vision and Pattern Recognition, pp. 1415–1424 (2017)
29. Montero, M.L., Ludwig, C.J., Costa, R.P., Malhotra, G., Bowers, J.: The role of disentanglement in generalisation. In: International Conference on Learning Representations (2020)
30. Leurent, E.: An environment for autonomous driving decision-making (2018). https://github.com/eleurent/highway-env
31. Treiber, M., Hennecke, A., Helbing, D.: Congested traffic states in empirical observations and microscopic simulations. Phys. Rev. E **62**(2), 1805 (2000)
32. Hidas, P.: Modelling vehicle interactions in microscopic simulation of merging and weaving. Transp. Res. Part C: Emerg. Technol. **13**(1), 37–62 (2005)

Visual Inertial Navigation Optimization Method Based on Landmark Recognition

Bochao Hou[1], Xiaokun Ding[1,2], Yin Bu[2], Chang Liu[1], Yingxin Shou[1], and Bin Xu[1](✉)

[1] Northwestern Polytechnic University, Xi'an 710129, China
smileface.binxu@gmail.com

[2] AVIC Xi'an Flight Automatic Control Research Institute, Xi'an 710065, China

Abstract. This study utilizes a low-cost, low-power, and highly accurate monocular visual-inertial odometry (VIO) as the navigation algorithm for unmanned aerial vehicles (UAV). However, VIO may experience positioning inaccuracies or even loss in cases of lighting changes and insufficient environmental textures. To mitigate the accumulation of errors in the algorithm, global localization information is introduced. In order to address the issue of positioning accuracy in the absence of Global Navigation Satellite Systems (GNSS), this paper proposes an optimized Visual-Inertial Navigation System (VINS) based on landmark recognition, considering the prior information of UAV flight missions. The system relies solely on visual and IMU information, employing the VINS-Mono algorithm for high-precision positioning. After flying a certain distance, the system recognizes landmarks from a landmark database and obtains their prior coordinate information. Finally, the pose graph and optimization algorithm are used to refine the pose of all keyframes. Experimental results indicate that this navigation system effectively reduces the accumulated errors of VINS and enhances the positioning accuracy in scenarios where GNSS signals are completely unavailable.

Keywords: GNSS-denied · VINS-Mono · Target Recognition

1 Introduction

Lightweight and small-sized UAV, represented by rotary-wing drones, possess agile and maneuverable flight characteristics in three-dimensional space. These attributes grant them significant application potential in areas such as reconnaissance, surveillance, and search operations. To enhance the intelligence of UAV and enable them to cope with unknown environments and complex tasks, accurate and real-time localization and mapping are crucial prerequisites [1]. GPS-based navigation systems demonstrate excellent positioning capabilities in outdoor environments by integrating inertial sensors or other sensors for combined navigation [2]. However, in urban areas or regions with signal interference, GPS-based navigation can suffer from positioning disturbances or even failure [3]. Visual-based navigation has attracted significant attention in the field of intelligent robotics due to its passive nature, low cost, and low power consumption. This

F. Sun and J. Li (Eds.): ICCCS 2023, CCIS 2029, pp. 212–223, 2024.
https://doi.org/10.1007/978-981-97-0885-7_18

is particularly crucial for rotary-wing UAV, which greatly benefit from visual navigation methods. Prone to scale estimation errors, Visual odometry (VO) [4–6] cannot be widely used in practical devices. Leveraging the complementary relationship between visual sensors and IMU, VIO [7] provides a low-cost, high-precision, and easily deployable visual navigation solution.

Simultaneous Localization and Mapping (SLAM) system PTAM [8] was proposed by Murray et al. in 2007. It utilized a monocular camera on an embedded platform to achieve parallel tracking and mapping, providing a comprehensive framework for subsequent research on visual navigation systems. In 2013, Google's Project Tango [9] introduced the MSCKF algorithm based on filtering methods, which employed Kalman filtering for data fusion between visual sensors and IMU, resulting in good localization accuracy. In 2015, Mur-Artal [10] drew inspiration from PTAM's multi-threading concept and proposed ORB-SLAM, which achieved state-of-the-art accuracy and stability in purely visual algorithms. Qin [11] from the Hong Kong University of Science and Technology proposed the VINS-Mono algorithm for monocular visual-inertial navigation systems, which fused visual and inertial data processing to achieve low-latency, precise, and robust localization capabilities. Gao [12] from Zhejiang University successfully implemented the visual-inertial navigation system on unmanned aerial vehicles, enabling perception and autonomy, thus significantly enhancing the intelligence level of drones.

VINS often suffers from inaccurate state estimation due to sensor noise. In the absence of global localization information, the accumulated state errors degrade the positioning accuracy. To address this issue, this study [13] posed a combination navigation solution that integrates GNSS, vision, and IMU. They introduced GNSS as ground truth to optimize the VINS, achieving promising results. He [14] focused on intermittent GNSS denial environments and proposed a method to assist VIO initialization using GNSS. They further solved the issue of sudden pose variations caused by GNSS connection/disconnection through pose smoothing techniques. Cao, based on the VINS-Mono algorithm, tightly coupled three types of data fusion for state estimation [15], significantly enhancing the positioning accuracy in areas with less prominent texture information. Jin introduced GNSS measurements into the ORB-SLAM algorithm and achieved accurate positioning on a vehicle platform [16]. In the case of strong GPS denial, Lynen [17] addressed the problem of VINS failure caused by rapid camera motion during high-altitude drone flights by incorporating barometric height measurements to calibrate the estimated height in the navigation system. Wang [18] aligned and fused magnetometer measurements with monocular visual-inertial data to mitigate attitude drift issues during pure rotation, thereby improving the algorithm's pose estimation accuracy. Chai [19] tackled the problem of GNSS signal loss during UAV power line inspection by combining a biomimetic sky polarization system and establishing a combined navigation model based on polarization and VINS. This approach reduced the cumulative error of VINS. Wookie [20] adopted a sliding window approach to fuse visual-inertial-wheel odometry measurements, which improved the accuracy and robustness of the VINS algorithm. Similarly, Zuo [21] incorporated laser radar (LiDAR) data into the MSCKF framework using a sliding window technique to achieve online spatiotemporal calibration of LiDAR-visual-inertial odometry, effectively handling 3-D point cloud data.

This study proposes a localization information optimization algorithm based on landmark recognition, specifically targeting the scenario where GNSS is completely denied and the visual-inertial navigation system accumulates errors over time. In the context of UAV flight missions, partial prior landmark images and coordinate information can be obtained. These landmarks are used to train a deep neural network, specifically the YOLOv4 [22] object detection network, to accurately recognize the landmarks. By obtaining global position information through accurate landmark recognition, the pose estimation of the VINS-Mono algorithm can be optimized, thereby improving the robustness of the navigation system and the intelligence level of the UAV.

2 The Algorithm Description

In this section, we first introduce the overall framework of the visual-inertial navigation localization optimization algorithm based on landmark recognition, as shown in Fig. 1. The framework consists of three main components: landmark recognition network module, visual-inertial navigation system, and localization optimization module. The algorithm is designed to address the scenario where a rotary-wing drone is flying autonomously in a GNSS-denied environment. Over time, the visual-inertial navigation system accumulates errors. To mitigate these errors and improve mapping accuracy, a deep neural network is trained in advance to recognize landmarks and obtain their position information. This information is then used to calculate the global coordinates. Finally, the estimated trajectory of the drone is corrected based on the obtained global coordinates. This correction process aims to reduce the cumulative errors of the visual-inertial navigation system and enhance mapping precision.

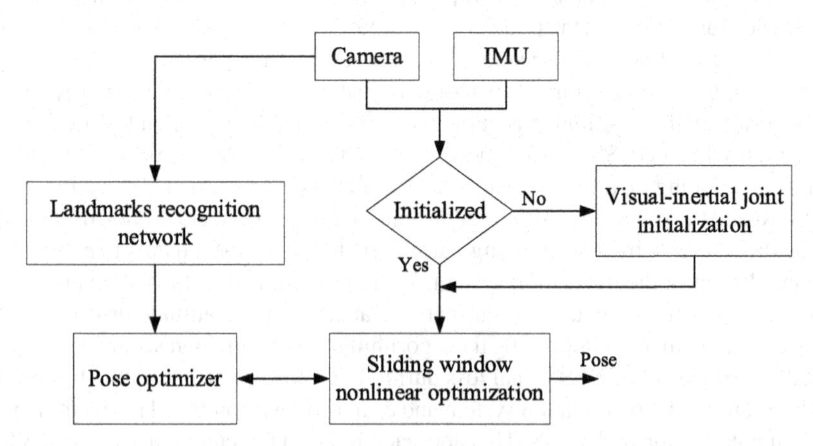

Fig. 1. The flowchart of VINS pose optimization algorithm based on landmark recognition.

2.1 Visual-Inertial Navigation System

This study utilizes the VINS-Mono monocular visual-inertial navigation system for real-time estimation of UAV pose. VINS-Mono integrates a monocular camera and a low-cost IMU, and through the fusion of camera and IMU data, it improves the robustness and positioning accuracy of the navigation system compared to pure visual navigation algorithms. VINS-Mono consists of three main components: front-end data preprocessing, visual-inertial odometry, and sliding window back-end optimization, as shown in Fig. 2. During data preprocessing, VINS-Mono performs pre-integration of IMU data, addressing data alignment issues and enhancing algorithm efficiency.

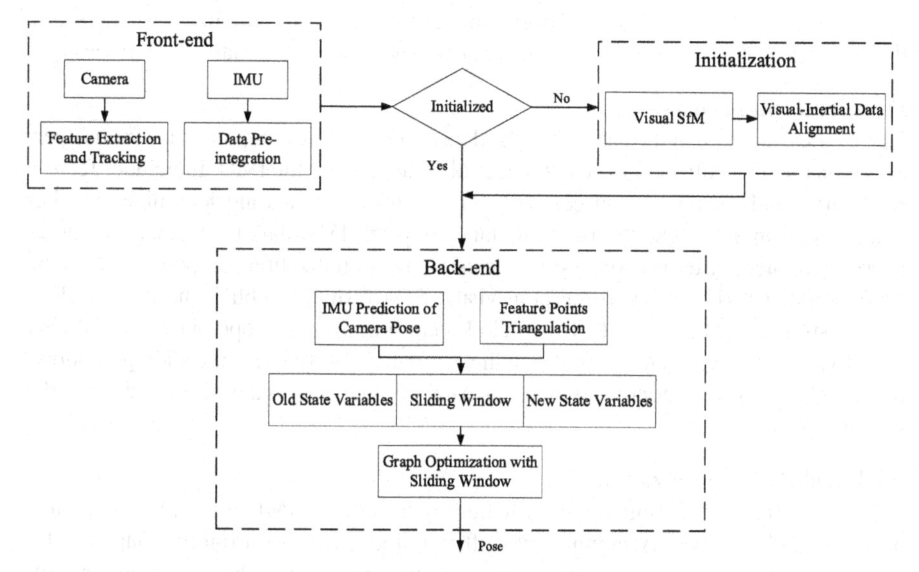

Fig. 2. The flowchart of VINS-Mono algorithm

IMU Pre-integration

When integrating the IMU measurements in the world coordinate system in VIO, the integration term includes the instantaneous rotation matrix between the body coordinate system and the world coordinate system. However, during pose optimization, the rotation matrix between the body coordinate system and the world coordinate system at the keyframe moments changes, requiring reintegration of the IMU measurements. Pre-integration is performed to avoid this repeated integration. Pre-integration changes the reference coordinate system to the body coordinate system of the previous frame, thus integrating the relative motion between two frames. The introduction of IMU pre-integration algorithm not only improves the state estimation performance of VINS (Visual-Inertial Navigation System) but also significantly enhances the efficiency and real-time capability of VINS initialization algorithms.

System Initialization

The initialization of VINS-Mono adopts a loosely coupled approach to obtain initial values. Firstly, the poses of all frames within the sliding window are estimated using Structure from Motion (SfM), with the first frame serving as the reference coordinate system. Additionally, the 3D positions of all landmarks are obtained. Then, the SfM results are aligned with the values from IMU pre-integration to correct the gyroscope biases. Subsequently, the velocity for each frame is solved, the direction of the gravity vector is estimated, and the scale factor of the monocular camera is recovered. It should be noted that during the initialization process, there is no calibration performed for the accelerometer biases. This is because the gravity is an unknown parameter in the initialization process, and the accelerometer biases are coupled with gravity. Moreover, the system's acceleration is relatively small compared to the gravitational acceleration, making it difficult to observe the accelerometer biases during initialization. Therefore, the correction of accelerometer biases is not considered in the initialization process.

Front-end Data Processing

During the execution of the algorithm, the data preprocessing component performs operations on the data collected from the monocular camera and the IMU. It includes feature recognition and tracking on consecutive frames of the image data and determines whether a new input image is a keyframe. Simultaneously, the IMU data from adjacent frames is pre-integrated. After preprocessing, the estimator initialization component aligns the pre-integrated IMU data with the pure visual SfM results to obtain the initial values of the estimator system's state. This includes calibrating gyroscope biases, initializing velocity, gravity, and scale factor. Once the estimator is initialized, the sliding window-based tightly-coupled monocular visual-inertial odometry estimates the variables of the system state.

Back-end Pose Optimization

It performs backend optimization on feature point depths, IMU errors, and keyframe positions and attitudes. Accumulative drift is mitigated through tightly-coupled relocalization. Furthermore, based on the relocalization results, the global pose graph optimization component achieves global consistency configuration of the historical poses, resulting in optimized pose data.

The performance of the visual-inertial odometry algorithm is shown in Fig. 3. The red bounding box represents the current camera's pose, the green line represents the estimated historical trajectory from odometry, the red line represents a portion of the trajectory after back-end optimization, white dots indicate the tracked feature points in the current frame, and green dots represent historical feature points. Through testing on datasets like EuRoC, the algorithm has demonstrated sub-decimeter-level localization accuracy, meeting the positioning requirements.

2.2 Landmarks Recognition Based on YOLOv4

In comparison to traditional object matching recognition methods, this study chooses the YOLOv4 network for landmark recognition. Deep neural networks offer better robustness, stability, and operational efficiency, enabling improved recognition accuracy even with limited sample data.

YOLO is a one-stage object detection algorithm based on deep regression models. It combines the classification and object localization regression problems using anchor boxes, which enhances the model's recognition capability and generalization performance. YOLOv4 can be divided into three parts: the backbone network, the neck network, and the prediction head network. The backbone network utilizes CSPDarkNet-53,

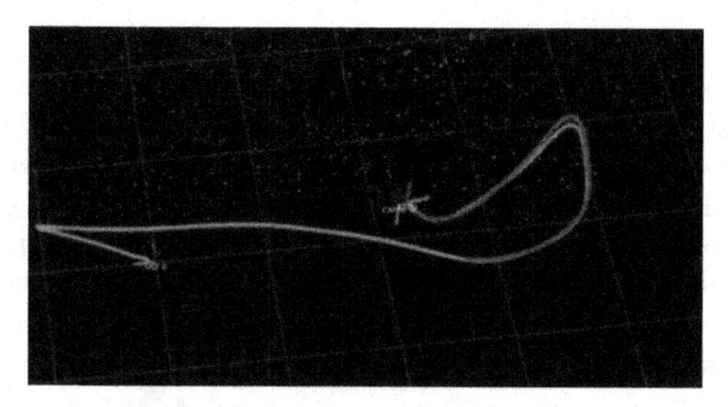

Fig. 3. Results of VINS-Mono on EuRoC

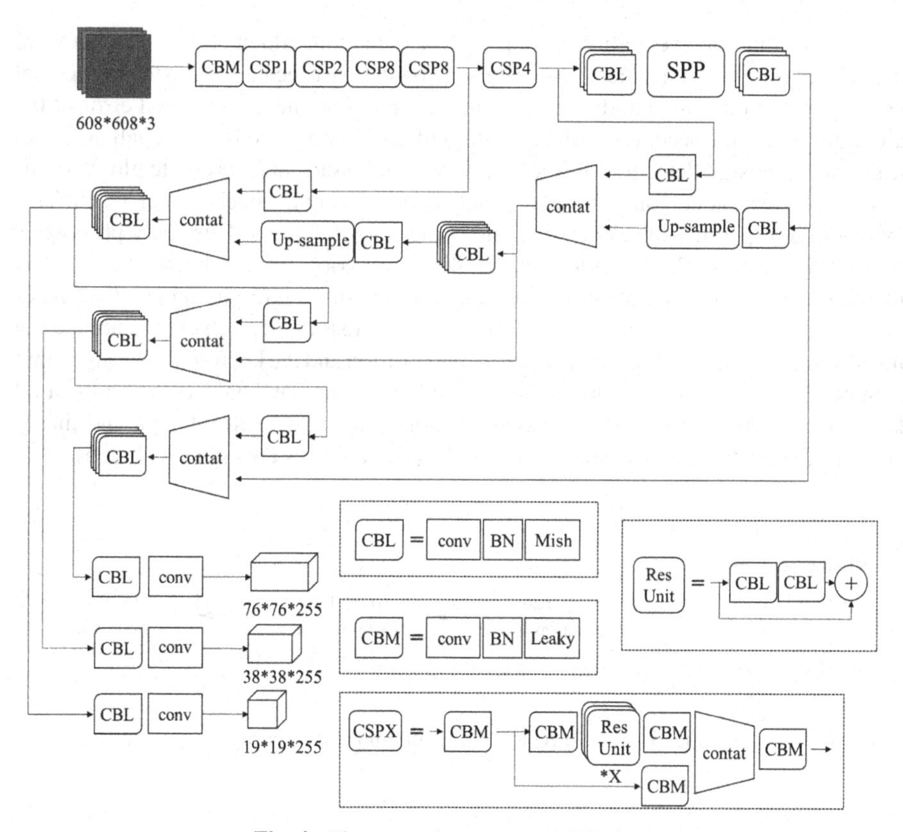

Fig. 4. The network structure of YOLOv4

which offers an excellent balance between speed and accuracy, as the feature extraction unit. The neck network fuses feature from different scales, and the prediction head network generates predicted bounding boxes for the detected objects. From Fig. 4, it can be observed that CBL (Conv2d-BN-Leaky) and CBM (Conv2d-BN-Mish) are fundamental deep convolutional network structures. They combine convolutional layers, batch normalization, and various activation functions. This combination effectively prevents overfitting, gradient vanishing, and gradient exploding during network training, ensuring the network's non-linear fitting capability. The input is an image with a resolution of 608 × 608 × 3. It first goes through the backbone neural network to extract features from the image. As the network depth increases, more detailed features of the target are extracted. Based on the principle of the Feature Pyramid Network (FPN), features from different scales are fused to generate more informative feature layers. Then, three prediction heads of different scales are used for regression to adapt to targets of different scales. After that, non-maximum suppression is applied to filter out overlapping bounding boxes, and the final positions of the targets in the image are outputted.

2.3 VINS with Relocation

VINS algorithm has the ability of loop closure detection, which allows it to compare the current pose with matched poses when revisiting past locations. It performs global pose optimization using bundle adjustment to ensure that the accumulated error of the algorithm does not become significant. In outdoor UAV missions, it is challenging to trigger loop closure detection. In such cases, it is necessary to incorporate global localization information to reduce the accumulated error during the operation of VINS. This involves the capability of relocalizing and optimizing the poses. This study proposes a VINS relocalization algorithm for outdoor flight scenarios. When a known landmark is identified, the algorithm obtains the current global position information and designates the current time as a keyframe in VINS. It utilizes a similarity solver to compute the transformation relationship between the global position and the keyframe. The algorithm then performs loop closure correction on the current frame and adjusts the poses of all keyframes. It conducts global pose optimization in the Covisibility Graph and finally applies global bundle adjustment to ensure the camera poses are updated (Fig. 5).

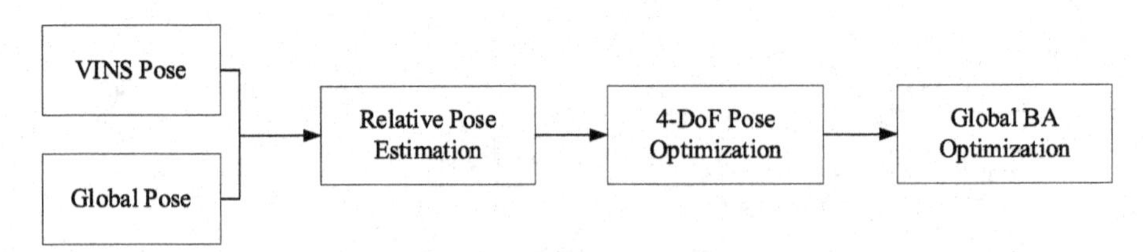

Fig. 5. The flowchart of pose optimization algorithm

3 Experiments and Results

3.1 Experimental Conditions

Model Training Platform

The Training utilized a GeForce 3060Ti graphics processor with 8GB of VRAM, an Intel Core i5-13490F CPU with a clock frequency of 3.4 GHz. The deep learning framework used was PyTorch, and the operating system was Windows 10. This section assumes the

Xi'an Giant Wild Goose Pagoda as the landmark for network training and presents the results.

In general, when there is a limited number of landmark images, there are two common approaches to address the high data requirements of deep learning:

(1) Utilizing pre-trained networks: By leveraging the concept of transfer learning, pre-trained networks can significantly reduce the demand for large datasets. Pre-trained models that have been trained on large-scale datasets can be used as a starting point and fine-tuned on the limited landmark dataset, thus benefiting from the learned features and reducing the need for extensive data.
(2) Applying data augmentation: Data augmentation techniques can be employed to artificially increase the diversity and quantity of the available data. By applying transformations such as rotation, scaling, flipping, or adding noise to the images, it is possible to generate additional variations of the existing dataset. This helps to address the fundamental issue of limited data by providing a larger and more diverse training set for the deep learning model.

VINS-Mono Platform

This study utilized the DJI M100 UAV as the platform for the visual-inertial fusion navigation system. The ZED 2 stereo camera was employed to capture images and inertial information. The UAV was equipped with an NVIDIA-NX microcomputer as the onboard computer, and all the software code was designed to run on the NX platform. Power for all devices on the UAV was supplied either directly from the UAV's power source or through DC-DC boost modules. The flight control system and the onboard computer communicated with each other through a TTL-to-USB module. In practical experiments, a local area network platform was set up, and Virtual Network Console (VNC) was used for remote control of the onboard computer (Fig. 6).

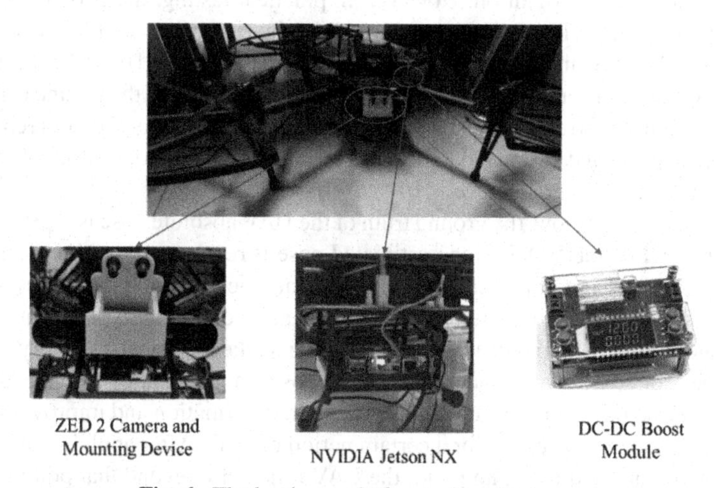

Fig. 6. The hardware platform schematic image

3.2 Experimental Results

Object Recognition Results

The training process involved a total of 120 epochs with an initial learning rate of 0.0003. This learning rate setting helps maintain convergence while preventing gradient instability during model training. The learning rate exponentially decays as the number of iterations increases. The performance of the trained model on the test set is shown in Fig. 7, demonstrating its effectiveness.

Fig. 7. The landmark recognition

Relocalization Results

The study proposes an optimization scheme for VINS (Visual-Inertial Navigation System) under landmark recognition. However, in practical testing, the two systems were not integrated together. The following results show the performance of VINS with relocalization on a UAV (Unmanned Aerial Vehicle). It can be observed that after running for some time, there is a discrepancy between the estimated results and the ground truth. This discrepancy includes errors in coordinate conversion, as the dataset did not record UAV attitude information and estimated values were used for calculations during coordinate transformation.

In the two figures below, the ground truth of the UAV absolute pose is represented by the red curve, while the VINS-Mono estimated pose is represented by the green curve. In Fig. 8, the processing point indicates the moment when the global pose information is introduced. Despite the potential errors introduced by coordinate transformation, the VINS-Mono estimation quickly adjusts its position and the historical trajectory becomes closer to the ground truth after incorporating the position information of the real points. This helps reduce the errors caused by coordinate transformation and improves the subsequent VIO positioning results for a certain period of time. After applying the relocalization function at the processing point, the UAV trajectory beyond that point becomes more consistent with the ground truth compared to the trajectory without relocalization. This demonstrates that introducing absolute real positions into the VIO for a certain

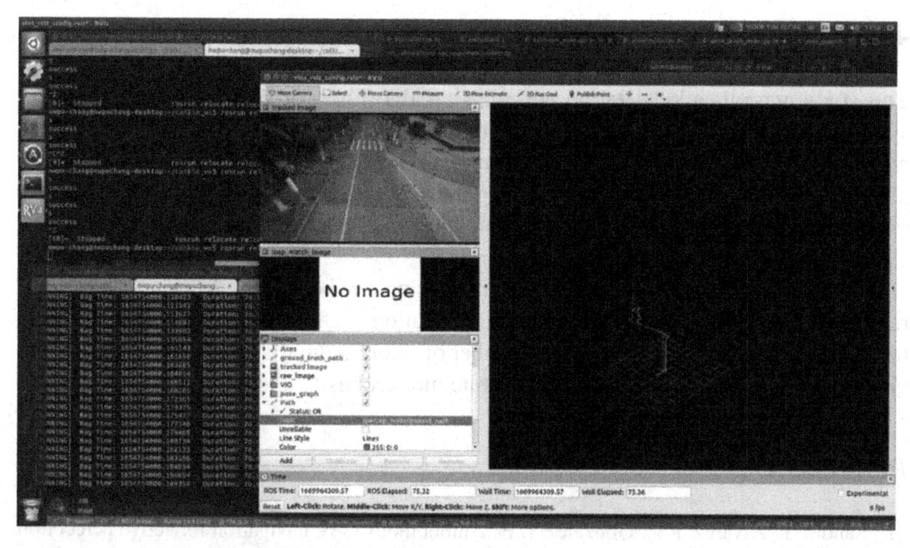

Fig. 8. Unoptimized UAV trajectory vs. ground truth

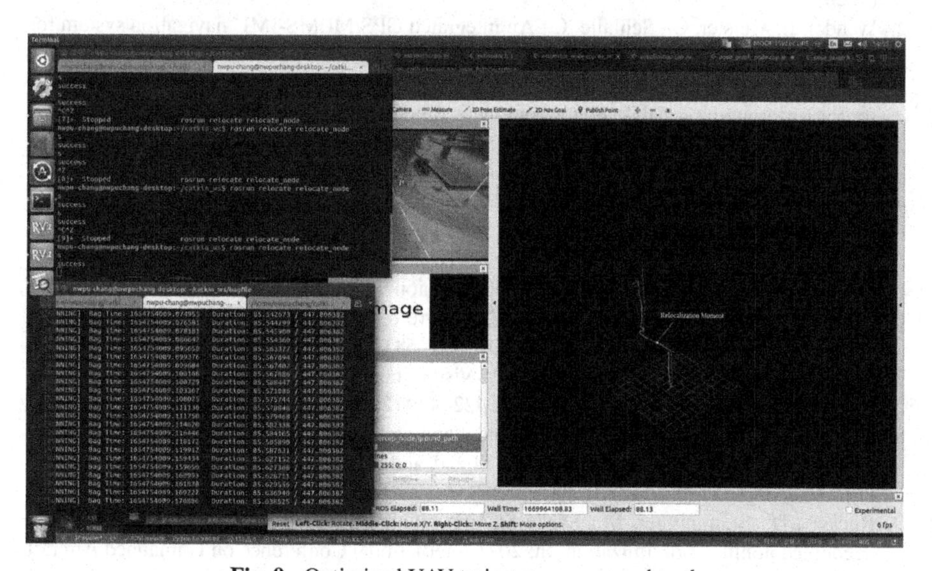

Fig. 9. Optimized UAV trajectory vs. ground truth

period of time effectively mitigates the accumulation of errors in subsequent estimation processes (Fig. 9).

4 Conclusion

This study presents a localization optimization algorithm for visual-inertial navigation systems based on landmark recognition in the scenario of complete GNSS denial. The proposed algorithm enables UAV to accomplish flight missions solely relying on visual information and low-cost IMU, enhancing the autonomy of UAV. However, one limitation is that it lacks comprehensive experimental validation of the entire system, which should be addressed using existing platforms.

In future work, not only can position information be obtained through landmark recognition, but also the current attitude information of the flying vehicle can be calculated. Furthermore, by increasing the number of prior landmarks in the environment, the visual-inertial navigation system can operate more robustly and reliably.

References

1. Sandio, J., Alvarez, F.V., Gonzalez, L.F.: Autonomous UAV navigation for active perception of targets in uncertain and cluttered environments. In IEEE Aerospace Conference. IEEE (2020)
2. Wendel, J., Meister, O., Schlaile, C.: An integrated GPS/MEMS-IMU navigation system for an autonomous helicopter. Aerosp. Sci. Technol. **10**(6), 527–533 (2006)
3. Hausman, S., Weiss, R., Brockers, R., Matthies, L., Sukhatme, G. S.: Self-calibrating multi-sensor fusion with probabilistic measurement validation for seamless sensor switching on a UAV. In: 2016 IEEE International Conference on Robotics and Automation (ICRA), pp. 4289–4296 (2016). https://doi.org/10.1109/ICRA.2016.7487626
4. Engel, J., Koltun, V., Cremers, D.: Direct sparse odometry. IEEE Trans. Pattern Anal. Mach. Intell. **99**, 1–17 (2017)
5. Mei, C., Sibley, G., Cummins, M., Newman, P.: RSLAM: a system for large-scale mapping in constant-time using stereo. Int. J. Comput. Vision **94**(2), 198–214 (2011)
6. Gutierrez-Gomez, D., Mayol-Cuevas, W., Guerrero, J.J.: Dense RGB-D visual odometry using inverse depth. Robot. Auton. Syst. **75**, 571–583 (2016)
7. Usenko, V., Demmel, N., Schubert, D.: Visual-inertial mapping with non-linear factor recovery. IEEE Robot. Autom. Lett. **5**(2), 422–429 (2020)
8. Klein, G., Murray, D.: Parallel tracking and mapping for small AR workspaces. In: 6th IEEE and ACM International Symposium on Mixed and Augmented Reality, 225–234 (2007). https://doi.org/10.1109/ISMAR.2007.4538852
9. Rückert, D., Stamminger, M.: Snake-SLAM: efficient global visual-inertial SLAM using decoupled nonlinear optimization. In: 2021 International Conference on Unmanned Aircraft Systems (ICUAS), pp. 219–228. IEEE (2021)
10. Mur-Artal, R., Tardós, J.D.: ORB-SLAM2: an open-source SLAM system for monocular, stereo, and RGB-D cameras. IEEE Trans. Robot. **33**(5), 1255–1262 (2017)
11. Qin, T., Li, P., Shen, S.: VINS-Mono: a robust and versatile monocular visual-inertial state estimator. IEEE Trans. Rob. **34**(4), 1004–1020 (2018). https://doi.org/10.1109/TRO.2018.2853729
12. Ren, Y., et al.: Bubble planner: planning high-speed smooth quadrotor trajectories using receding corridors. In: 2022 IEEE/RSJ International Conference on Intelligent Robots and Systems (IROS), Kyoto, Japan, pp. 6332–6339 (2022). https://doi.org/10.1109/IROS47612.2022.9981518

13. He, P., Jiao, J., Li, N.: Seamless indoor-outdoor localization system based on multi-sensor fusion in GNSS intermittent denial environment: GNSS/VIO localization system with nonlinear optimization. In Proceedings of the 13th China Satellite Navigation Conference - S09PNT System and PNT New Technology, p. 9. Academic Exchange Center of China Satellite Navigation System Management Office, Beijing Economic and Information Bureau, People's Government of Shunyi District, Beijing (2022)
14. Qin, T., Cao, S., Pan, J., Shen, S.: A general optimization-based framework for global pose estimation with multiple sensors (2019)
15. Cao, S., Lu, X., Shen, S.: GVINS: tightly coupled GNSS–visual–inertial fusion for smooth and consistent state estimation. IEEE Trans. Robot. **38**(4), 2004–2021 (2022). https://doi.org/10.1109/TRO.2021.3133730
16. Jin, R.H., Liu, J.N., Zhang, H.P.: Fast and accurate initialization for monocular vision/INS/GNSS integrated system on land vehicle. IEEE Sens. J. **21**(22), 26074–26085 (2021)
17. Lynen, S., Achtelik, M.W., Weiss, S.: A robust and modular multi-sensor fusion approach applied to MAV navigation. In: IEEE/RSJ International Conference on Intelligent Robots and Systems, pp. 3923–3929 (2013)
18. Wang, J.Z., Li, L.L., Yu, H.: VIMO: a visual-inertial-magnetic navigation system based on non-linear optimization. Sensors **20**(16), 4386 (2020)
19. Chai, P., Yang, J., Liu, X.: Design and implementation of a polarization/VINS integrated navigation system for power line inspection UAV. In Proceedings of the 2018 Chinese Automation Congress (CAC2018), organized by the Chinese Association of Automation, 2018, p. 6 (2018)
20. Lee, W., Eckenhoff, K., Yang, Y., Geneva, P., Huang, G.: Visual-inertial-wheel odometry with online calibration. In: 2020 IEEE/RSJ International Conference on Intelligent Robots and Systems (IROS), pp. 4559–4566 (2020). https://doi.org/10.1109/IROS45743.2020.9341161
21. Zuo, X.X., Yang, Y.L., Geneva, P., Lv, J.J., Liu, Y., Huang, G., Pollefeys, M.: LIC-Fusion 2.0: LiDAR-Inertial-camera odometry with sliding-window plane-feature tracking. In: 2020 IEEE/RSJ International Conference on Intelligent Robots and Systems (IROS), Las Vegas, NV, USA, pp. 5112–5119 (2020). https://doi.org/10.1109/IROS45743.2020.9340704
22. Bochkovskiy, A., Wang, C.Y., Liao, H.Y.M.: YOLOv4: optimal speed and accuracy of object detection (2020). https://doi.org/10.48550/arXiv.2004.10934

A Novel Approach to Trajectory Situation Awareness Using Multi-modal Deep Learning Models

Dai Xiang(✉), Cui Ying, and Lican Dai

Southwest China Institute of Electronic Technology, Chengdu 610036, China
daixiang81@gmail.com, licand@mail.ustc.edu.cn

Abstract. This paper presents a novel multi-modal deep learning framework, TSA-MM, for accurate trajectory situational awareness in transportation and military systems. The proposed framework integrates multiple sources of data, including trajectory data, USNI NEWS data, and track situation maps, and effectively addresses the challenges of data heterogeneity, data sparsity, and model interpretability. The proposed approach combines visual, textual, and numerical information to predict the future trajectory of aircraft and achieves state-of-the-art performance. Notably, the proposed framework balances the importance and relevance of different tasks, demonstrating its flexibility and adaptability to various scenarios. The experiments conducted in this study demonstrate the accuracy and real-time performance of the proposed framework, highlighting its potential for trajectory situational awareness. Furthermore, this paper emphasizes the importance of developing more effective and interpretable models that can handle the complexity and heterogeneity of trajectory data and integrate domain knowledge into the model. Overall, the proposed multi-modal deep learning framework provides a promising approach for trajectory situational awareness and contributes to the development of more advanced and practical TSA methods.

Keywords: multi-modal deep learning framework · trajectory situational awareness · trajectory data

1 Introduction

1.1 Background and Motivation

Trajectory Situation Awareness (TSA) is a critical task in many industries, including aviation, maritime transportation, and military operations. TSA involves the interpretation of trajectory data to understand the current and future situation of a moving object, such as an aircraft, a ship, or a missile. Accurate TSA is essential for ensuring the safety, efficiency, and effectiveness of transportation and military systems.

Traditionally, TSA has relied on manual analysis of trajectory data, which is time-consuming and prone to errors. With the advent of advanced technologies such as Automatic Dependent Surveillance-Broadcast (ADS-B) and Automatic Identification System

F. Sun and J. Li (Eds.): ICCCS 2023, CCIS 2029, pp. 224–232, 2024.
https://doi.org/10.1007/978-981-97-0885-7_19

(AIS), vast amounts of trajectory data can now be collected in real-time. This has led to an urgent need for automated TSA methods that can process and interpret the large volumes of trajectory data.

Multi-modal Deep Learning Models have emerged as a promising approach for TSA. By integrating multiple sources of data, such as trajectory data, weather data, and text data, these models can provide a more comprehensive understanding of the situation. However, the development of effective multi-modal deep learning models for TSA is still in its early stages, and there are many challenges to be addressed, such as data heterogeneity, data sparsity, and model interpretability.

In this paper, we focus on the application of multi-modal deep learning models to TSA in aviation, maritime transportation, and military operations. Specifically, we use aircraft and ship trajectories as well as text data, such as flight plans and ship manifests, as examples to demonstrate the effectiveness of our proposed approach. We aim to develop a framework that can accurately interpret multi-modal trajectory data and provide real-time situation awareness for transportation and military systems. Our proposed approach has the potential to greatly enhance the safety, efficiency, and effectiveness of these industries.

1.2 Related Work

Trajectory situation awareness is an important issue in air traffic management, attracting the attention of numerous researchers both domestically and abroad. In recent years, significant progress has been made in this field. In China, researchers have focused on processing and analyzing trajectory data, as well as trajectory prediction and decision support. For instance, Liu et al. (2019) proposed a machine learning-based trajectory prediction method that accurately predicts flight arrival times. Meanwhile, Wang et al. (2019) developed a deep learning-based real-time air traffic control system that can simultaneously process multiple data sources.

In foreign countries, researchers have focused on developing more advanced algorithms and technologies to improve the accuracy and real-time performance of trajectory prediction and decision support. For example, Dequaire et al. (2018) proposed a method for trajectory prediction using convolutional neural networks, achieving promising results in experiments. Additionally, Li et al. (2019) developed a multi-task learning method for simultaneous trajectory prediction and flight delay prediction.

Overall, trajectory situation awareness is a complex problem that requires the comprehensive use of multiple algorithms and technologies. In recent years, multi-modal deep learning models have become a hot research topic, as they can simultaneously process multiple data sources and achieve good results in trajectory prediction and decision support. In addition, technologies such as multi-task learning and convolutional neural networks have also been widely applied in the field of trajectory situation awareness.

In this paper, our research focus is on developing a framework that accurately interprets trajectory data and provides real-time situation awareness. We will comprehensively use the above algorithms and technologies, and combine them with the actual needs of air traffic management for research and experimentation.

1.3 Research Objectives

The main objective of this paper is to develop a multi-modal deep learning framework that can accurately interpret trajectory data and provide real-time situation awareness for transportation and military systems. Specifically, we aim to:

1. Integrate multiple sources of data, such as trajectory data, USNI NEWS data, and track situation maps, to provide a more comprehensive understanding of the situation.
2. Develop effective multi-modal deep learning models that can address the challenges of data heterogeneity, data sparsity, and model interpretability, and balance the importance and relevance of different tasks to improve model performance and generalization.
3. Use aircraft and ship trajectories as well as text data, such as flight plans and ship manifests, as examples to demonstrate the effectiveness of the proposed approach.
4. Conduct experiments and evaluations to demonstrate the accuracy and real-time performance of the proposed framework.

By achieving these research objectives, the paper aims to contribute to the development of automated TSA methods that can enhance the safety, efficiency, and effectiveness of transportation and military systems, and to address the challenges of constructing multi-modal datasets and developing effective multi-modal deep learning models.

2 Methodology

2.1 Data Collection and Preprocessing

Our research method utilizes the ADS-B dataset for trajectory analysis. This dataset provides real-time location and speed information for global flights, including flight identifiers, longitude, latitude, altitude, speed, and direction. We preprocessed the data using standard techniques such as data cleaning, deduplication, and normalization. We also used specialized techniques to handle unique properties of the ADS-B dataset, such as sorting the data by timestamp, smoothing flight trajectories, and labeling flight states.

In addition to the ADS-B dataset, we collected a large amount of text reports from USNI NEWS on aircraft and ship trajectory situations, such as those in the South China Sea. We associated these text reports with trajectory situations based on timestamps and target names, resulting in more comprehensive and accurate situational information. After preprocessing the data, we trained our models using common machine learning and deep learning algorithms, including convolutional neural networks and recurrent neural networks. Our model achieved high accuracy and efficiency in trajectory analysis (Table 1).

In this example, each row represents the ADS-B data of a flight, including message type, ADS-B message type, message number, transponder ID, aircraft ICAO address, receiver ID, date, time, UTC date, UTC time, position, altitude, ground speed, heading, vertical speed, distance, signal strength, and transponder track. The position information is empty in this example.

The following is an example of the content of text data on aircraft situation interpretation collected from the internet: "The USNI indicates that the amphibious assault ship

Table 1. Examples of ADS-B data

Message Type	ADS-B Message Type	Message Number	Transponder ID	Aircraft ICAO Address	Receiver ID	Date	Time	UTC Date	UTC Time
MSG	5	111	11111	4CAE6E	111111	2021/08/26	04:25:40.988	2021/08/26	04:25:40.970

Table 2. Examples of ADS-B data

Position	Altitude (ft)	Ground Speed (knots)	Heading	Vertical Speed (ft/min)	Distance (nmi)	Signal Strength (dBm)	Transponder Track
	401.7	164.2	180°	0	10	0	

(LHD-8) has entered the central South China Sea for activity, while the AIS signal shows that the Anchorage amphibious dock landing ship (LPD-23) has arrived in Sri Lanka for maritime readiness and training/exercises (CARAT/MAREX) with the Sri Lankan Navy, and the Musa (LPD-26) is still docked in Sabang." (Fig. 1)

In this paper, the USNI NEWS and ADS-B datasets were used. Firstly, the data was sorted based on timestamps, and the flight trajectories were smoothed and the flight states were labeled. Secondly, named entity recognition was used to extract target names, times, and activity areas from the USNI NEWS text content, and key turning points were marked on the situational map. By combining these three elements, a training dataset of 10,000 flight situational interpretations was constructed, and 3,003 of them were selected for training the model proposed in this paper. This process can be referred to as data collection and preprocessing.

2.2 Multi-modal Deep Learning Models for Trajectory Situation Awareness

We proposed a multi-modal deep learning model for trajectory situation awareness (TSA-MM). The model combines visual, textual, and numerical information to predict the future trajectory of aircraft. The experimental results show that the proposed model outperforms the baseline models and achieves state-of-the-art performance.

Modality embedding layer and joint training: The modality embedding layer takes input data from different modalities (e.g., trajectory data, flight information, USNI NEWS text content data) and performs feature extraction and transformation for each modality. The output of the modality embedding layer is a modality-specific representation. To leverage the joint information across multiple modalities, we use joint training strategy, where all modalities are processed concurrently within a single network. This helps the network to learn features that are discriminative for multiple modalities simultaneously.

Modality embedding matrix and deep complementary learning: After the modality embedding layer and joint training, we have multiple modality-specific representations.

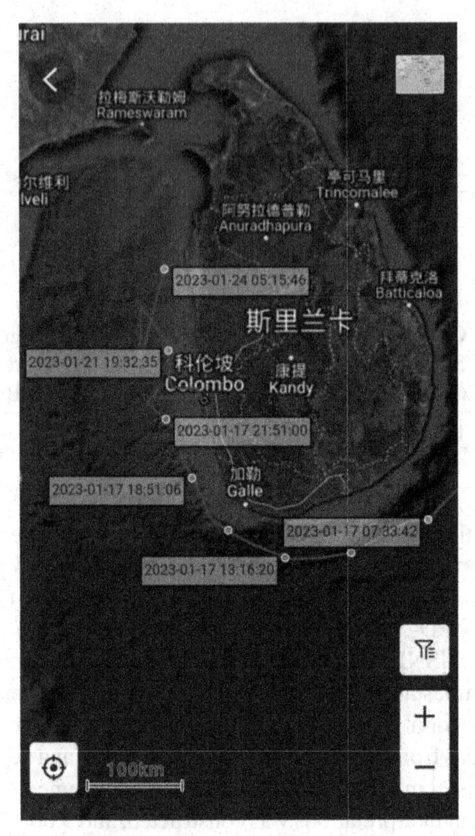

Fig. 1. Track situation map:the Anchorage-class amphibious dock landing ship (LPD-23) arrives in Sri Lanka.

The modality embedding matrix is used to concatenate these modality-specific representations to get a multimodal representation. The modality embedding matrix could be a vector or a matrix depending on the modality types. If two modalities have the same number of features, then the modality embedding matrix is a matrix. Otherwise, it is a vector. To learn complementarity among modalities, we use deep complementary learning strategy, which enhances the correlation and coherence among different modalities.

Multimodal prediction model: The multimodal prediction model takes the multimodal representation as input, and applies a deep learning algorithm to build the prediction model. The output of the multimodal prediction model is the predicted activities, location, and tasks of the target aircraft.

In this process, we need to consider how to balance the importance and correlation of different tasks. This paper proposes a multi-task joint training strategy, which refers to learning multiple related tasks in a single model to improve the model's performance and generalization ability. The main steps include the following:

1. Define the loss function

Firstly, we need to define the loss function to measure the model's performance on different tasks. This paper proposes a uncertainty-based loss function, which combines the loss functions of scene geometry and semantic segmentation tasks, and dynamically adjusts the weights based on the uncertainty of each task. Specifically, the loss function can be represented as:

$$L = \lambda gLg + \lambda sLs \tag{1}$$

where Lg and Ls represent the loss functions of scene geometry and semantic segmentation tasks, respectively, and λg and λs represent the weights of the two tasks, which can be dynamically adjusted based on the uncertainty of each task.

2. Calculate uncertainty

Next, we need to calculate the uncertainty of each task to dynamically adjust the weights. This paper uses a Bayesian neural network to estimate the uncertainty of each task. Specifically, an additional noise node is added to the output layer of the network to represent the model's uncertainty. By maximizing the likelihood function, we can obtain the posterior distribution of the network parameters and the noise node, and then calculate the uncertainty of each task.

3. Dynamically adjust weights

Finally, we need to dynamically adjust the weights based on the uncertainty of each task. This paper uses an exponential distribution-based method to represent the weights of each task as:

$$\lambda_i = \frac{1}{1 + e^{-\alpha\sigma^i}} \tag{2}$$

where σ^i represents the uncertainty of task i, and α is a hyperparameter used to control the impact of uncertainty on weights. Through this method, we can reduce the impact of tasks with larger uncertainties on the entire model by assigning smaller weights to them.

Through the uncertainty-based multi-task joint training strategy, we can effectively solve the mutual influence and conflict problems between scene geometry and semantic segmentation tasks. The key of this method lies in dynamically adjusting the weights of tasks based on their uncertainties, which improves the model's performance and generalization ability.

3 Experimental Results

In this paper, we used precision, mAP, and F1-Score as evaluation metrics. Precision refers to the proportion of correctly predicted samples among all samples predicted as a certain category, while mAP refers to the average precision of all categories. To calculate mAP, we need to first calculate the precision and recall for each category, and then calculate the average precision. Therefore, mAP is a comprehensive indicator of precision and recall, which can more comprehensively evaluate the performance of the model. If a model has a high precision for a certain category, then the mAP for that category is also likely to be high.

Precision is the ratio of true positive samples to all samples predicted as positive. Specifically, precision measures how many of the predicted positive samples are truly positive. The formula for calculating precision is:

$$precision = TP/(TP + FP) \tag{3}$$

where TP represents the number of true positive samples, and FP represents the number of samples that are actually negative but predicted as positive.

Recall is the ratio of the number of samples correctly predicted as positive to the total number of actual positive samples. Specifically, recall measures how many positive samples are correctly predicted as positive. The formula for calculating recall is:

$$recall = \frac{TP}{TP + FN} \tag{4}$$

where TP represents the number of true positive samples, and FN represents the number of samples that are actually positive but predicted as negative.

F1-Score is the harmonic mean of precision and recall, which comprehensively evaluates the advantages and disadvantages of precision and recall. The formula for calculating F1-Score is:

$$F1 = \frac{-2 * precision * recall}{precision + recall} \tag{5}$$

The value of F1-Score ranges from 0 to 1, and the closer it is to 1, the better the performance of the model (Table 3).

Table 3. Comparison of experimental results

Method	Heading level	Precision	mAP (Mean Average Precision)	F1-Score
TSA-MM	target states	89.02%	87.78%	86.23%
	activity areas	91.76%	91.09%	90.34%
	key turning points	93.15%	91.12%	89.78%
SVM	target states	81.37%	78.58%	79.32%
	activity areas	80.96%	79.56%	78.46%
	key turning points	78.63%	77.13%	76.77%
RNN	target states	87.84%	87.72%	86.57%
	activity areas	88.53%	87.23%	87.12%
	key turning points	89.53%	89.54%	89.08%
LSTM	target states	87.21%	86.32%	86.11%
	activity areas	90.32%	89.19%	88.54%
	key turning points	91.23%	90.82%	89.48%

In addition, we conducted experiments and compared the results of three tasks based on the same dataset using LSTM, RNN, SVM, and the proposed TSA-MM model (Table 2). In the trajectory situation awareness prediction task, TSA-MM, LSTM, and RNN performed better than SVM. This is because LSTM and RNN can handle sequence data and learn long-term dependencies in sequence data, while SVM is usually used to handle non-sequence data and is less suitable for trajectory prediction tasks. TSA-MM performed better than LSTM and RNN because it adopted a multi-task joint training and deep complementary learning strategy.

4 Conclusion

Trajectory situational awareness is an important task in aviation safety and air traffic management. With the increasing amount of data generated by various sensors and sources, multi-modal deep learning models have become a promising approach to solving the challenges of trajectory situational awareness. One of the challenges is the heterogeneity and complexity of multi-modal trajectory datasets, which makes it difficult to construct a training dataset. In this paper, we constructed a training dataset and a test dataset using USNI NEWS, ADS-B datasets, and trajectory situational map data, and conducted various task experiments, achieving good results. Another challenge addressed in this paper is balancing the importance and correlation of different tasks. We propose a multi-task joint training strategy, which learns multiple related tasks in a single model to improve the performance and generalization ability of the model. We propose a multi-modal deep learning model (TSA-MM) for trajectory situational awareness, which combines visual, textual, and numerical information to predict the future trajectory of aircraft. Through the uncertainty-based multi-task joint training strategy, we can effectively solve the mutual influence and conflict problems between scene geometry and semantic segmentation tasks. Experimental results show that the proposed model outperforms the baseline model and achieves state-of-the-art performance.

In conclusion, multi-modal deep learning models show great potential in trajectory situational awareness. Future research should focus on developing more effective and interpretable models to handle the complexity and heterogeneity of trajectory data and integrate domain knowledge into the model.

References

SarahBolton · RichardDill· MichaelR.Grimaila· DouglasHodson. ads b classification using mu1tivariate 1ong shorttenn memory fu1ly convo1utiona1 networks and data reduction techniques. J. Supercomput. **79**(2), 2281–2307 (2023)

Gers, F.A., Schmidhuber, J., Cummins, F.: Learning to forget: continual prediction with LSTM. Neural Comput. **12**(10), 2451–2471 (2000)

Jia, Z., Sheng, M., Li, J., Niyato, D., Han, Z.: LEO-satellite-assisted UAV: joint trajectory and data collection for internet of remote things in 6g aerial access networks. IEEE Internet Things J. **8**(12), 9814–9826 (2019)

Gao, J., Li, P., Chen, Z., Zhang, J.: A survey on deep learning for multimodal data fusion. Neural Comput. **32**(5), 829–864 (2020)

Yu, B., Yin, H., Zhu, Z.: Spatio-temporal graph convolutional networks: a deep learning framework for traffic forecasting. In: Proceedings of the Twenty-Seventh International Joint Conference on Artificial Intelligence (2018)

Liu, J., Mao, X., Fang, Y., Zhu, D., Meng, M.Q.-H.: A survey on deep-learning approaches for vehicle trajectory prediction in autonomous driving, arXiv preprint arXiv:2110.10436 (2021)

Sun, J., Li, Y., Fang, H.-S., Lu, C.: Three steps to multimodal trajectory prediction: modality clustering, classification and synthesis. In: 2021 IEEE/CVF International Conference on Computer Vision (ICCV), pp. 13230–13239 (2021)

Cui, H., et al.: Multimodal trajectory predictions for autonomous driving using deep convolutional networks, arXiv preprint arXiv:1809.10732

Huang, Y., Du, J., Yang, Z., Zhou, Z., Zhang, L., Chen, H.: A survey on trajectory-prediction methods for autonomous driving. IEEE Trans. Intell. Veh. **7**(3), 652–674 (2022)

Xue, H., Huynh, D.Q., Reynolds, M.: SS-LSTM: A hierarchical LSTM model for pedestrian trajectory prediction. In: 2018 IEEE Winter Conference on Applications of Computer Vision (WACV), pp. 1186–1194 (2018)

Yi, S., Li, H., Wang, X.: Pedestrian behavior understanding and prediction with deep neural networks. In: Leibe, B., Matas, J., Sebe, N., Welling, M. (eds.) ECCV 2016. LNCS, vol. 9905, pp. 263–279. Springer, Cham (2016). https://doi.org/10.1007/978-3-319-46448-0_16

Besse, P.C., Guillouet, B., Loubes, J.-M., Royer, F.: Destination prediction by trajectory distribution-based model. IEEE Trans. Intell. Transp. Syst. **19**(8), 2470–2481 (2018)

Wu, Y., Yu, H., Du, J., Liu, B., Yu, W.: An aircraft trajectory prediction method based on trajectory clustering and a spatiotemporal feature network. Electronics **11**(21), 345 (2022)

Pinto Neto, E.C., Baum, D.M., Almeida, J.R.D., Jr., Camargo, J.B., Jr., Cugnasca, P.S.: Deep learning in air traffic management (ATM): a survey on applications, opportunities, and open challenges. Aerospace **10**(4), 358 (2023)

Cong, F., He, X., Xuewu, W., Ji, X.: Multi-modal vehicle trajectory prediction based on mutual information. IET Intell. Transp. Syst. **14**(3), 148–153 (2020)

Pathiraja, B., Munasinghe, S., Ranawella, M., De Silva, M., Rodrigo, R., Jayasekara, P.: Class-aware attention for multimodal trajectory prediction, arXiv preprint arXiv:2209.00062 (2022)

Gong, C., McNally, D.: A methodology for automated trajectory prediction analysis. In: AIAA Guidance, Navigation, and Control Conference and Exhibit (2004)

Du, S., Li, T., Gong, X., Horng, S.J.: A hybrid method for traffic flow forecasting using multimodal deep learning, arXiv preprint arXiv:1803.02099 (2018)

A Framework for UAV Swarm Situation Awareness

Weixin Han(✉), Yukun Wang, and Jintao Wang

School of Automation, Northwestern Polytechnical University, Xi'an 710129, China
hanweixin@nwpu.edu.cn

Abstract. This paper proposes a framework for situation awareness in a multi-agent environment. In our work, the framework is composed of 3 key components: situation elements identification, situation comprehension, and situation prediction. Taking the example of a UAV swarm engaged in search and rescue operations. Firstly, we study the UAV swarms from a holistic perspective and define situation elements as the formation and motion characteristics of swarm. The Multi-Layer Perceptron (MLP) neural network is designed to identificate the swarm collective behavior formation. Then, this paper proposes a fuzzy inference system to comprehend the current situation of the swarm, deducing the current behavioral intentions. Finally, the Long-Short Term Memory (LSTM) neural network is employed to predict the future situation of swarm. And the advantage assessment function is given to quantitatively analyze the swarm's advantage in the search and rescue operation.

Keywords: Multi-agent · Situation awareness · Situation elements identification · Advantage assessment

1 Introduction

In recent years, multi-agent systems have garnered extensive application across diverse domains due to their autonomy, fault tolerance, scalability, and cooperative distribution [1]. These systems have found utility in domains spanning robotics, autonomous vehicles, swarm intelligence, social networks, and even sophisticated infrastructures such as smart grids.

With the wide application of multi-agent systems, the situation awareness of multi-agent systems has gradually become a significant research field. Through situation awareness methods, multi-agent systems can effectively extract environmental elements and understand the surrounding environmental situation, so

This work was partially supported by the Fundamental Research Funds for the Central Universities (HYGJZN202303), Key Research and Development Program of Shaanxi (2021GXLH-01-13), Aeronautical Science Foundation of China 2020000505053004, Young Elite Scientist Sponsorship Program by SNAST (20220125) and CAST (2022QNRC001), Natural Science Basic Research Plan in Shaanxi Province (2022JQ-610).

F. Sun and J. Li (Eds.): ICCCS 2023, CCIS 2029, pp. 233–246, 2024.
https://doi.org/10.1007/978-981-97-0885-7_20

as to promote the overall action of the system. Situation awareness, as defined by the American psychologist Endsley M.R [2], is the perception, comprehension, and prediction of situation elements within a specific time and space environment, and it has a significant impact on the behavior of pilots and navigators.

In order to accurately identify situation information, scholars have done a lot of research. The traditional situation recognition technology adopts Bayesian network [3], template matching [4,5], expert system [6] and other methods. With the rapid development of artificial intelligence technology, deep learning is gradually being applied to situation recognition problems [7–9]. In [10], a feature representation method of air target situation recognition based on fuzzy inference method was proposed. [11] proposed a target intention recognition model based on radial basis function neural network. However, the situation awareness methods proposed in [10,11] are all for a single target, and the main work focuses on the design and processing of situational features and situational classification results.

For the research on situation awareness of swarm, [12] designed an identification method for unmanned swarm adversarial situation elements using Transformer. In [13], a multi-entity hierarchical Bayesian network-based method was proposed to analyze swarm intentions. And a model based on analytic hierarchy process (AHP) method is proposed to evaluate advantages of target clusters in [14]. However, these studies are also based on the identification of individual situation elements in swarm or the prediction of individual situation awareness of swarm, in the face of large-scale swarm situation awareness, there will be the problem of complex calculation.

In this paper, a UAV swarm situation awareness scheme is proposed. The main contributions of this paper are outlined as follows: (1) A situation awareness framework of UAV swarm is established. The situation of UAV swarm are discussed from three layers, including the identification of situation elements, situation comprehension, and advantage assessment for UAV swarm. (2) The UAV swarms are studied in a holistic perspective, and several novel situation elements are proposed. (3) A advantage assessment function is designed combining present situation and futuren states.

The rest of this paper is structured as follows: In Sect. 2, the problem of uav swarm situation awareness is introduced and the established framework is presented. Section 3 presents the methods for situation elements identification in UAV swarm and how to derive situation comprehension and identify swarm intentions based on the recognized elements. In Sect. 4, we discuss situation prediction for UAV swarm, including formation prediction and trajectory prediction, as well as advantage assessment for the swarm. In Sect. 5, a series of simulation are conducted to demonstrate the rationality of the framework.

2 The Framework for Swarm Situational Awareness

Endsley proposed a three-level model of situation awareness in 1995 [15]. Level 1 situation awareness involves the perception of situation elements; Level 2 situation awareness entails the comprehension of the current situation by recognizing the meaning of situation elements; Level 3 situation awareness involves the prediction of future situations and gives an advantage assessment of the target.

Based on the Endsley SA model, we divide the UAV swarm situation awareness into 3 levels: situation elements identification, present situation comprehension and advantage assessment, as shown in Fig. 1.

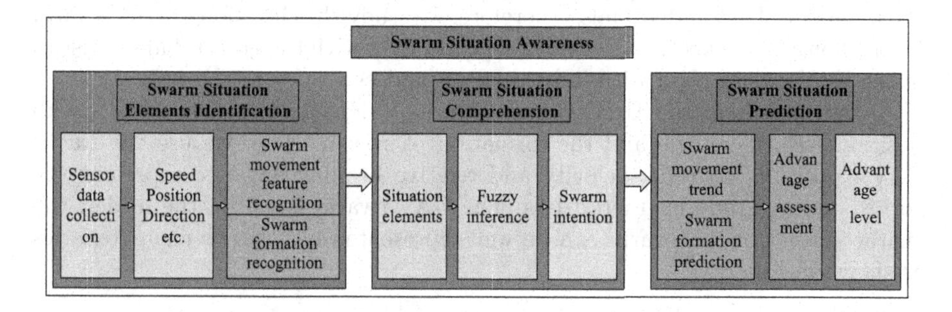

Fig. 1. The framework of swarm situation awareness.

In our research, UAV swarm situation elements can be divided into two categories: dynamic elements and static elements. We study the UAV swarm from a holistic perspective. In our study, we use the leader agent's position, altitude, velocity and relative heading angle to represent the corresponding group characteristics. Besides, we introduce group polarization and group momentum to describe the formation of UAV swarm. We design a Multi-Layer Perception (MLP) neural network to identificate the swarm collective behavior formation from p_{group} and m_{group}.

Situation comprehension involves interpreting the current situation based on the identification of situation elements and assess the situation of UAV swarm. We propose a method based on fuzzy inference to achieve the intention recognition of swarm. The fuzzy inference model is established using the identified situation elements. The fuzzy inference rules are then formulated by combining expert knowledge.

Advantage assessment refers to the quantitative description of the advantage level of the UAV swarm. We design Long-Short Term Memory (LSTM) networks to predict swarm trajectories and formation in the future. In our research, advantage assessment integrates current states and the futuren states by multiple elements, including angles, positions, velocities and group formation.

3 Situation Elements Identification and Situation Comprehension

The identification of swarm situation elements and comprehension of the present situation are introduced in this section. Firstly, several situation elements are defined and defined based on group characteristics. Then we design fuzzy rules to infer the intention and comprehend the present situation of the swarm based on the identified situation elements.

3.1 Situation Elements Identification

The computational complexity of UAV swarm perception increases greatly with the number of UAV individuals. Therefore, we study the situation of UAV swarm from a holistic perspective. We are more concerned with the group characteristics of the UAV swarm than with individual properties.

In our research, the UAV swarm situation elements are defined as group position, velocity, angle and the formation of swarm. Here, we use the leader UAV's position, altitude, velocity and relative heading angle to represent the corresponding group characteristics. The UAV swarm is led by the leader, the characteristics of the swarm can be well represented by leader's characteristics as shown in Fig. 2.

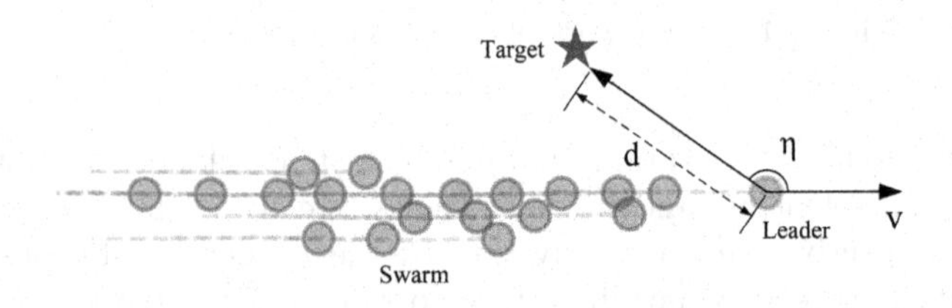

Fig. 2. Group characteristics. Wherein, d represents the distance between the swarm and the target, v represents the velocity of the swarm and η represents the relative heading angle.

As for swarm formation, some concepts in hydrodynamics are refered to and we introduced the following two indicators to describe the swarm formation.

$$\begin{aligned} p_{group}(k) &= \frac{1}{N}\left\| \sum_{i=1}^{N} \overline{v}_i(k) \right\|^2 \\ \overline{v}_i(k) &= \frac{\mathbf{v}_i(k)}{\sqrt{\sum_{i=1}^{N} \|\mathbf{v}_i(k)\|^2}} \end{aligned} \tag{1}$$

$$\begin{aligned} m_{group}(k) &= \frac{1}{N}\left| \sum_{i=1}^{N} r_{ic}(k) \times v_i(k) \right| \\ r_{ic} &= c_i - c_{group} \\ c_{group}(k) &= \frac{1}{N}\sum_{i=1}^{N} c_i(k) \end{aligned} \tag{2}$$

where N is the number of agents, v_i is the velocity of agent i in swarm, $\bar{v}_i$ is the normalized velocity, c_i represents the position of individual i and c_{group} is the mean position of the swarm.

Group polarization, p_{group}, measures the degree of alignment among individuals within the group and is also a value between 0 and 1. The group polarization of a swarm reaches a maximum value of 1 when all the agents head in the same direction. The group polarization of a swarm reaches a maximum value of 1 when all the agents head in the same direction. Group angular momentum, m_{group}, is a measure of the degree of rotation of the group about the group centroid and is a value between 0 and 1. The group angular momentum of a swarm reaches a maximum value of 1 if all the agents rotate around the group centroid in the same direction.

It has been indicated that a bio-inspired swarm model exhibits two stable collective behaviors: a flock and a torus. From Fig. 3 we can see a clear spatial difference. In a flock the UAVs are tightly packed while in a torus has a void in the center. A flock can be characterized by p_{group} close to 1 and m_{group} close to 0. A torus can be characterized by p_{group} close to 0 and m_{group} close to 1. In summary, based on the two indicators of swarm polarization and group angular momentum, it is easy to classify the formation of collective behavior into flocking formation and torus formation.

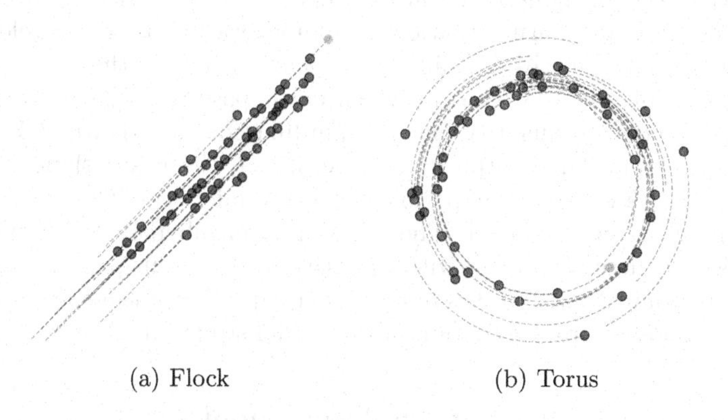

(a) Flock (b) Torus

Fig. 3. UAV swarm collective formation

To identificate the swarm collective behavior formation from p_{group} and m_{group}, a Multi-Layer Perception (MLP) neural network is designed. The MLP network consists of three layers as shown in Algorithm 1: input layer, hidden layer and output layer, and the different layers of the MLP neural network are fully connected to each other.

Algorithm 1 MLP Classification

Require: Formation features: p_{group}, m_{group}
Ensure: Labels: Flocking, Torus or Swarm
1: **function** FORWARD(*input*)
2: $z_1 = Dense(input)$
3: $z_2 = Dense(z_1)$
4: $output = ReLU(z_2)$
return *output*

3.2 Present Situation Comprehension

In this section, a swarm intention recognition method based on fuzzy inference is proposed, which can realize the inference of target intention without extensive prior knowledge. Utilizing the present state of the swarm, i.e., the recognized situation elements, we construct a fuzzy inference system to infer the swarm's intention. When using fuzzy inference for target intention recognition, it is essential to construct a fuzzy inference system from two aspects: fuzzy inference model and rules. And it is necessary to establish corresponding models and rules according to the actual situation.

Within this framework, we employ a multiple-input and single-output Mamdani fuzzy inference model. The inputs of the model consist of the identified elements of the UAV swarm situation, including swarm formation, velocity, altitude, distance, and relative heading angle. The outputs of the model represent the intentions of the UAV swarm, which encompass rescue, search and leave. The input and output quantity membership functions are shown in Fig. 4.

Once the inputs and outputs are determined, the establishment of corresponding rules is essential for performing fuzzy inference. The fuzzy rules for the system are given in Table 1. The selection of membership functions and the formulation of these rules align with objective facts. In this paper, the design of the fuzzy reasoning system takes into account expert knowledge and experience, ensuring a more accurate reflection of the actual situation.

4 UAV Swarm Advantage Assessment

In this section, the 3rd level of situation awareness, advantage assessment, is introduced. Firstly, a LSTM network is designed to predict the future states. Then, combining with the present situation and the future swarm states, a advantage assessment function is established to quantify the advantage level of the UAV swarm.

4.1 Prediction of Swarm Movement Trends and Collective Motion Formation

The original situation sequence information contains multi-dimensional situation information with multiple agents and multiple moments. To effectively obtain

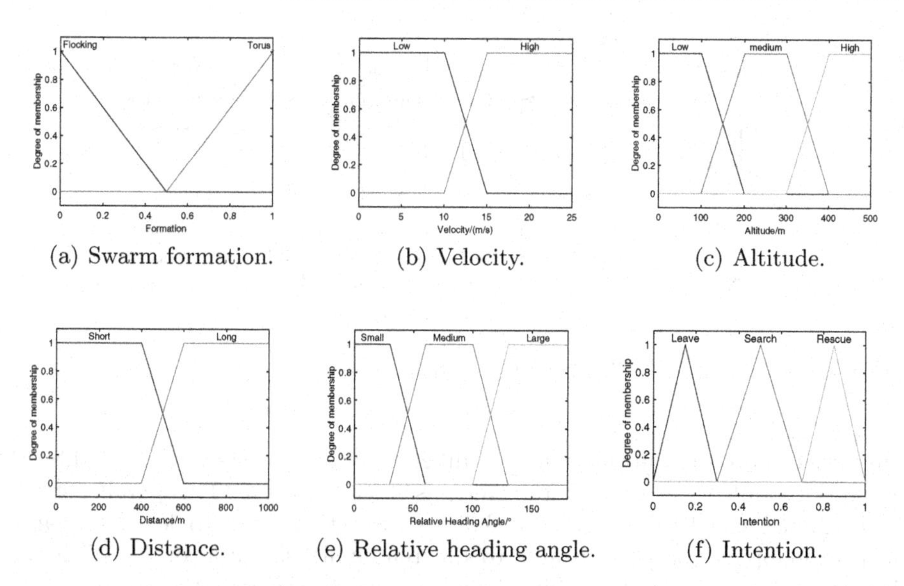

(a) Swarm formation. (b) Velocity. (c) Altitude.

(d) Distance. (e) Relative heading angle. (f) Intention.

Fig. 4. Membership functions

Table 1. The Fuzzy Rules.

Formation	Velocity	Altitude	Distance	Angle	Intention	Formation	Velocity	Altitude	Distance	Angle	Intention
Flocking	Low	Low	Short	Small/ Medium	Rescue	Flocking	High	Medium	Short	Small/ Medium	Rescue
Flocking	Low	Low	Long	Small	Rescue	Flocking	High	Medium	Long	Small/ Medium	Rescue
Flocking	Low	Low	Long	Medium	Search	Flocking	High	Medium/High	Short/Long	Large	Leave
Flocking	Low	Low	Long	Large	Leave	Flocking	High	High	Short	Small/ Medium	Rescue
Flocking	Low	Medium	short	small/ Medium	Rescue	Flocking	High	High	Long	Small/ Medium	Search
Flocking	Low	Medium	Long	Small	Rescue	Torus	Low	Low	Short	/	Rescue
Flocking	Low	Medium	Long	Medium	Search	Torus	Low	Low/ Medium	Long	/	Search
Flocking	Low	Medium/ High	Short	Large	Search	Torus	Low	Medium	Short	/	Rescue
Flocking	Low	Medium/ High	Long	Large	Leave	Torus	Low	High	Long	/	Search
Flocking	Low	High	Short	Small/ Medium	Search	Torus	Low/High	High	Short	/	Rescue
Flocking	Low	High	Long	Small/ Medium	Search	Torus	High	Low	Short	/	Search
Flocking	High	Low	Short	Small/ Medium	Rescue	Torus	High	Low/ Medium	Long	/	Search
Flocking	High	Low	Long	Small/ Medium	Rescue	Torus	High	Medium	Short	/	Rescue
Flocking	High	Low	Long	Large	Leave	Torus	High	High	Long	/	Search

the future situation information, it is necessary to process the original situation information in time dimension.

In this paper, the LSTM recurrent neural network module is employed to perform trajectory prediction and swarm formation prediction. The spatial-temporal information of each unit in the swarm is parallelly fed into the LSTM temporal feature extraction layer for computation. The modular network struc-

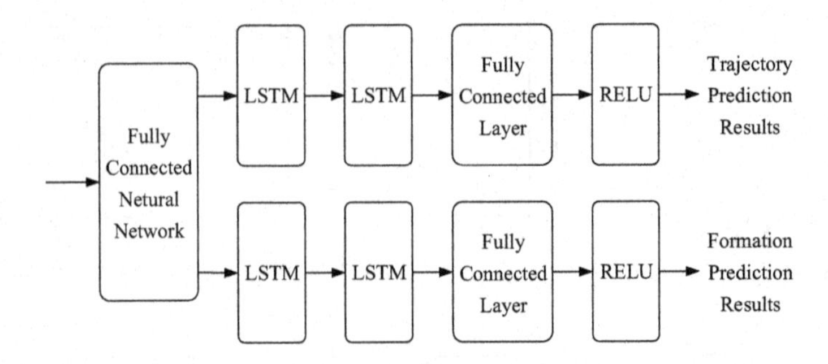

Fig. 5. LSTM prediction network

ture used to process information from different units consists of multiple LSTM network modules that share parameters, as shown in Fig. 5. This parallel structure enables the LSTM temporal feature extraction layer to learn a shared feature mapping space applicable to different units simultaneously. With the increase in swarm size, it exhibits the advantage of accelerated learning.

The training input of the network is sequences containing two types of characteristics, trajectory of the UAV swarm and the swarm collective motion formation features (p_{group} and m_{group}). And the output is sequences of prediction containing same states as the input.

The forward propagation algorithm of LSTM is shown in Algorithm 2. We designed two LSTM layers, where the first LSTM layer is responsible for extracting past time series information, and the second LSTM layer is used for predicting future information. During each forward propagation, we perform predictions for a total of $target\_length$ steps. For each prediction step, we append the result to the end of the input sequence and remove the first state of the input sequence. Then, we use the updated input sequence to predict the state for the next time step.

Algorithm 2 LSTM Forward Propagation

Input: Sequence of Group Characteristics

Output: Sequence of Future Characteristics

1: **function** FORWARD($input\_seq$)
2: $input = input\_seq$
3: $output =[\,]$
4: **for** $i \leftarrow 1$ to $target\_length$ **do**
5: $z_1, (hidden, cell) = LSTM(input)$
6: $z_2, \_ = LSTM(z_1, (hidden, cell))$
7: $z_3 = Dense(z_2[:, -1, :])$
8: $z_4 = ReLU(z_3)$
9: $input = input[:, 1, :]$
10: $input = concatenate(input, z_4, axis = 1)$
11: $output.append(z_4)$
12: $output = concatenate(output, axis = 1)$
return $output$

4.2 Swarm Advantage Assessment

To assess the advantage of swarm, it is necessary to pay attention to its characteristics. In this paper, we construct the angle advantage function, velocity advantage function, and distance advantage function based on the identified situation elements and predicted information to determine the overall advantage value of the swarm.

Angle Advantage Function. When the velocity direction of the swarm is closer, the easier it is to approach the target, that is, as the relative heading angle gets smaller, the swarm becomes more of a advantage to the target. Therefore, the angle advantage function can be constructed as:

$$T_{\varphi} = e^{sign(\Delta\eta)(\frac{\Delta\eta}{a\pi})^2} \tag{3}$$

wherein, a is a variable parameter, it can change as the distance d between the swarm and the target changes. The larger d is, the less influence $\Delta\eta$ has on the angle advantage function, so it is assumed that a is proportional to d. $\Delta\eta$ represents the difference between the predicted future relative heading angle of the swarm and its present relative heading angle.

Velocity Advantage Function. It can be set that the larger the swarm velocity, the greater the advantage to the target. The velocity advantage function of the UAV swarm to the target can be constructed as follows.

$$T_v = \frac{v - v_{\min}}{v_{\max} - v_{\min}} \tag{4}$$

wherein, $v_{\min}$ and $v_{\max}$ represent the maximum and minimum speed of the UAV swarm respectively.

Distance Advantage Function. To assess the distance advantage, we predict the swarm's next position at the subsequent moment and compare it with its current position. This calculation allows us to determine the change in distance. The distance advantage function of the UAV swarm to the target can be constructed as follows:

$$T_d = e^{sign(\Delta d)(\frac{\Delta d}{\sigma_d})^2} \tag{5}$$

where, σ_d is a variable determined by the relative heading angle of the swarm. Δd represents the difference between the predicted future position of the swarm and its present position.

Comprehensive Advantage Assessment Function. According to the above advantage functions, a comprehensive advantage assessment function between the swarm and the target can be constructed. In the case of a large distance advantage or velocity advantage, if the angle deviation is large, the comprehensive advantage degree will not be large. Therefore, the three advantage indexes are processed into a multiplicative relationship, and the comprehensive advantage assessment function is obtained.

$$T = \begin{cases} \alpha_1 \cdot T_\varphi \cdot T_d + \alpha_2 \cdot T_\varphi \cdot T_v & \text{Flocking} \\ \beta_1 \cdot T_d + \beta_2 \cdot T_v & \text{Torus} \end{cases} \tag{6}$$

Herein, α_1, α_2, β_1 and β_2 are weight coefficients respectively, the values are determined by the influence of each advantage index on the comprehensive advantage index.

$$\begin{aligned} 0 \leq \alpha_i \leq 1, \sum \alpha_i = 1 \\ 0 \leq \beta_i \leq 1, \sum \beta_i = 1 \end{aligned} \tag{7}$$

5 Simulation Result

To verify the effectiveness of the proposed UAV swarm situation awareness scheme, a series of simulation are performed. The identification and prediction of situation elements are firstly validated. Then the capabilities of the proposed framework in situation elements integration for situation awareness and advantage assessment are presented.

5.1 Situation Elements Identification and Prediction

At the beginning, we conduct simulations to identificate and predict situation elements, especially for swarm formation and swarm trajectory. We use Python and Pytorch to build neural networks analyze data. And the simulations are conducted based on a computer running Windows 11 with AMD Ryzen5500 and NVIDIA RTX3090. The training parameters are listed in Table 2.

Table 2. Parameters of neural networks for identification and prediction

Parameter	Identification (MLP)	Prediction (LSTM)
Number of agents	50	50
Epochs	100	500
Learning rate	0.001	0.001
Batch size	128	128
Hidden size	128	256
Dense layers	2	1
Sequence length	-	50
Prediction length	-	10
LSTM layers	-	2

Firstly, we use cross entropy as loss function to train a MLP neural network for swarm formation identification. The identification of swarm formation can be greatly simplified by using group polarization and group angular momentum.

The training results are presented in Fig. 6, the accuracy of UAV swarm formation identification increases with the number of iterations. Studying the UAV swarm from group characteristics makes it esay for situation elements recognition. After 100 iterations, the identification accuracy can reach 99%.

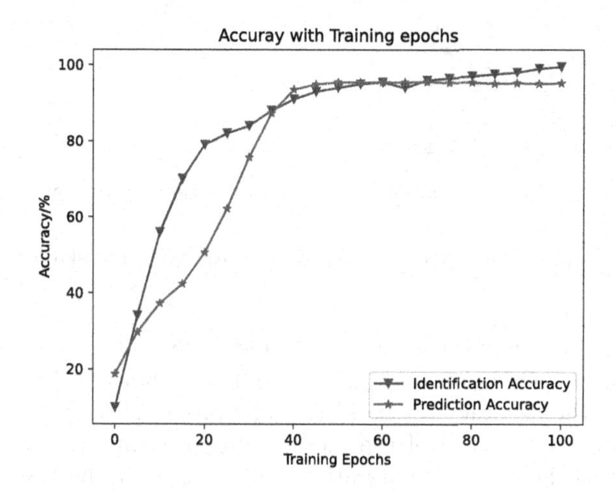

Fig. 6. Identification and Prediction Accuracy with Training Epochs.

Secondly, we use mse-square error as loss function to train a LSTM neural network for swarm formation and trajectory prediction.

In this simulation, we use the past 50 states to predict the future 10 states and each state includes position and formation characteristics of the UAV swarm. The states are sampled every 0.5 s, and a prediction of the states in the upcoming 5 s is made after 25 s of sampling.

The training process is also presented in Fig. 6, the accuracy of the swarm characteristics prediction increases with the number of iterations and converges to 96%.

In Fig. 7, two prediction precesses are presented. The trajectory and shape of the UAV swarm are predicted well in our simulation. The red solid line represents the input trajectory. The green solid line represents the actual trajectory in the future. The blue dashed line represents the predicted trajectory. The blue dots represents individual UAVs. The transparent blue dots represents individuals in the future. The transparent dots in the figure is only for the convenience of the shape prediction, and does not mean making accurate predictions of each individual in the future.

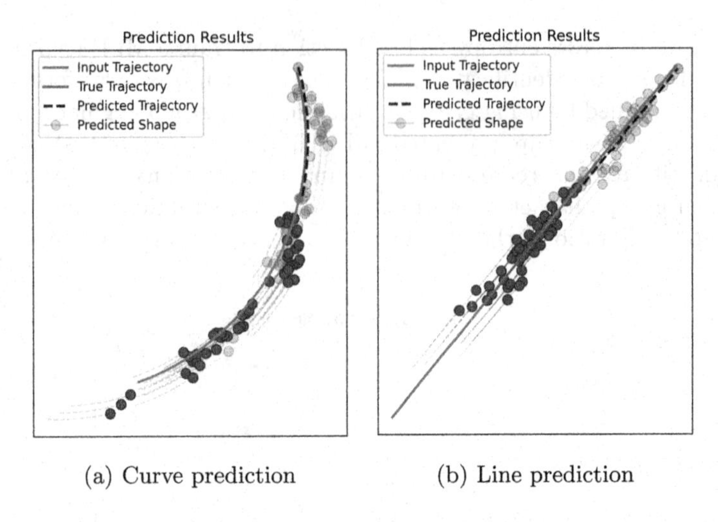

(a) Curve prediction (b) Line prediction

Fig. 7. Prediction of UAV swarm trajectory and shape

The comparison simulations are conducted in the same situation identification and prediction environment and the results are shown in Fig. 8. The yellow line represents the accuracy of the IAT model proposed in [12], while the blue line represents the accuracy of the method proposed in this paper. It can be observed that as the number of iterations increases, both methods achieve prediction accuracy of over 95% for future swarm trends, including trajectories and formations. However, for the current swarm formation identification, the method proposed in this paper achieves an accuracy of 99%, indicating a relatively higher recognition accuracy.

5.2 Situation Awareness and Advantage Assessment

Finally we conducted simulations to validate the situational awareness framework of UAV swarm. The simulation scenario is set up as follows: in our area, the UAV swarm appear. The swarm are reconnoitred at high altitudes at low speeds, search and rescue, then leave. The results of the situation awareness and advantage assessment of drone swarm are shown in the Fig. 9, where the light blue dots indicate the UAV swarm at different moments and the red stars indicate our position.

During the simulation, the UAV swarm first searched at high altitude, then found us and circled at low altitude to rescue, and finally leave. The simulation results show that the proposed method accords with the hypothesis and is effective for the intention recognition and advantage assessment of swarm.

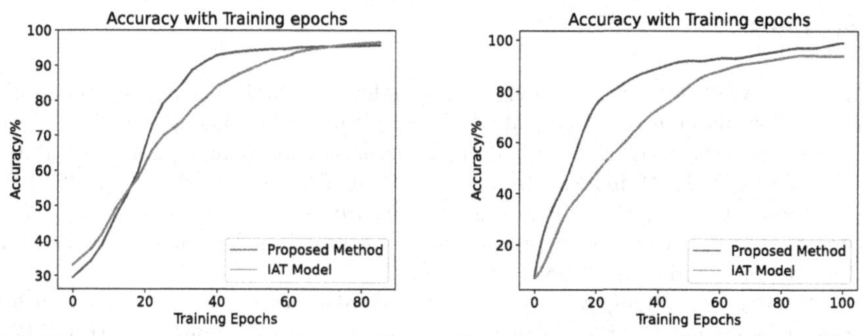

(a) The comparison curves of different models in situation prediction.

(b) The comparison curves of different models in formation recognition.

Fig. 8. Comparison results of different models in the same situation identification and prediction environment.

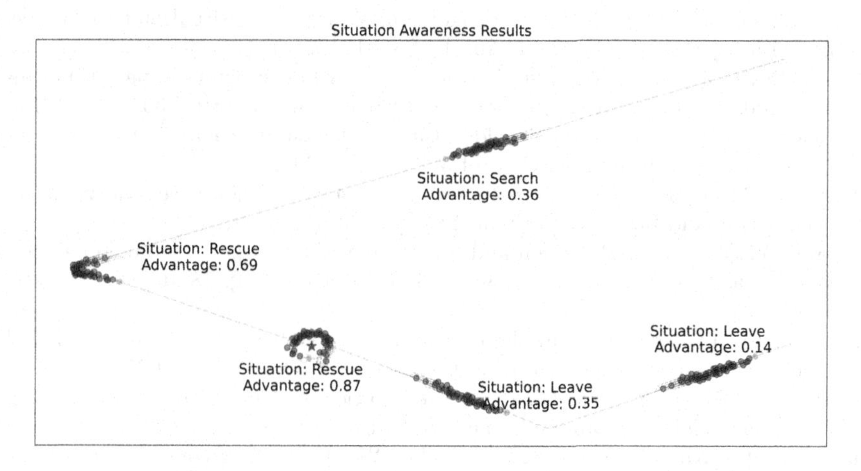

Fig. 9. Situation awareness and advantage assessment results.

6 Conclusion

In summary, this paper establishes a framework for UAV swarm situation awareness in complex environments. The framework includes swarm situation elements identification, swarm situation comprehension, and swarm situation prediction. In addition, unlike other studies that rely on individual UAVs for situation elements identification, this paper takes a holistic approach by considering the swarm as a whole, leveraging advantages in the situation awareness of large-scale UAV swarms. Finally, the proposed framework is validated through simulations, demonstrating its rationality and completeness.

References

1. Antonyshyn, L., Silveira, J., Givigi, S., Marshall, J.: Multiple mobile robot task and motion planning: a survey. ACM Comput. Surv. **55**(10), 1–35 (2023)
2. Endsley, M.R.: Design and evaluation for situation awareness enhancement. In: Proceedings of the Human Factors Society Annual Meeting, vol. 32, pp. 97–101. Sage Publications Sage CA: Los Angeles, CA (1988)
3. Luo, W., Li, M., Zhang, X.: Situation awareness based on Bayesian networks. Fire Control Command Control **35**(3), 89–92 (2010)
4. Li, M., Feng, X., Zhang, W.: Template-based inference model and algorithm for situation assessment in information fusion. Fire Control Command Control **35**(6), 64–66 (2010)
5. Li, W., Wang, B.: A template matching algorithm for implementing situation assessment. Comput. Sci. 33(5), 229–230 (2006). 249
6. Wang, C.: Research and implementation on non-cooperative target recognition expert system. Fire Control Command Control **35**(5), 132–134 (2010). 148
7. Li, G., Ma, Y.: Feature extraction algorithm of air combat situation based on deep neural networks. J. Syst. Simul. 29(z1), 98–105 (2017). 112
8. Ou, W., Liu, S., He, X., Guo, S.: Tactical intention recognition algorithm based on encoded temporal features. Command Control Simul. **38**(6), 36–41 (2016)
9. Liao, Y., Yi, Z., Hu, X.: Battlefields situation elementary comprehension based on deep learning. J. Command Control **3**(1), 67–71 (2017)
10. Xu, J., Zhang, L., Han, D.: Air target intention recognition based on fuzzy inference. Command Inf. Syst. Technol. **11**(3), 44–48 (2020)
11. Wei, W., Wang, G.: Detection and recognition of air targets by unmanned aerial vehicle based on RBF neural network. Ship Electron. Eng. **38**(10), 37–40, (2018). 110
12. Li, Y., et al.: Recognition method for unmanned swarm adversarial situation elements using transformer. J. Harbin Inst. Technol. **54**(12), 1–9 (2022)
13. Qiao, D., Liang, Y., Ma, C., et al.: Recognition and prediction of group target intention in multi-domain operations. Syst. Eng. Electron. 1–12 (2021)
14. Jiahui, H., Xin, S., Xiaodong, C., Jiabo, W.: Research on air objective threat assessment technology based on cluster analysis. Tactical Missile Technol. **02**, 96–104 (2023)
15. Endsley, M.R.: Toward a theory of situation awareness in dynamic systems. Hum. Factors: J. Hum. Factors Ergonomics Soc. **37**(1), 32–64 (1995)

The Weapon Target Assignment in Adversarial Environments

Rui Gao[1,2], Shaoqiu Zheng[2], Kai Wang[2], Longlong Zhang[1], Zhiyuan Zhu[1(✉)], and Xiaohu Yuan[3]

[1] College of Electronic and Information Engineering, Southwest University, Chongqing 400715, China
846008321@qq.com

[2] Key Lab of Information System Requirement, Nanjing Research Institute of Electronics Engineering, Nanjing 210007, China

[3] Department of Computer Science and Technology, Tsinghua University, Beijing 100084, China

Abstract. In context of the current weapon target assignment (WTA) problem, consideration of target capability is often disregarded. To address this issue, we propose an optimized model for weapon target assignment in complex environments. Our model provides a comprehensive description and conducts a detailed analysis of constraints and objective functions in the context of adversarial weapon target assignment (AWTA). To efficiently tackle the assignment problem, we employ the multi-particle swarm optimization algorithm (MOPSO). During the optimization process, we incorporate learning factors and inertia weights to update particle positions, seamlessly implementing the model with a set of adjustable parameters. Experimental results demonstrate that our model effectively captures the concept of adversarial scenarios. Furthermore, by utilizing the MOPSO optimization algorithm, we achieve a superior Pareto front even for small-scale adversarial scenarios, successfully resolving the adversarial WTA problem. Our research contributes an effective optimization approach to address the weapon task assignment problem and offers valuable inspiration for future investigations into multi-objective research. In conclusion, our study reveals the potential of multi-objective optimization research in the context of weapon target assignment.

Keywords: AWTA problem · MOPSO · Pareto front · penalty parameter

1 Introduction

Weapon-target assignment (WTA) is one of the core issues in the field of military operations research. Optimal Weapon Assignment Strategy for Expected Value Maximization [1, 2]. Such problems belong to the general case of resource allocation, but also involve intelligent information processing in cognition and computation.

After decades of research, the WTA problem has made great progress in terms of models, algorithms, and applications. The problem can be expressed as a nonlinear integer programming problem known to be NP-complete [3]. WTA problems can be divided

F. Sun and J. Li (Eds.): ICCCS 2023, CCIS 2029, pp. 247–257, 2024.
https://doi.org/10.1007/978-981-97-0885-7_21

into single-objective WTA and multi-objective WTA according to different decision-making objectives. According to the different ways of combat confrontation, it can be separated into direct confrontation WTA and indirect WTA. In direct confrontation WTA, both sides choose the form of direct confrontation; while in indirect confrontation WTA, one party attacks while the other passively defends [4]. When classifying WTA models, they can be divided into static WTA models and dynamic WTA models according to the time dimension. At the same time, according to the choice of objective function, it can be divided into the maximum hit probability model, the maximum expected penetration number model and the minimum expected cost model [5].

Although there are many ways to find the optimal solution, there is a lack of consideration from the perspective of the opponent's capabilities [6]. With the widespread application of artificial intelligence technology and unmanned system technology, WTA issues are also facing new challenges and development opportunities, such as swarm operations and cross-domain unmanned swarm coordinated operations. Therefore, on the basis of existing research, exploring new problem types and technical means will become a research hotspot in the future of intelligent and unmanned WTA problems.

This paper proposes a new weapon-target assignment model based on the confrontation environment, which fully considers the ability of the target.And the multi-objective particle swarm optimization algorithm (MOPSO) is applied to solve the Pareto optimal solution set of the AWTA problem.At the same time, we verified the effectiveness of the new weapon-target assignmen model and multi-objective particle swarm optimization algorithm, and finally obtain the conclusion of this paper.

2 Related Work

2.1 MOWTA Problem

Due to the gradual increase in the scale of operations, it is necessary to face more complex scenarios, and it is necessary to find a suitable weapon to distribute to multiple targets so that the total expected value of all targets is maximized. Computationally more difficult.Li et al.transformed the WTA problem into a multi-stage WTA problem (MWTA) and further the multi-objective optimizer: the adaptive mechanism of NSGA-II (based on domination) and MOEA/D-AWA (based on decomposition). This operation extends the WTA problem into a multi-objective optimization problem (MOP), maximizing the efficiency of destroying targets and minimizing the consumption of weapons [7].

Aimning at the WTA problem with multiple optimization objectives, Fu proposed a multi-objective particle swarm optimization (MOPSO) algorithm based on multi-population co-evolution to achieve rapid search for the global optimal solution. This algorithm constructed a master-slave population co-evolution model [8]. Shi proposed an evolutionary algorithm (SparseEA) for the asset-based weapon target assignment (ABWTA) problem, and proposed a new sparse large-scale ABWTA problem framework evolutionary algorithm SpareEA-ABWTA to solve the problem in different large-scale under the weapon target assignment problem [9]. Considering the uncertainty in the dynamic environment, Wang proposed an adaptive MA algorithm, and adaptive replacement strategy and restart strategy were used to maintain the genetic diversity of the population [10].

2.2 Adversarial Elements

In adversarial decision-making, as the risk of failure increases, the adversarial factor become more important in task assignment [11]. In order to highlight the capabilities of the target defense system, researchers such as LUO conducted research on the problem of missile target assignment to counter defense systems. They set up a deep Q-network to learn optimal policies from historical data [12]. Zhang utilizes gray correlation analysis methods for threat assessment of targets to be attacked during the target selection process for aviation force attacks on operational targets [13]. Reily, on the other hand, focuses on the impact of passive team changes caused by adversarial attacks or disturbances [14]. Similarly, Rudolph solves the task assignment decision problem with nondeterministic agency capabilities to ensure various task constraints [15]. However, researchers such as Li fully consider the intelligent decision-making of the target. In the adversarial decision-making, all targets can hide themselves by jamming the sensors of a multi-agent system (MAS), preventing agents using the same type of sensors from detecting them [16].

2.3 MOPSO

Multi-Objective Particle Swarm Optimization (MOPSO) is an extension of PSO algorithm to multi-objective optimization problems, aiming at solving optimization problems with multiple objective functions.

Chen et al. designed an improved multi-objective particle swarm optimization algorithm (MOPSO) to optimize the allocation, taking the solution of the top 5% of the crowding distance as the global optimal value, and the obtained pareto frontier solution has a higher accuracy, which conforms to The principle of the least probability of failure and the least use of weapons [17]. Wei et al. took advantage of the maximum damage probability of the single-target particle swarm optimization algorithm and reduced ammunition consumption. The experimental comparative analysis results show that the improved multi-target particle swarm optimization algorithm converges faster and more effectively than NSGA-II. The gap is more obvious when the scale of the target assignment problem is larger [18]. Gu et al. used a multi-objective discrete particle swarm optimization-orbital search hybrid algorithm based on the minimum division of weapon consumption and the total expected remaining threat, which has global search capabilities and fast convergence speed, which meets the experimental requirements [19]. Kong et al. improved the multi-objective particle swarm algorithm, repeatedly learned and evolved the dominant and non-dominant solutions of the algorithm, and adopted a dynamic file maintenance strategy to solve the problems of maximizing combat benefits and minimizing weapon costs [20].

3 Problem Formulation

In the problem of weapon-target tasking in an adversary-based environment, an adequate mathematical description and modeling analysis is required in order to fully explore and solve the problem.

3.1 AWTA Modeling

In the AWTA problem, first element considered is the capability of the weapon platform and target. It is first necessary to define the types and properties of the elements of the AWTA problem. The set of weapon platforms is $S_w = \{W_1, W_2, ..., W_m\}$, in which m is the number of weapon platforms, and the load capacity of the i-th weapons platform W_i is l_i. Weapons platform capacity is the maximum amount of ammunition that can be carried by each weapons platform W_i, and

$$L = [l_1 \ l_2 \ ... \ l_m] \tag{1}$$

The cost of ammunition carried by each weapons platform is c_i, and

$$C = [c_1 \ c_2 \ ... \ c_m] \tag{2}$$

The set of targets is $S_T = \{T_1, T_2, ..., T_n\}$, where n is the number of targets and the value of the j-th target is V_j, and

$$V = [v_1 \ v_2 \ ... \ v_m] \tag{3}$$

The probability of destruction of the j-th target T_j by each munition in the weapon platform W_i is p_{ij}, and

$$P = \begin{bmatrix} p_{11} & \cdots & p_{1n} \\ \vdots & \ddots & \vdots \\ p_{m1} & \cdots & p_{mn} \end{bmatrix} \tag{4}$$

Restrictive Condition. To characterize the AWTA problem, we introduce the matrix of decision variables X, and

$$X = \begin{bmatrix} X_{11} & \cdots & X_{1n} \\ \vdots & \ddots & \vdots \\ X_{m1} & \cdots & X_{mn} \end{bmatrix} \tag{5}$$

where x_{ij} is denoted as the value of the element in row i and column j of the ATWA decision variable matrix. The rows represent the number of weapon platforms, the columns represent the number of targets, and each weapon in an element is capable of attacking only one target. $x_{ij} = \{0, 1, ..., min(w_i, s_i)\}$, the i-th weapon platform carries a maximum number of weapons W_i, and

$$\sum_{j=1}^{n} x_{ij} \leq w_i, i = 1, 2, ..., m \tag{6}$$

At most s_j weapons can be used to strike the j-th target, and

$$\sum_{i=1}^{m} x_{ij} \leq s_j, j = 1, 2, ..., n \tag{7}$$

The sum of the ammo caps assigned to all targets is W, and

$$W = [w_1\ w_2\ ...\ w_n] \tag{8}$$

W needs to be lower than the sum L of the ammunition loads of all weapon platforms, and

$$\sum_{j=1}^{n} w_j \leq \sum_{i=1}^{m} l_i, i = 1, 2, ..., m, j = 1, 2, ..., n \tag{9}$$

Objective Function. In order to comprehensively consider the confrontation capability and the consumption of weapon resources in the AWTA problem, we introduce two objective functions: the maximization of the target damage benefit and the minimization of the consumption of weapon resources. At the same time, we use the value of the target to measure the target's strike revenue, that is, the higher the value of the target hit, the better the benefit effect.

The damage probability of the i-th weapon platform w_i to the j-th target T_j is:

$$p(j) = 1 - \prod_{i=1}^{m} (1 - p_{ij})^{x_{ij}} \tag{10}$$

The benefit of the weapon platform to strike the target as:

$$f(x) = \sum_{j=1}^{n} v_j \cdot p(j) = \sum_{j=1}^{n} v_j \cdot \left[1 - \prod_{i=1}^{m} (1 - p_{ij})^{x_{ij}}\right] \tag{11}$$

When attacking the target, the target will carry out targeted defense. We use a matrix D to represent the defense capability of each target, and

$$D = \begin{bmatrix} d_{11} & \cdots & d_{1m} \\ \vdots & \ddots & \vdots \\ d_{n1} & \cdots & d_{nm} \end{bmatrix} \tag{12}$$

where d_{ji} represents the weapon defense strength of the j-th target against the enemy's i weapon platform. Additionally, the defensive performance of a target typically degrades as the number of weapons platforms assigned increases. This is because the target needs to face more attacks, causing its defense capabilities to be scattered and difficult to resist large-scale attacks.

The defense performance of the target decreases with the increase of the allocation number of weapon platforms, then the defined weapon cost loss is:

$$g(x) = \sum_{i=1}^{m} \sum_{j=1}^{n} c_i \left(1 - d_{ji}\right) \cdot x_{ij} \tag{13}$$

Combining the constraints and optimization objectives of the ATWA problem, the mathematical model can be established as:

$$\min f(x) = \sum_{j=1}^{n} v_j \cdot \left[1 - \prod_{i=1}^{m} (1 - p_{ij})^{x_{ij}}\right]$$

$$\min g(x) = \sum_{i=1}^{m} \sum_{j=1}^{n} c_i \cdot (1 - d_{ji}) \cdot x_{ij}$$

$$s.t. \begin{cases} \sum_{j=1}^{n} x_{ij} \leq w_i, \\ \sum_{i=1}^{n} x_{ij} \leq s_j, \\ \sum_{i=1}^{n} w_i \leq \sum_{i=1}^{m} s_j \end{cases} \tag{14}$$

When optimizing the AWTA problem, we not only need to maximize the damage benefit of the target, but also need to minimize the weapon cost loss. Such an optimization goal will help to comprehensively consider attack effects and resource consumption in task allocation to obtain a balanced and efficient task allocation scheme.

3.2 Multi-objective Optimization Problem

The AWTA model is a typical multi-objective optimization problem (MOP). In this problem, there are multiple optimization objectives, which are not comparable to each other and may conflict. This means that it is impossible to find a single solution that optimizes all objective functions simultaneously. Therefore, we need to balance these goals to be optimized so that each objective function can get a better solution, and finally get an optimal solution set, calling Pareto optimal solution set or non-dominated solution set.

If the given decision-making $Y = (y_1, y_2, ..., y_Q)$ vector satisfies the constraints $h_k \geq 0, k = 1, 2, ..., Q$ and there are s conflicting optimization objectives $f(Y) = (f_1(y), f_2(y), ..., f_s(y))$, then $Y^* = (y_1^*, y_2^*, ..., y_Q^*)$ is the Pareto optimal solution that makes the objective function reach Pareto optimal under the constraints.

Pareto Dominance. Only two levels of headings can be numbered. Lower level headings are still unnumbered; they are initialized to insert headings.

Assume that the goal of a system is to minimize a series of objective functions $\min f(Y)$, $Y \in R^m$ is a set of its variable constraints, and both Y_1 and Y_2 are Pareto optimal solutions. If it holds for all ff, then Y1 is said to dominate Y2. Pareto efficiency refers to the state in which one objective cannot be improved without affecting other objectives in multi-objective optimization.

Pareto Front. The concept of Pareto efficiency builds on the definition of dominance, Dominant solutions are a subset of the Pareto optimal solutions, representing the solutions that outperform all others in all objectives. The relationship between Pareto optimal and dominant solutions is essential for understanding the behavior of multi-objective optimization algorithms and finding the most suitable solutions for decision-making in complex problems with multiple conflicting objectives. Relationship between dominant solution and Pareto is shown in Fig. 1.

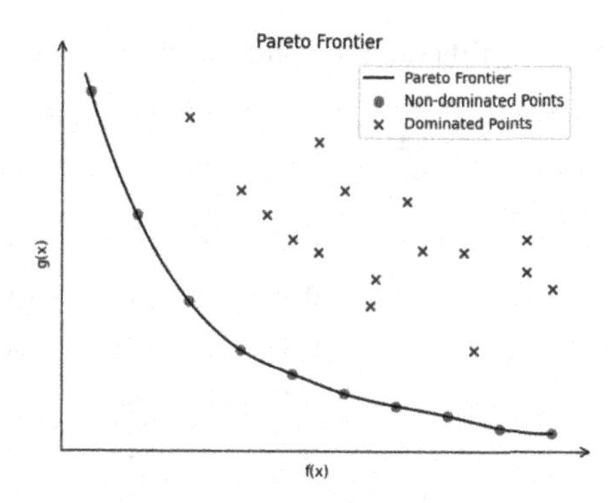

Fig. 1. Schematic diagram of the Pareto front

In the optimization problem of the two objective functions, the solutions represented by the red solid points are all Pareto optimal solutions, that is, there are no other solutions in the objective function space that can simultaneously outperform them on all objectives. The first set of Pareto optimal solutions from the Pareto optimal solution set, and mapping this set of solutions to the objective function, forms the Pareto optimal front or Pareto front of the problem. In short, the objective function value corresponding to the Pareto optimal solution is the Pareto optimal front [21].

4 Experiment and Results Analysis

In order to verify the effectiveness of the model and algorithm in this paper, we use Python and NumPy libraries to implement the MOPSO optimization algorithm for the AWTA problem to solve the Pareto optimal solution set. This algorithm will help us find a solution with an optimal balance between multiple objectives.

4.1 Scene Setting

In this study, we construct a small-scale confrontation scenario, where the number of weapon platforms is set to 6 and the number of targets is set to 5. The amount of ammunition mounted on each weapon platform and the maximum amount of ammunition that can be allocated to each target during strikes are preset values. It is worth noting that the same type of weapon is configured on the same weapon platform, and the hit probability of the target is consistent in this scenario. The specific parameters are shown in Table 1 and Table 2. This scenario aims to verify the effectiveness of our proposed AWTA model and MOPSO optimization algorithm in a small-scale confrontation environment.

4.2 Algorithm Process

In order to solve the constraint limitation and simplify the computational complexity in the weapon target assignment problem in adversarial environments, we use the penalty function method to transform the constraints into a series of unconstrained optimization

Table 1. Probability of hits.

Weapons platform number	Target				
	1	2	3	4	5
1	0.47	0.82	0.69	0.61	0.34
2	0.34	0.28	0.77	0.61	0.67
3	0.26	0.83	0.75	0.38	0.36
4	0.36	0.43	0.56	0.51	0.42
5	0.62	0.33	0.43	0.47	0.52
6	0.72	0.37	0.56	0.61	0.28

Table 2. Target's defense.

Assigned to the same target number	Target				
	1	2	3	4	5
1	0.5	0.4	0.5	0.5	0.5
2	0.32	0.3	0.23	0.29	0.28
3	0.15	0.19	0.12	0.08	0.15

problems. The specific method is to add the constraint inequality into the objective function after being penalized, so as to transform the constrained optimization problem into an unconstrained optimization problem. When the particle's solution does not satisfy the constraints, we regard it as an infeasible solution and add corresponding penalties to it. In order to ensure the effectiveness of the solution, we introduce a large penalty factor φ_{i} to punish infeasible solutions during the optimization process.

$$\varphi_1 \sum_{i=1}^{m} \left[\min\left(0, s_j - \sum_{i=1}^{m} x_{ij}\right) \right]^2 + \varphi_2 \sum_{j=1}^{n} \left[\min\left(0, w_i - \sum_{j=1}^{n} x_{ij}\right) \right]^2 \tag{15}$$

We will apply the penalty function based MOPSO algorithm to solve the problem of weapon target assignment in adversarial environments. The flow chart is shown in the figure. The algorithm will use the search ability of particle swarms and consider the penalty of constraints in the objective function to find the optimal solution. The flow chart of the MOPSO algorithm for solving the weapon target assignment problem in the confrontation environment is shown in Fig. 2.

By introducing the penalty function method and designing the corresponding optimization algorithm, we will be able to better deal with constraints and optimization objectives, and provide a more effective solution to the problem of weapon target assignment in adversarial environments.

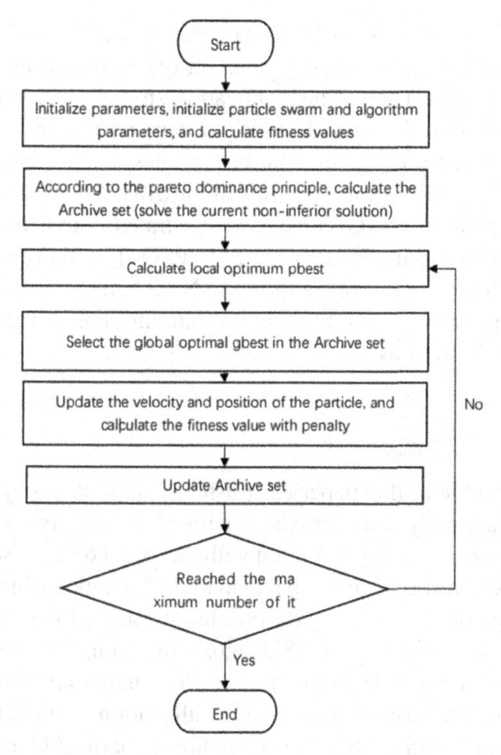

Fig. 2. Flowchart of MOPSO algorithm

4.3 Experiment and Results

In order to better analyze the improved MOPSO algorithm, we set the number of iterations of the algorithm to 100, and used 100 particles. In order to further verify our proposed method, we also introduced NSGA-II (non-dominated sorting genetic algorithm) in the experiment for comparison, The experimental results are shown in Fig. 3.

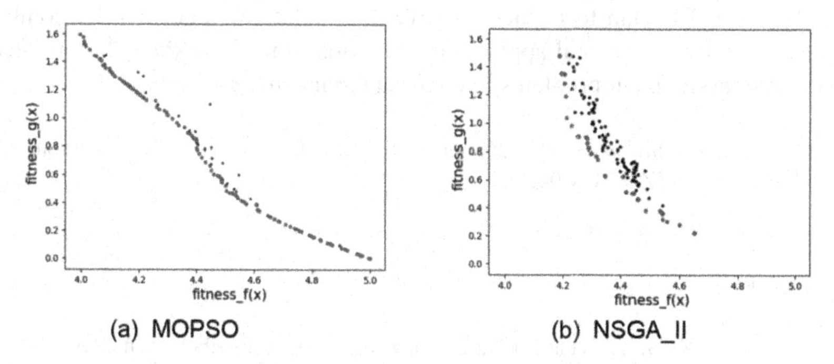

Fig. 3. Performance comparison between MOPSO and NSGA-II

Upon introducing objectives f(x) and g(x), the Allocation of Weaponry and Tasks Assignment (AWTA) problem evolves into a multi-objective optimization scenario. Our primary aim is to identify a set of task allocation strategies that optimize the objective of target destruction while simultaneously minimizing weapon resource consumption, taking into account both adversarial capabilities and resource utilization. Both the Multi-Objective Particle Swarm Optimization (MOPSO) and the Non-dominated Sorting Genetic Algorithm (NAGA) have demonstrated commendable proficiency in addressing our research inquiry. However, the MOPSO algorithm exhibits a more extensive exploration of the search space compared to NAGA, thereby enhancing its capacity to discover a greater number of Pareto-optimal solutions, consequently yielding a more diverse set of optimal solutions.

5 Conclusion

Aiming at AWTA problem, this paper establishes a static dual-objective optimization model that comprehensively considers the benefit of the weapon platform to strike the target and the cost of the strike ammunition with the target defense scheme. Constraints of the model include constraints on the payload of each weapon platform, constraints on the target distribution of ammunition, and constraints on the total ammunition. In order to solve the model, we adopt the MOPSO algorithm and use the constraint processing mechanism to optimize the calculation process. Through simulation experiments, we analyzed the iterative situation of weapon target allocation in small-scale confrontation scenarios, and we can obtain the Pareto front in this scenario. At the same time, we also verified the performance of the MOPSO algorithm and the NSGA algorithm in the above model, and compared their solution effects.

Currently, our static bi-objective optimization model has proven effective in the problem of weapon target assignment in adversarial environments. However, in order to better adapt to the needs of dynamic environments, we will further build a model that is more in line with dynamic performance. In future work, we will focus on dealing with dynamic timeliness issues and explore better convergence performance. Dynamic timeliness refers to the situation that the state of the target and the weapon platform changes with time, which puts forward higher requirements for the decision-making of task assignment. We plan to further improve the model and algorithm by combining the dynamic features in actual application scenarios to improve the solution effect of weapon target assignment problems in confrontational environments.

Acknowledgments. This work was supported by Key Laboratory of Information System Requirement, NO: LHZZ2021-M04.

References

1. Li, M., Chang, X., Shi, J., et al.: Advances in weapon target distribution: models, algorithms and applications. Syst. Eng. Electron. **45**(4), 1071–1123 (2023)
2. Xue, X., Wang, Z., Huang, X.: Weapon target assignment based on expected utility maximization. Command Inf. Syst. Technol. **9**(6), 61–65 (2018)

3. Ni, M., Yu, Z., Ma, F., et al.: A Lagrange relaxation method for solving weapon-target assignment problem. Math. Probl. Eng. **2011**, 1–10 (2011)
4. Wang, N., Dai, J.: Weapons-target allocation method based on INSGA-II algorithm. J. Yunnan Univ. (Nat. Sci. Edn.) **44**(6), 1155–1165 (2022)
5. Kline, A., Ahner, D., Hill, R.: The weapon-target assignment problem. Comput. Oper. Res. **105**, 226–236 (2019)
6. Lee, Z.J., Su, S.F., Lee, C.Y.: Efficiently solving general weapon-target assignment problem by genetic algorithms with greedy eugenics. IEEE Trans. Syst. Man Cybern. Part B (Cybern.) **33**(1), 113–121 (2003)
7. Li, J., Chen, J., Xin, B., Dou, L.: Solving multi-objective multi-stage weapon target assignment problem via adaptive NSGA-II and adaptive MOEA/D: a comparison study. In: 2015 IEEE Congress on Evolutionary Computation (CEC), Sendai, Japan, pp. 3132–3139 (2015). https://doi.org/10.1109/CEC.2015.7257280
8. Fu, G., Wang, C., Zhang, D., et al.: A multiobjective particle swarm optimization algorithm based on multipopulation coevolution for weapon-target assignment. Math. Probl. Eng. **2019** (2019)
9. Shi, X., Zou, S., Song, S., et al.: A multi-objective sparse evolutionary framework for large-scale weapon target assignment based on a reward strategy. J. Intell. Fuzzy Syst. **40**(5), 10043–10061 (2021)
10. Wang, Y., Xin, B., Chen, J.: An adaptive memetic algorithm for the joint allocation of heterogeneous stochastic resources. IEEE Trans. Cybern. **52**(11), 11526–11538 (2021)
11. Cacciamani, F., Celli, A., Ciccone, M., et al.: Multi-agent coordination in adversarial environments through signal mediated strategies. arXiv preprint arXiv:2102.05026 (2021)
12. Luo, W., Lü, J., Liu, K., et al.: Learning-based policy optimization for adversarial missile-target assignment[J]. IEEE Trans. Syst. Man Cybern.: Syst. **52**(7), 4426–4437 (2021)
13. Zhang, H., Wang, D., Gou, Q.: Method for attacking target assignment and priority sequencing based on threat assessment. Command Inf. Syst. Technol. **9**(6), 55–60 (2018)
14. Reily, B., Zhang, H.: Team assignment for heterogeneous multi-robot sensor coverage through graph representation learning. In: 2021 IEEE International Conference on Robotics and Automation (ICRA), Xi'an, China, pp. 838–844 (2021)
15. Rudolph, M., Chernova, S., Ravichandar, H.: Desperate times call for desperate measures: towards risk-adaptive task allocation. In: IEEE/RSJ International Conference on Intelligent Robots and Systems (IROS), Prague, Czech Republic, pp. 2592–2597 (2021). https://doi.org/10.1109/IROS51168.2021.9635955
16. Li, Y., Liu, H., Sun, F., et al.: Adversarial decision making against intelligent targets in cooperative multi-agent systems. IEEE Trans. Cogn. Dev. Syst. (2023)
17. Chen, M., Zhou, F., Zhang, C.: Improvement of MOPSO's combined fire targeting distribution. Firepower Command Control (2019)
18. Xia, W., Liu, X., Fan, Y., et al.: Weapon-target assignment with an improved multi-objective particle swarm optimization algorithm. Acta Armamentarii **37**(11), 2085 (2016)
19. Gu, J., Zhao, J., Yan, J., et al.: Cooperative weapon-target assignment based on multi-objective discrete particle swarm optimization-gravitational search algorithm in air combat. J. Beijing Univ. Aeronaut. Astron. **41**(2), 252–258 (2015)
20. Kong, L., Wang, J., Zhao, P.: Solving the dynamic weapon target assignment problem by an improved multiobjective particle swarm optimization algorithm. Appl. Sci. **11**(19), 9254 (2021)
21. Lin, X., Chen, H., Pei, C., et al.: A pareto-efficient algorithm for multiple objective optimization in e-commerce recommendation. In: Proceedings of the 13th ACM Conference on Recommender Systems, pp. 20–28 (2019)

A Distributed Vehicle-Infrastructure Cooperation System Based on OpenHarmony

Jingda Chen[1,2], Hanyang Zhuang[1](✉), and Ming Yang[3]

[1] University of Michigan-Shanghai Jiao Tong University Joint Institute, Shanghai Jiao Tong University, Shanghai 200240, China
zhuanghany11@sjtu.edu.cn

[2] Artificial Intelligence Institute, Shanghai Jiao Tong University, Shanghai 200240, China

[3] Department of Automation, Shanghai Jiao Tong University, Shanghai 200240, China

Abstract. The rapid advancement of transportation has sparked interest in Vehicle-Infrastructure Cooperation System (VICS), a novel transportation system that emphasizes the coordination of transportation facilities. However, this system consists of multiple heterogeneous devices, including roadside sensors, autonomous vehicles, traffic lights, cellphones, etc. The disparate communication capabilities among these devices result in poor development efficiency and compatibility, necessitating the establishment of a unified and standardized communication mechanism. Distributed operating systems, which operate across multiple devices and manage inter-device communication, offer a promising solution to this challenge. OpenHarmony, an emerging distributed operating system, has been applied in various interconnection scenarios to successfully resolve the communication and development complexities, but not has limited use cases in VICS. Therefore, this paper addresses this gap by constructing a miniaturized hardware platform of VICS and utilize OpenHarmony as the universal operating system to develop and deploy self-driving algorithms. The main target is to explore and demonstrate the capability of OpenHarmony in the application area of VICS. The miniatured hardware consists roadside perception systems and self-driving vehicles at smaller scale to simulate the real-world scenario. Initially, the paper proposes a system framework based on OpenHarmony, leverages the communication capabilities of OpenHarmony's distributed soft bus to enable data exchange between vehicles and roadside perception systems, and designs a quality measurement index for evaluating multi-perception results. Subsequently, the roadside perception system with vehicle localization algorithm and the self-driving vehicle with planning and control algorithms are developed. In the end, this paper utilizes this comprehensive VICS system to explore the capability of OpenHarmony and discuss the potential of OpenHarmony in this area.

Keywords: Vehicle-Infrastructure Cooperation System · OpenHarmony · Distributed Operating System · Intelligent Transportation System

F. Sun and J. Li (Eds.): ICCCS 2023, CCIS 2029, pp. 258–271, 2024.
https://doi.org/10.1007/978-981-97-0885-7_22

1 Introduction

With the rapid expansion of transportation systems, the requirement of management and operation are incurring escalating human resource costs. Consequently, there is a growing demand for Intelligent Transportation Systems (ITS) to address this challenge. The Vehicle-Infrastructure Cooperation System (VICS) emerges as a technical solution for implementing ITS. Leveraging wireless communication technology and the Internet of Things (IoT), ITS facilitates dynamic information exchange between vehicles and between vehicles and infrastructure, enabling collaborative interactions and enhancing the safety and efficiency of the transportation system [1]. By treating vehicles and infrastructures as an integrated entity, this collaboration enhances the overall efficiency and safety of the global transportation system.

Nevertheless, the development of VICS encounters certain unresolved issues that impede their progress. Notably, the communication and coordination problem among multiple heterogeneous devices within the system remains a key challenge. The system encompasses diverse devices, including roadside perception systems, autonomous vehicles, and others, each characterized by disparate hardware resources and communication capabilities such as WiFi, Bluetooth, ZigBee, and more [2]. The divergent nature of these devices necessitates individualized development efforts for seamless integration into a unified system. Consequently, system development becomes intricate, hampered by inadequate data interaction between devices, thus undermining the coordination of the VICS and limiting its scalability. Therefore, there is a need for a mechanism that allows developers to overlook the differences between devices, enabling more efficient development and achieving seamless and efficient communication between devices. This mechanism would fulfill the development requirements of VICS.

Distributed operating systems are commonly built on multiple devices and manage inter-device communication to achieve the characteristics of distribution. Additionally, operating systems can provide developers with a unified application development interface for developing applications. Therefore, during development, developers only need to focus on adapting the operating system to the hardware and developing the application, while the specific implementation mechanisms are handled by the operating system itself. From the characteristics of distributed operating systems, it is possible that they can serve as one of the solutions to enhance cooperativity in VICS by abstracting device types and communication methods.

OpenHarmony, as an emerging distributed operating system, features a multi-kernel design that can adapt to devices with different resource types. Moreover, OpenHarmony's distributed soft bus function can integrate various communication interfaces, managed by the operating system itself, and provide a unified interface to the upper layers [3]. Furthermore, it facilitates automatic device discovery, networking, and data transmission, thereby improving the efficiency of information exchange between devices. Based on these characteristics of OpenHarmony, it may provide a more unified and convenient communication mechanism for VICS. However, there is currently limited research in this area. Therefore, this paper aims to explore the application of OpenHarmony in VICS with following contributions:

1. We accomplished the construction and validation of a software and hardware system for VICS using OpenHarmony.
2. We investigated the role and impact of a distributed software bus on VICS data exchange and conducted autonomous driving verification on a miniatured hardware platform.

In this paper, we propose a VICS architecture based on OpenHarmony. We explore the communication capabilities of OpenHarmony's distributed soft bus and validate them through corresponding miniaturized vehicle experiments. In Sect. 2, we briefly review the relevant work on VICS and VICS operating systems. In Sect. 3, we present the overall system framework, followed by an exploration of the communication mechanism of the VICS using OpenHarmony and its distributed soft bus. Subsequently, we implement the functionalities of perception, localization, data fusion, and system control within the proposed system. In Sect. 4, we conduct corresponding experiments using miniaturized system to evaluate the implemented system. In Sect. 5, we provide a summary of the paper as a whole and discuss the future plan.

2 Related Works

2.1 Vehicle-Infrastructure Cooperation System

Vehicle Infrastructure Cooperation Systems (VICS) represents the next generation of transportation system. Leveraging wireless communication and Internet of Things (IoT) technologies, VICS enables dynamic information exchange among vehicles, roads, and pedestrians, fostering collaborative efforts to enhance safety and efficiency within the transportation system. The collaboration among devices is often characterized by interoperability, which refers to the ability of devices to exchange information and utilize the exchanged information. Current research focuses on improving the collaboration among VICS devices, and a brief overview is provided below.

Hussain et al. [4] analyzed interoperability strategies in the context of vehicular networks, highlighting the trade-offs such as increased execution time and decreased processing speed. They also emphasized the lack of unified interconnection standards for VICS and suggested exploring middleware solutions from the IoT domain, without specifying the research details [5]. Datta et al. [6] identified the absence of unified standards as a hindrance to interoperability among vehicles, roads, and pedestrians. They proposed leveraging open standards like SenML and oneM2M to enhance interoperability in VICS. However, these approaches did not consider the impact of device heterogeneity and diverse communication methods on interoperability.

In terms of network architecture and protocols, Chekkouri et al. [7] proposed an architecture that integrates multiple heterogeneous networks, combining the characteristics of VANET and 4G LTE-A networks. Wang et al. [8] utilized 5G technology to introduce a low-latency, low-power consumption routing protocol for vehicle-road cooperation. However, they did not address the issue of heterogeneous networks.

Kaiwartya et al. [9] proposed a five-layer network architecture for VICS, comprising perception, coordination, artificial intelligence, applications, and services layers, to enhance interoperability. However, this approach may introduce additional network latency. While these studies enhance the quality and interoperability of vehicular networks from a specific network structure perspective, they do not incorporate an operating system. As a result, developers are required to perform multiple independent developments on individual devices, leading to increased development costs.

2.2 VICS Operating System

It must be mentioned in the beginning that there are very limited operating systems specific for VICS. But it is promising to utilize and extend single-agent vehicular operating system into VICS. Moreover, some researchers also focus on develop operating system designated for VICS. Therefore, in this section, both vehicular and VICS operating system will be surveyed.

Currently, vehicle operating systems mainly consist of embedded real-time operating systems for low-level control and middleware for providing development support. For embedded operating systems, there are several options available. QNX is a microkernel-based operating system known for its high security and low latency [10]. Real-time Linux systems offer stable performance, open-source nature, and a wide range of mature applications [11]. OpenHarmony that based on the LiteOS kernel, is a cross-platform, distributed system that enhances collaboration in vehicle-infrastructure cooperation system through the use of a software bus [12].

As for middleware, ROS (Robot Operating System) [13] and ROS 2.0 [14] are widely used in automotive systems. They rely on a publish-subscribe architecture for message passing [13]. However, using ROS on smaller embedded systems can be challenging, and it requires running on top of the underlying operating system, which may introduce compatibility issues and development difficulties. Additionally, Liu et al. [15] developed Zero, a middleware solution with high performance, reliability, and security. Zero achieved a 41% reduction in communication latency compared to ROS 2.0 and significantly reduced CPU core usage. Testouri et al. [16] proposed FastCycle, a lightweight middleware solution that meets the demands of modular, real-time, and secure communication for autonomous driving, although it still exhibits some message jitter. In OpenHarmony, the software bus mechanism is similar to ROS, making it a potential middleware solution for vehicle-infrastructure cooperation operating systems. Furthermore, OpenHarmony allows cross-platform and cross-communication-method communication, which can further enhance information transfer rates.

With regard to VICS operating systems, there are also integrated operating systems such as AIR-OS [17]. AIR-OS is an operating system designed for intelligent roads, featuring a vehicle-infrastructure-cloud integrated architecture. It comprises a real-time kernel layer, hardware abstraction layer, and middleware layer, while also providing rich upper-layer applications. AIR-OS shows great potential for development and advancement in this field.

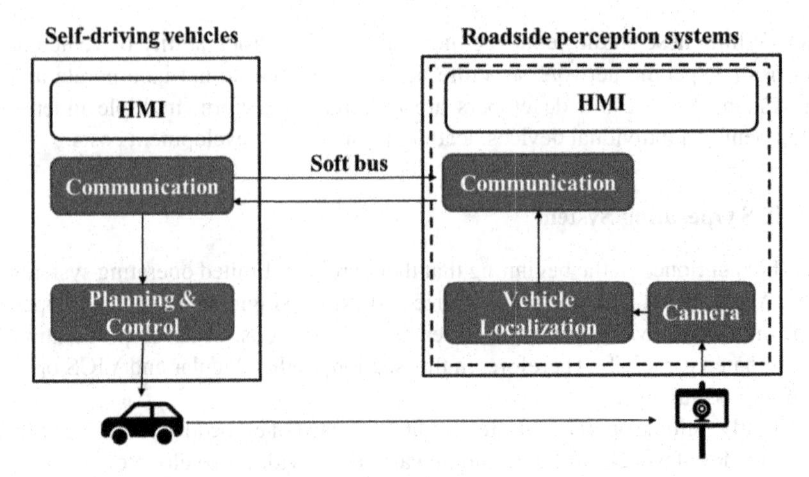

Fig. 1. The overall system framework

3 Methodology

3.1 System Framework

The overall system framework is depicted in Fig. 1. The roadside perception system consists of a camera module, vehicle localization module, and communication module based on distributed soft bus. The self-driving vehicle includes a communication module, and planning and control module.

For the roadside perception system, its primary purpose is to detect the vehicle's pose from the roadside. Firstly, the camera module captures images and stores them in the device. The storage address of the image is then transmitted to the vehicle localization module. This module analyzes the images using a program based on the OpenCV library to obtain the car's position, heading angle, and detection confidence. The data is packaged and sent to the autonomous vehicle via the communication module based on distributed soft bus.

For the self-driving vehicle, it first receives data from the roadside perception system through the communication module based on distributed soft bus. If there are multiple roadside perception units, the data is fused based on confidence levels and timeliness. The fused data is then sent to the control core board. The control core board analyzes the received position and heading angle data to determine the vehicle's desired motion. The motion is then controlled using the PWM peripheral.

By employing the communication and data exchange between the roadside perception system and the self-driving vehicle equipment, the system achieves vehicle pose detection and motion control.

3.2 Distributed Soft Bus for Data Communication

In the context of the Internet of Things (IoT), there are diverse requirements for data communication between devices. OpenHarmony, being a distributed operating system designed for various application scenarios, offers a range of data communication methods based on a software bus in its current version. These include RPC communication, distributed data objects, and distributed databases. However, the current versions of RPC and distributed data objects have limitations on the number of connected devices. Considering the scenario of VICS, where multiple roadside perception systems serve multiple vehicles, this design chooses to utilize the distributed database functionality provided by OpenHarmony as the communication module.

This approach involves creating a Key-Value (KV) type database to facilitate data communication between two devices. The operational mechanism is illustrated in Fig. 2. It allows for searching and modifying data entries using key-value pairs. After modification, the system can automatically synchronize the data and perform data maintenance. The upper-layer application only needs to access and manipulate data through the corresponding distributed database interface.

In this system, the road-side perception system and self-driving vehicles collaborate to maintain a distributed database. The road-side perception system continuously updates the values of perception data in the database, while the self-driving vehicles promptly retrieve the data after it changes.

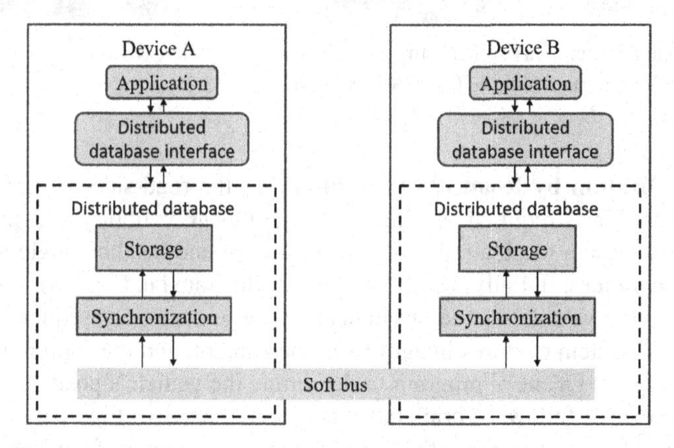

Fig. 2. The mechanism of the distributed database in OpenHarmony

3.3 Vehicle-Infrastructure Cooperation System

Hardware Platform. This paper uses a miniatured hardware platform to simulate real-world scenario and validate the capability, but both the hardware boards and algorithms were selected and designed based on the architecture of a real-size vehicle-infrastructure cooperation system.

For the self-driving vehicle, a scaled-down car equipped with an HI3516 device was utilized, as depicted in Fig. 3(a). The HI3516 board features a dual-core Cortex-A7 MP2 CPU clocked at 900 MHz, 1 GB DDR3, 8 GB eMMC, and includes a 2.4 GHz WIFI communication chip. The version of OpenHarmony is 3.1 Release in the HI3516 device. The primary mode of communication between devices is via 2.4 GHz WIFI. The scaled-down car utilizes PWM signals generated by the board to control the DC motors for motion.

Regarding the hardware for the road-side perception system, its physical representation is shown in Fig. 3(b) and mainly consists of an HI3516 device. The hardware specifications are the same as those used on the self-driving vehicle, and it is additionally equipped with a SONY Sensor IMX335 for image input, which has been calibrated by us.

For the experimental test site, the system employed a layout depicted in Fig. 3(c). The white circular frame within the site serves as the lane marker, enabling the vehicle to traverse along the lane as accurately as possible.

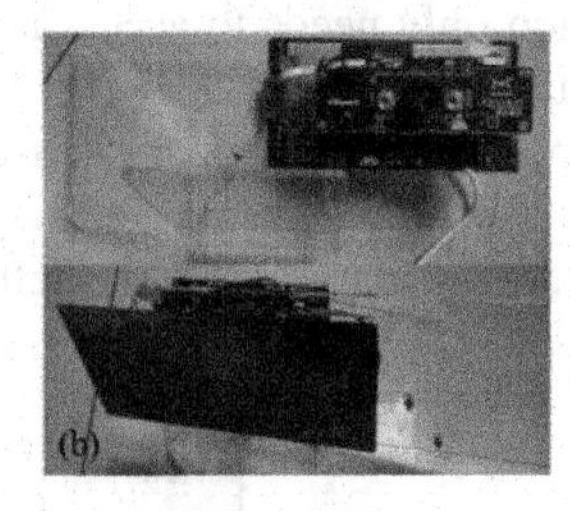

Fig. 3. Hardware system, (a) self-driving vehicle with controller, (b) one roadside perception system with processor and camera, (c) experiment site.

Vehicle Localization by Roadside Perception. In the road-side perception system, vehicle localization is required. In this study, the OpenCV library was ported to the OpenHarmony system to obtain the vehicle's position and heading angle using image processing techniques. Initially, a color marker, as illustrated in Fig. 4(a), was displayed on the self-driving vehicle's screen to indicate its pose status. Subsequently, the road-side perception system captures images from the camera, and the captured images are processed using the OpenCV program to determine the vehicle's position and heading angle. The program follows a brief flowchart, as depicted in Fig. 4(b). It begins by applying color filtering to separate the red and green components of the image. Gaussian filtering is then performed on each component, followed by edge detection. Finally, based on the geometric relationship of the patterns, the vehicle's pose and the confidence level of the detection results are computed.

This study considers the scenario where multiple road-side perception units jointly detect the pose information of the same self-driving vehicle. To address this, a data fusion scheme is designed. The scheme considers two key criteria: detection accuracy and real-time performance.

For detection accuracy, this paper utilizes a similarity measure between the captured image of the vehicle's screen display and the image shown in Fig. 4(a) to assess the accuracy of the detection results. The program calculates the Hu moments of the captured image using the built-in libraries of OpenCV, and calculates the Hu moments of the image shown in Fig. 4(a). The distance between the two sets of Hu moments is then calculated to obtain an indicator of detection accuracy. The formula for calculating this indicator

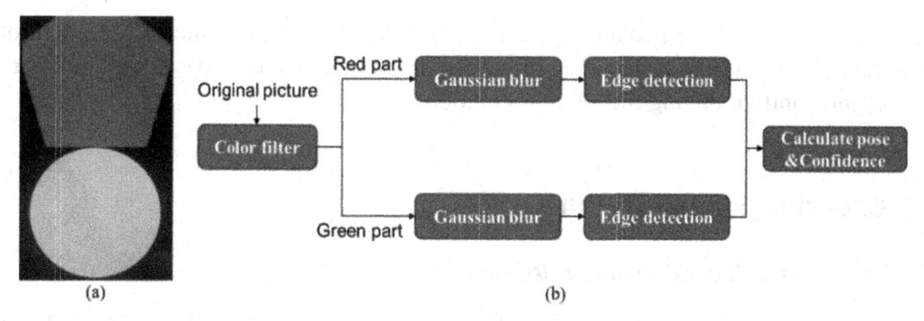

Fig. 4. Vehicle Localization by Roadside Perception program (a) is the picture displayed on the vehicle's screen, (b) is the flowchart of OpenCV pose detection program.

is shown in Eq. (1):

$$C^{pic} = \frac{1}{\sum_{i=1}^{7}\left|\frac{1}{sign(h_i^A)\cdot log(h_i^A)} - \frac{1}{sign(h_i^A)\cdot log(h_i^A)}\right|} \tag{1}$$

where C^{pic} is the confidence level of the geometric accuracy, h_i^A and h_i^B represent the values of the Hu moments of the standard image and the detected image, respectively.

For real-time performance, this paper considers the received information as the starting point. Initially, a weight value of 5 is assigned. After each utilization of the data, the weight value is reduced by 1 until it reaches 0 or until the data from that specific device is updated. This process yields another confidence level C^{time} that represents the real-time performance.

The overall data fusion weight formula is represented as Eq. (2):

$$D_{fuse} = \frac{\sum_{i=1}^{m} D_i \cdot C_i^{pic} \cdot C_i^{time}}{C_{max}^{time} \sum_{i=1}^{m} C_i^{pic}} \tag{2}$$

where D_{fuse} represents the fused result, m represents the total number of devices, D_i represents the perception result, C_i^{pic} represents the geometric accuracy confidence level, C_i^{time} represents the real-time performance confidence level, and C_{max}^{time} represents the maximum value of the real-time performance confidence level, which is set to 5 in this paper.

Vehicle Motion Planning and Control. After receiving data from multiple road-side perception units and performing fusion, the self-driving vehicle begins calculating the difference between its current pose and the target pose. This difference is used to determine the desired heading angle, which is then compared to the vehicle's current heading angle. Based on this comparison, decisions are made regarding the vehicle's actions, such as turning left or right, or moving forward.

In this study, the road-side perception units only handle the perception tasks, because the remaining computational resources are limited. Therefore, the system only transfers perception data, while the planning and control aspects are handled by the self-driving vehicle itself. However, if the road-side perception system has abundant computational resources, it can perform planning and control of self-driving vehicles. The control instructions can then be transmitted through a distributed database based on a software bus, significantly reducing the vehicle's burden.

4 Experiments and Results

4.1 Perspective Transformation Results

To ensure the fusion of perception results from multiple road-side perception units, it is crucial to align the images onto a common plane in a consistent direction. Due to practical limitations encountered in this experiment, we adopted a method that involves detecting the vertices of reference markers and applying perspective transformation to the images. This technique effectively aligns the images and facilitates accurate fusion. The visualization of this process is presented in Fig. 5.

The results depicted in Fig. 5 indicate the program's ability to successfully detect and extract the four vertices of the blue quadrilateral. Moreover, by accurately repositioning the vertices based on the orange points, the program effectively applies perspective transformation to the images, converting them from an oblique view to a bird's-eye view perspective. These outcomes provide a solid basis for subsequent stages of the program, enabling further processing and analysis.

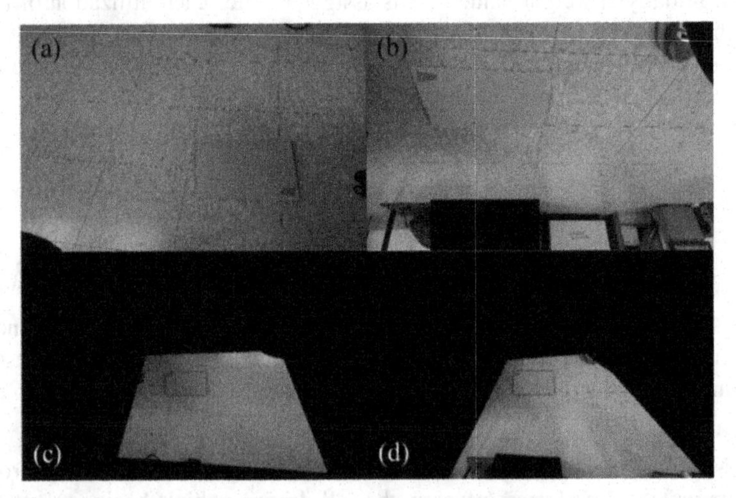

Fig. 5. The result of perspective transformation, (a) and (b) represent the raw images acquired from distinct road-side perception units, whereas (c) and (d) depict the transformed outcomes following the application of perspective transformation.

4.2 Perception Results

Based on the perception system mentioned earlier, we initially validated the program on a personal computer. The results of this validation are presented in Fig. 6, which shows the detection of the vehicle's x-coordinate as 359, the y-coordinate as 265, and the heading angle as 90°.

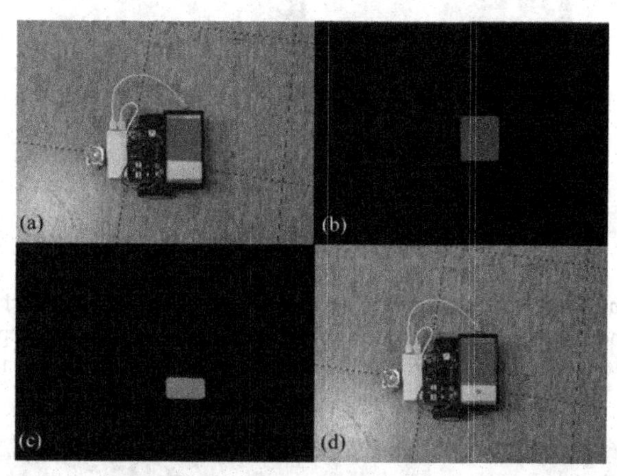

Fig. 6. Perception results of vehicle on the personal computer, (a) the original image, (b) the detected green region, (c) the detected red region, and (d) the center points of the detected red and green regions separately.

Then, we proceeded to deploy it on the road-side perception system for real-world validation. The outcomes of this practical testing are depicted in Fig. 7. The detection results were shown in the second row on the screen. Figure 7(a) indicate that the pose of the vehicle is located at x-coordinate 452, y-coordinate 164 (in pixel coordination), with a heading angle of 148°, while Fig. 7(b) indicate that the pose of the vehicle is located at x-coordinate 294, y-coordinate 207, with a heading angle of 184°.

To improve the integrity of the system, we proceeded to detect the lane lines and vehicle path. To conserve space, we divided the lane lines and path into multiple line segments. The detection results, as shown in Fig. 8, indicate relatively satisfactory performance.

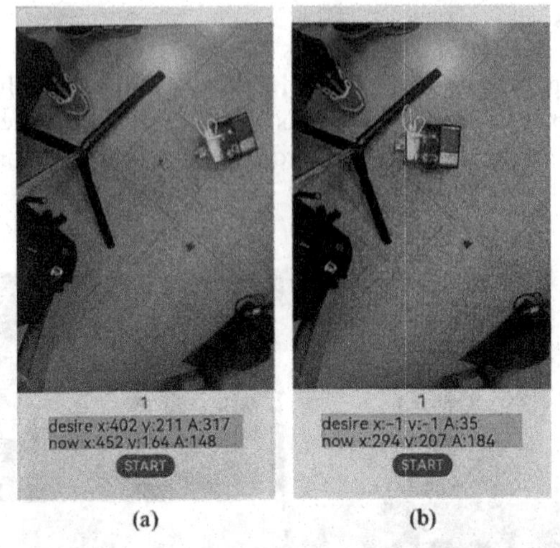

Fig. 7. Perception results of vehicle on the road-side perception system, which is a device with OpenHarmony. The results are shown in the second row on screen. (a) and (b) represent the screen photos of HI3516 on the road-side perception system, which display perception results.

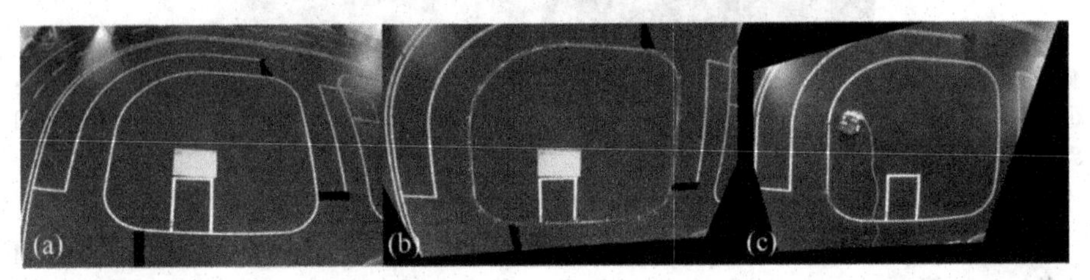

Fig. 8. Perception results of lane lines and vehicle path on the road-side perception system, (a) the original image captured at the site, (b) the detected results of the circular lane lines, and (c) the vehicle's planned motion path.

4.3 Motion Planning and Control Results

After completing the perception experiments, we proceeded with the motion planning and control experiments. The self-driving vehicle was instructed to follow the trajectory depicted in Fig. 8(c). The resulting trajectory of the vehicle is illustrated in Fig. 9(a), where the red line is the planned trajectory, the green dots indicate the target points, and the yellow curve is the actual movement of the vehicle. The variations in the vehicle's distance from the target points are shown in Fig. 9(b), where the blue curve is the distance between the vehicle and the subsequent target point, and the orange indicator points to the distance threshold for switching to the next target point. Once the vehicle reaches a close proximity to the target point, as indicated by the orange line in Fig. 9(b), the vehicle will transition to the next target point.

From the result, it can be observed that although the vehicle eventually gets close to the target point, the trajectory during the process is not particularly ideal. Through research, the prominent issues highlighted by this system are latency and computational

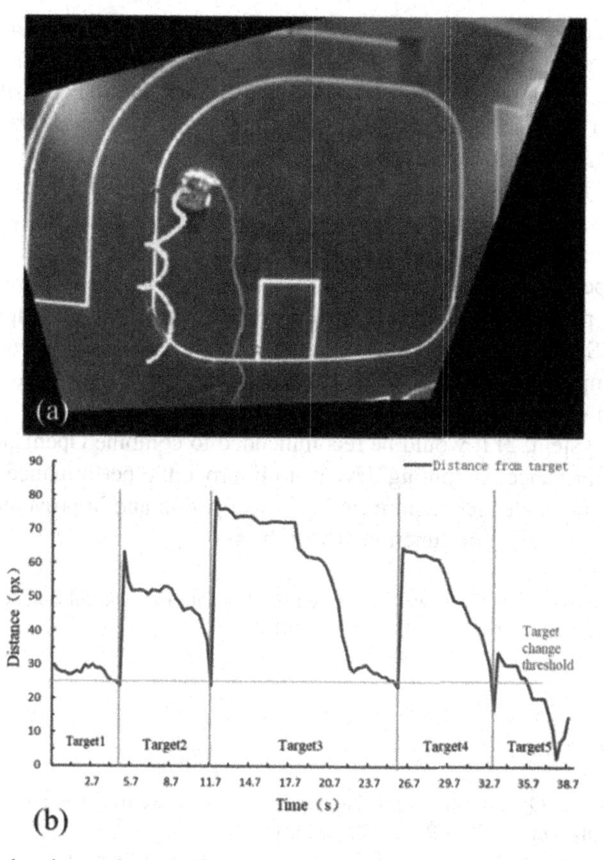

Fig. 9. Motion planning and control results. (a) the actual trajectory of the vehicle. (b) the variations in the vehicle's distance from the target points.

resource constraints. Regarding latency, measurements indicate an average communication delay of 44.15 ms. This delay may be attributed to the hardware limitation of the system, which only supports 2.4 GHz WiFi, thereby restricting the communication speed. In terms of computational resources, the system is equipped with a dual-core 900 MHz CPU without an image processor, resulting in significant processing time for running OpenCV-related programs. Testing has shown an average processing time exceeding 300ms. These two factors contribute to insufficient real-time performance of the overall system, necessitating further hardware improvements in subsequent iterations.

5 Conclusions

Vehicle-Infrastructure Cooperation System (VICS) is gaining increasing attention, and the issues of device heterogeneity and device interoperability are also receiving more focus. The adoption of distributed operating systems, represented by OpenHarmony, provides a systematic solution to address these problems.

In this paper, we have implemented OpenHarmony for demonstration in a miniatured setup of VICS and propose a comprehensive system framework based on OpenHarmony. We investigate a communication method using the OpenHarmony distributed software bus and design the communication module of the system based on the OpenHarmony distributed database. Furthermore, we utilize visual methods with OpenCV to implement perception and localization functionalities in the system. Additionally, we determine the overall control scheme for the system. Based on these contributions, we also conduct separate tests on each module, resulting in the overall system achieving a basic, complete, and smooth operation.

The focal point of this paper is the successful application attempt of OpenHarmony on VICS, where the entire system runs smoothly. However, there are still areas that require improvement: 1) It would be beneficial to port OpenHarmony to devices equipped with efficient communication chips to further enhance the real-time performance of the system. 2) It would be recommended to combine OpenHarmony devices with high-performance computing devices to improve the performance of the system. These remaining challenges require further investigation and improvement to enhance the overall performance and functionality of the system.

Acknowledgements. This work was supported by the National Natural Science Foundation of China under Grant 62203294, and Grant U22A20100.

References

1. Shladover, S.E.: Opportunities and challenges in cooperative road vehicle automation. IEEE Open J. Intell. Transp. Syst. **2**, 216–224 (2021)
2. Bloom, G., Alsulami, B., Nwafor, E., Bertolotti, I.C.: Design patterns for the industrial internet of things. In: 2018 14th IEEE International Workshop on Factory Communication Systems (WFCS), pp. 1–10. IEEE (2018)
3. OpenHarmony. https://www.openharmony.cn. Accessed 11 Aug 2023
4. Hussain, S.M., Yusof, K.M., Hussain, S.A.: Interoperability in connected vehicles—a review. Int. J. Wireless Microw. Technol. **9**(5), 1–11 (2019)
5. Hussain, S.M., Yosof, K.M., Hussain, S.A.: Interoperability issues in internet of vehicles- a survey. In: 2018 3rd International Conference on Contemporary Computing and Informatics (IC3I), pp. 257–262 (2018)
6. Datta, S.K., Harri, J., Bonnet, C., Da Costa, R.P.F.: Vehicles as connected resources opportunities and challenges for the future. IEEE Veh. Technol. Mag. **12**(2), 26–35 (2017)
7. Chekkouri, A.S., Ezzouhairi, A., Pierre, S.: A new integrated VANET-LTE-A architecture for enhanced mobility in small cells HetNet using dynamic gateway and traffic forwarding. Comput. Netw. **140**, 15–27 (2018)
8. Wang, X., Weng, Y., Gao, H.: A low-latency and energy-efficient multimetric routing protocol based on network connectivity in VANET communication. IEEE Trans. Green Commun. Netw. **5**(4), 1761–1776 (2021)
9. Kaiwartya, O., et al.: Internet of vehicles: motivation, layered architecture, network model, challenges, and future aspects. IEEE Access **4**, 5356–5373 (2016)
10. Hildebrand, D.: An architectural overview of QNX. In: USENIX Workshop on Microkernels and Other Kernel Architectures, pp. 113–126 (1992)

11. Sivakumar, P., Devi, R.S., Lakshmi, A.N., Vinoth Kumar, B., Vinod, B.: Automotive grade Linux software architecture for automotive infotainment system. In: 2020 International Conference on Inventive Computation Technologies (ICICT), pp. 391–395. IEEE (2020)
12. Cao, Q., Abdelzaher, T., Stankovic, J., He, T.: The LiteOS operating system: towards unix-like abstractions for wireless sensor networks. In: 2008 International Conference on Information Processing in Sensor Networks (IPSN 2008), pp. 233–244. IEEE (2008)
13. Quigley, M., et al.: ROS: an open-source Robot Operating System. In: ICRA Workshop on Open Source Software, Kobe, Japan, p. 5 (2009)
14. Macenski, S., Foote, T., Gerkey, B., Lalancette, C., Woodall, W.: Robot operating system 2: design, architecture, and uses in the wild. Sci. Robot. **7**(66), eabm6074 (2022)
15. Liu, W., Jin, J., Wu, H., Gong, Y., Jiang, Z., Zhai, J.: Zoro: a robotic middleware combining high performance and high reliability. J. Parallel Distrib. Comput. **166**, 126–138 (2022)
16. Testouri, M., Elghazaly, G., Frank, R.: FastCycle: a message sharing framework for modular automated driving systems. arXiv preprint arXiv:2211.15495 (2022)
17. AIROS. https://www.apollo.auto/zhiluos. Accessed 11 Aug 2023

A Practical Byzantine Fault Tolerant Algorithm Based on Credit Value and Dynamic Grouping

Haonan Zhai and Xiangrong Tong(✉)

College of Computer and Control Engineering, Yantai University, Yantai, China
txr@ytu.edu.cn

Abstract. The Practical Byzantine Fault Tolerance (PBFT) consensus algorithm plays a pivotal role in the efficiency of blockchain networks. Despite several enhancements, optimized PBFT still grapples with issues, notably incomplete credit assessments, and uneven node groupings. While existing trust models predominantly focus on objectively evaluating nodes within the system, the subjective assessment of inter-node interactions remains relatively underexplored. Furthermore, post-grouping, the reduction in nodes within each group leads to a diminished capacity for Byzantine nodes. This poses a notable challenge to the algorithm's effectiveness. In this paper, a practical Byzantine fault-tolerant algorithm (CDGPBFT) based on credit value and dynamic grouping is proposed to solve these problems. First, credit attributes are extended for nodes. The design of the node model introduces a dual judgment mechanism, including dynamic evaluation of node interactions and static system evaluation. Subsequently, nodes are categorized into three groups based on their credit scores: high, medium, and reserve node groups. Next, formulas are designed to enhance the master node election scheme and simplify the consensus process. Finally, a reward function based on node computational resources, network bandwidth, and storage capacity is proposed. Nodes that effectively complete the consensus process receive tokens as rewards, thus promoting accurate and rapid participation. Experimental validation shows that the enhanced algorithm has lower latency and communication overheads compared to PBFT, while also improving message throughput.

Keywords: PBFT · Dynamic · Group · Credit · Reward

1 Introduction

Blockchain technology, characterized by its reliability, decentralization, tamper resistance, and anonymity, provides a robust solution to the persistent chal-

Supported by the National Natural Science Foundation of China (62072392,61972360), the Major Innovation Project of Science and Technology of Shandong Province (2019522Y020131), the Natural Science Foundation of Shandong Province (ZR2020QF113) and the Yantai Key Laboratory: intelligent technology of high-end marine engineering equipment.

F. Sun and J. Li (Eds.): ICCCS 2023, CCIS 2029, pp. 272–291, 2024.
https://doi.org/10.1007/978-981-97-0885-7_23

lenge of Byzantine attacks [1]. To enhance user privacy and data monitoring, Vitalik Buterin, the creator of Ethereum, introduced the concept of a federated blockchain [2,3]. In a federated blockchain, access is restricted to authorized nodes, thereby fortifying the network against potential attacks [4,5]. Crucial to blockchain technology are consensus algorithms, with PoW, PoS, DPoS, BFT, and PBFT emerging as prominent mechanisms [6,7].

While PBFT remains a pivotal consensus algorithm widely embraced by federated blockchains, it encounters certain problems. The efficacy and security of the PBFT algorithm hinge on the integrity and precision of participating nodes, necessitating an elevated degree of node honesty. The susceptibility of certain nodes to furnish deceptive information or execute spoofing attacks poses a credible threat to the consensus process. Instances of node failure or malicious incursions precipitate data incongruity or impede consensus attainment. Consequently, several researchers have endeavored to introduce a trust model that enhances the PBFT algorithm, effecting a resolution through the calibration of node scores via a static trust model. The PBFT approach, lacking the mechanism of grouping, entails a scenario wherein all nodes must engage in comprehensive communication upon each consensus iteration. As the node count escalates, this communication overhead experiences a dramatic surge, thereby conferring complexity and protraction upon the consensus process. In this context, the introduction of grouping to the PBFT framework emerges as a seminal enhancement, particularly within expansive and high-throughput systems. Traditional PBFT lacks a predetermined or fixed reward function. The purpose of a reward mechanism is to motivate nodes to actively participate in the consensus process by providing them with appropriate incentives or rewards. This ensures that their participation is meaningful and valuable.

The introduction of trust mechanisms and grouping techniques into PBFT pose significant challenges. Existing trust evaluation models, reliant on static attributes derived from historical node behavior, fall short in capturing dynamic interactions crucial for consensus performance. To address this, a more comprehensive trust model must be developed, encompassing both static system assessment and dynamic inter-node evaluations. The grouping implementation introduces the challenge of load balance, particularly in distributing Byzantine nodes. This uneven distribution can lead to performance disparities among groups and increased latency. Thus, the grouping design prioritizes allocating potential Byzantine nodes to a specialized reserve node group based on trustworthiness, ensuring optimal consensus performance and efficiency in the remaining groups. Concurrently, the reward function's architecture requires meticulous consideration to ensure fairness and equity. It should seamlessly align with the overarching goals of the consensus algorithm and system, factoring in resource utilization and system sustainability.

To address the above problems and challenges, we propose an improved practical Byzantine fault-tolerant algorithm based on credit values and dynamic grouping(CDGPBFT). Initially, we introduce a novel node evaluation model, amalgamating dynamic inter-node assessments with static system evaluations.

This model categorizes nodes into three distinct groups based on their credit values. Following grouping, the algorithm is refined to streamline and enhance the consensus process. Additionally, we devise a novel primary node election scheme exclusively within the high-credit node group, thereby mitigating the likelihood of selecting Byzantine nodes as primary nodes. Lastly, we implement a reward function to incentivize nodes, allocating tokens to those that execute consensus accurately and swiftly. This allocation is contingent on considerations of computational resources, network bandwidth, and storage capacity. The main contributions of this research are listed below:

1. Our approach introduces a credit attribute, which involves both static and dynamic evaluations, accounting for nodes' subjectivity within the group. This enables the categorization of nodes into three groups based on trust. The credit model's implementation facilitates the timely detection of Byzantine behavior among nodes, effectively resolving the issue of node consistency in distributed environments.
2. Enhancements in primary node election involve a refined process exclusively within high credit groups. This selection amalgamates node latency, historical behavior, and transaction completion, markedly reducing the likelihood of Byzantine nodes assuming primary roles.
3. A pioneering reward function is introduced to recognize nodes' contributions in computational resources, network bandwidth, and storage capacity during the consensus process. This design not only ensures precise consensus execution by nodes but also cultivates greater engagement, fortifying the security of the system-wide consensus mechanism.

It is experimentally verified that CDGPBFT outperforms PBFT, GPBFT [8] and CPBFT [9] algorithms in terms of delay, message throughput and communication overhead.

2 Related Works

The Byzantine Fault Tolerance Algorithm represents a pivotal avenue of inquiry within the domain of distributed systems. Its core objective revolves around achieving consensus even in the presence of malfunctioning nodes. The Practical Byzantine Fault Tolerance (PBFT) algorithm underscores the fundamental tenets of trustworthiness and security within distributed systems. This bears profound significance for applications spanning the realm of blockchain technology and beyond. In the context of blockchain technology, the introduction of a consensus mechanism serves to confront the intricate challenges pertaining to trust and security inherent in decentralized networks. With the rapid development of blockchain in recent years, Blockchain can be categorized into public chain, coalition chain, and private chain [10]. These three kinds of blockchains have their own consensus algorithms, and public chain consensus algorithms can be categorized into PoW, PoS, and improvements based on PoW and PoS. For example, POA [11], POSV [12]. Coalition chain consensus, in general, can be divided into two categories: byzantine fault-tolerant consensus and non-byzantine fault-tolerant consensus, Non-byzantine fault-tolerant consensus is also known as fault-tolerant type of consensus, which refers to the ability to guarantee the reliability of the system in the event of a system downtime problem, but when there is a byzantine node, it is not possible to guarantee the reliability of the blockchain system, for example, PAXOS [13], RAFT [14]; byzantine fault-tolerant consensus refers to the fact that the nodes forge and respond maliciously to broadcasted messages, but the system is able to allow the existence of a certain number of Byzantine nodes, for example, PBFT. Coalition chains, unlike the other two chains, are used only within organizations and institutions, and most of them use consensus such as PBFT and RAFT.

However, they still have some problems in terms of energy consumption and security. In contrast, the improved PBFT algorithm solves the problems of consensus and security in a different way in terms of improving the consensus structure, literature [15] improves the accuracy of node classification by applying the C4.5 grouping method and introduces the integral mechanism, which effectively alleviates the inefficiency of the original PBFT consensus algorithm consensus in the case of a large scale of the number of nodes. Literature [16] proposes an arbitrarily scalable distributed model that combines the PBFT consensus algorithm with the P2P trust computation model, which is able to improve the fault tolerance and trustworthiness of the system. In addressing the communication complexity in the system literature [17] evaluates the performance of Hyperledger fabric v0.6 with PBFT consensus, this consensus algorithm is able to contribute a large amount of throughput to the system, but with the increase in the number of nodes, the performance of the system tends to decrease. Literature [18] enhanced the fault tolerance performance of the algorithm by exceeding the maximum fault tolerance by one-third. Literature [8] proposed GPBFT, which utilizes the geographic information of fixed IoT devices to reach consensus, reducing the overhead of verifying and recording transactions. Literature [9] proposes CPBFT, which introduces credit attributes, changes the architecture, and reduces the consensus steps, thereby reducing the amount of data transfer and increasing throughput.

3 CDGPBFT Algorithm Design and Implementation

3.1 Grouping Strategy

The nodes are categorized into high credit group, medium credit group, and standby node group according to their credit value, where the high credit group is the group where the nodes with the highest credit value are located. The nodes in the reserve node group do not participate in the consensus and their consensus results are not adopted, but the credit scores are updated in each round of consensus [19,20]. When the number of nodes increases, the system triggers the detection mechanism and if the total number of nodes can be divided by 3, nodes are distributed equally in each group. If the total number of nodes

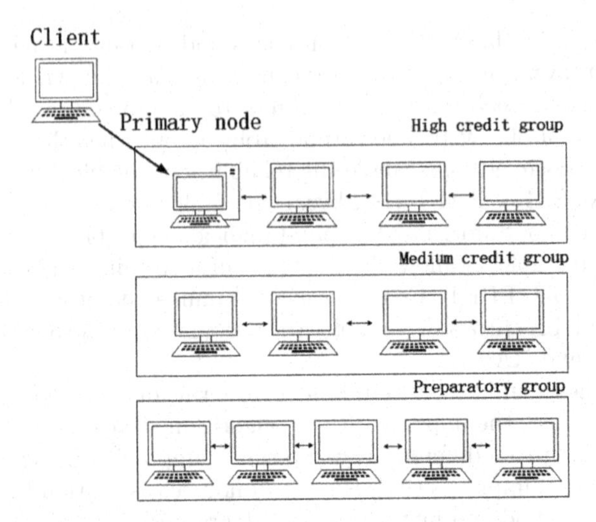

Fig. 1. CDGPBFT grouping architecture diagram

cannot be divided by 3, the extra nodes will be placed in the spare node group [17,21].

The model schematic of CDGPBFT is shown in Fig. 1. The nodes are arranged in descending order based on the trust value and are sequentially placed in three groups. Newly added nodes are stored in the reserve node group. Figure 2 simulates the case of 13 nodes, 4 nodes in the high credit node group and 4 nodes in the medium credit node group, The nodes cannot be divided equally, so there are 5 nodes in the reserve node group.

3.2 Criteria for Assessing Creditworthiness

The credit value, an essential metric for assessing a node's reliability within the blockchain network [22], is derived from a combination of the node's historical consensus performance and its initial configuration. Variables such as network bandwidth, memory, and processor settings during node initialization influence this value. Typically initialized between 60 to 75, it directly mirrors the robustness of the server configuration and network environment. Consequently, a more potent server setup and favorable network conditions translate to a higher initial credit value for the node. It's worth noting that these credit values undergo periodic updates throughout the ongoing consensus process.

The key determinant for node grouping lies in their respective credit values. Nodes endowed with higher credit values are identified as non-Byzantine entities and are subsequently assigned to either the high or medium credit groups. Conversely, nodes with relatively lower credit values fall into the Byzantine category and are directed to the backup node group. Byzantine nodes hold the potential to deliberately introduce false consensus outcomes, thereby disrupting the network's regular operations. As a result, the process of primary node selection and the ensuing consensus protocol exclusively involve nodes with elevated credit values [23].

Node evaluation comprises two pivotal facets. The first involves a static assessment, entailing an evaluation of nodes based on their historical consensus behavior and their level of involvement in the consensus process. The second facet pertains to dynamic evaluation, where nodes within the same group are subject to ongoing evaluation of each other throughout the consensus process.

The trust value evaluation formula (1) for node i is shown below:

$$Trust_i = \alpha TS_i + \beta TD_i^t \tag{1}$$

where $Trust_i$ denotes the current trust value of node i. The trust value consists of two parts: the static evaluation and the dynamic evaluation of node i. TS_i is the historical consensus behavior and participation of node i to evaluate it. TD_i^t is the interaction behavior between node i and other nodes to evaluate it. α and β are the weights corresponding to the static and dynamic evaluations, respectively. It is important to note that the scope of dynamic evaluation of nodes is within the subgroup in which they are divided.

The static trust value of node i is calculated based on the historical behavior of the node during the consensus process, which can be expressed by the following Eq. (2):

$$TS_i = \gamma B_i + \theta P_i \tag{2}$$

where B_i is the historical behavior factor of node i in the formula process,P_i is the historical participation degree factor of node i in the consensus process, γ and θ are the corresponding weights of each.

The historical behavior factor is the degree of influence of node i historical behavior on its trust value. The historical behavior factor can be expressed as Eq. (3):

$$B_i = \frac{\varphi R_{is} - \varpi R_{if}}{R_i} \tag{3}$$

where R_{is} denotes as the number of times node i has shown honest behavior in the ever consensus process.R_{if} denotes the number of times that node i had malicious behavior in the consensus process. φ and ϖ is the corresponding weights, $\varphi + \varpi = 1$. R_i denotes the number of times node i has ever participated in the consensus process, $R_i = R_{is}+R_{if}$. From this formula, it can be seen that the more times a node has made honest behavior, the greater the impact on increasing the trust value of the node.

The degree of node participation is to evaluate whether the node participates in the consensus or not, expressed as Eq. (4):

$$P_i = \frac{R_i}{R} \tag{4}$$

where R denotes the total number of consensus. R_i denotes the number of times node i participates in the consensus process.

The dynamic evaluation of node i is the mutual evaluation between nodes. The dynamic evaluation of node i can be expressed as Eq. (5):

$$TD_i^t = \begin{cases} \dfrac{\sum_{J=1}^{\frac{N}{3}} \dfrac{\sum_{m=l}^{l} f(m) TD_{ij}^{tm}}{l}}{\frac{N}{3}} & l \neq 0 \\ 0 & l = 0 \end{cases} \tag{5}$$

where node i is not equal to node j, N is the total number of nodes in the system, $N/3$ denotes the number of nodes in a single subgroup, which indicates that the dynamic evaluation is only performed within the subgroup, t denotes the period in which the client sends a request message, l denotes the number of consensus in period t, m is the mth consensus, TD_{ij}^{tm} is the rating of node j on node i after node j and node i interact in the mth consensus. $f(m)$ denotes the time decay factor, which can be expressed as Eq. (6):

$$f(m) = \sigma^{1-m}(0 < \sigma < 1, 1 \leq m \leq l) \tag{6}$$

where, $f(m)$ represents the distance between the consensus count and the current consensus round, the farther away from the current consensus round, the greater the time decay and the smaller the impact on the dynamic evaluation of the node.

3.3 Primary Node Rotation Mechanism

The election of the primary node takes place only in the high credit node group because the nodes in the high credit node group have high reliability [17]. We elect the primary node by a combination of the node's latency, transaction completion rate, and historical credit value.

The delay in the exponential response of a node to various information processes is expressed as Eq. (7):

$$D(i) = \frac{\sum_{l=1}^{m} \sum_{j=1}^{g} \left[1 - (\frac{d_{ijl}}{d_{max}})^3\right]}{gm} \tag{7}$$

where d_{ijl} denotes the delay of the lth transaction from node i to node j; d_{max} denotes the maximum delay allowed by the exchange. m is the number of transactions that node i has made, and g is the number of nodes that node i has interacted with. PBFT is an asynchronous Byzantine fault tolerant algorithm, the communication delays between the nodes can be different, so the different delays between the nodes are considered comprehensively.

Transaction completion rate is the percentage of nodes successfully participating in each transaction after entering the network and is expressed as Eq. (8):

$$T(i) = \frac{\sum_{i=1}^{m} \mu_i}{n} \tag{8}$$

where n is the total number of system transactions and m denotes the number of transactions completed by node i. μ is the identifier of whether the transaction is successful or not, with a successful transaction μ of 1 and a failed transaction μ of -1. Both the promotion effect of successfully completed transactions on the node and the adverse effect on the node by affecting the normal conduct of transactions are considered, which can better distinguish the credit value of the node. Influence of historical credit values. The credit status of the current node is influenced by the historical credit value, and is expressed as Eq. (9):

$$C(i) = zC(i-1) \tag{9}$$

The coefficient z indicates the degree of historical state influence.

The combined final score for the primary node election is shown below in Eq. (10):

$$P(i) = \frac{1}{3}(xD(i) + yT(i) + C(i)) \tag{10}$$

x is the weight of the node's transaction latency, y is the weight of the node's own completed transactions, and $x + y + z = 1$. The credit model intuitively reflects the node's performance in the consensus. If a node has small latency, a high transaction completion rate, and good historical credit value, it is highly credible; conversely, if a node has large latency, a low transaction completion rate, and poor historical credit value, it is less credible.

3.4 Improved Consensus Process

1. Request Stage
   Client c sends an execution request <REQUEST,m,c,t >to the primary node of each group. REQUEST denotes the message sent by the client, m is the specific message, c is the client address, and t is the timestamp.
2. Pre-preparation stage
   After the client sends the request, the pre-preparation phase is executed. The primary node sends a pre-preparation message to the rest of the nodes in the group. The message format is <PRE-PREPARE,v,n,d,m,p >, where v is the view number, n is the sequence number of the requested message, d is the hash value of m, m is the specific message, and p denotes the primary node. When the node accepts the message an acknowledgement will be made:
   (a) Whether the view number and primary node number are consistent with local.
   (b) Whether the client's request is consistent with the preparation phase.
   (c) Whether the hash values of d and m are the same.
   (d) Whether the message number n of the prep phase is in range.
   (e) Whether the message number n received by the node is duplicated.
3. Preparation Stage
   All nodes in the group receive a pre-preparation message broadcast by the primary node and check it. After the verification is complete, each node forwards the message to the other nodes in the group. and writes the readiness

message to the local message log. The nodes will enter the preparation phase. The message format <PREPARE, v, n, d, i>, where v is the current view number, n is the serial number of request m in the current view, d is the summary of m, i is the node serial number, and p is the primary node.

4. Response Stage
When the node checks that the ready message is true, it enters the reponse phase. At the beginning of the reponse phase, the node verifies the message and returns the result to the primary node.
5. Reply Stage
Consensus is considered reached when the master node collects a sufficient number of confirmations from the group. Primary nodes will send the confirmation results to the client and update the scores and trustworthiness of all nodes at the same time.

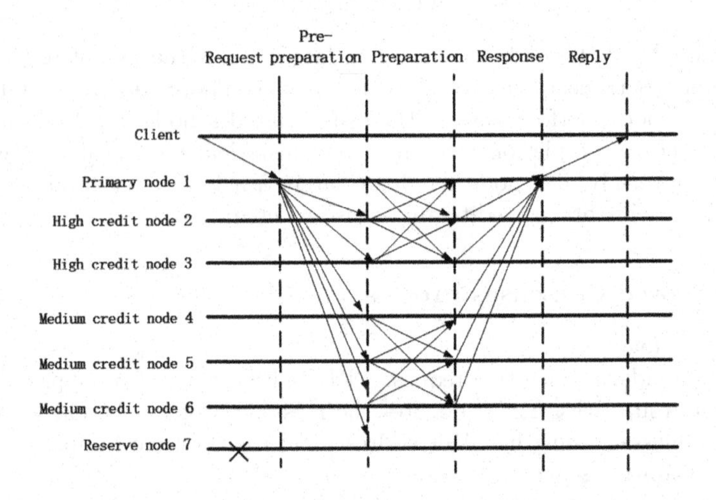

Fig. 2. CDGPBFT consensus process

Figure 2 shows the CDGPBFT consensus process, with nodes 1, 2, and 3 in the high credit group, nodes 4, 5, and 6 in the medium credit group, and node 7 in the standby node group. Commonly used symbols as well as shown in Table 1 Algorithm 1.

Table 1. Table of common symbols

Symbols	Definition
o	Client request operation
t	Timestamp
c	Client
m	Client request
i	Node number
v	View Number
n	Request Serial Number
d	Request hash value
f	Number of Byzantine nodes
σ_c	Signing of messages by the client
σ_p	Signature of the message by the primary node
σ_i	Node signature on messages

Algorithm 1 CDGPBFT algorithm

1: While <REQUEST,o,t,c >= TRUE do
2: broadcast <<PREPREPARE,v,n,d,>σ_p,m>
3: **if** <PREPREPARE>= TRUE **then**
4: broadcast <PREPARE,v,n,d,i>σ_i;
5: receive <PREPARE,v,n,d,i>σ_n;
6: **else** do nothing;
7: **if** prepared(m,v,n,i)=TRUE **then**
8: $result\{\} = result\{\} \cup result\{i\}\ \sigma_m$
9: **else** do nothing;
10: **if** $\exists \sum (result\{\text{i}\} \in result\{\})>2f+1$ **then**
11: Primary nodes broadcast the result to client.
12: **else** do nothing;
13: UpdateConNodes() and ReGroup();
14: END

3.5 Group Adjustments

In this study, we present a dynamically adaptable grouping scheme. This scheme involves updating nodes' credit values upon concluding each round of consensus and subsequently adjusting the grouping structure based on these values. To elaborate, the scheme encompasses the subsequent two steps:

1. Following the completion of each consensus round, nodes' credit values undergo updates. Within their respective groups, if a node's credit value surpasses that of all nodes in the current group, the node is promoted to a higher grouping level. However, if the node already resides within the highest

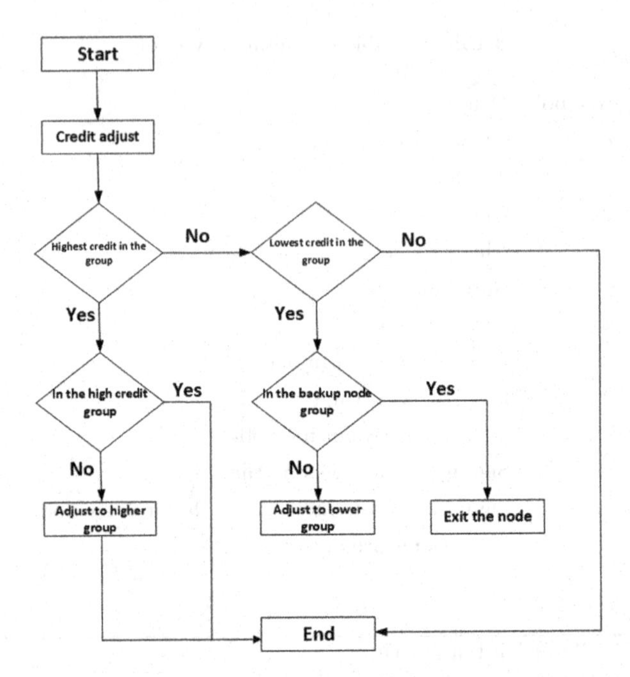

Fig. 3. Group adjustments

credit group, no further adjustments are enacted. This dynamic process facilitates an upward shift of nodes with superior credit values to higher levels of grouping (Fig. 3).

2. Lowering the node with the least credit value to the subsequent grouping tier is performed as follows: Within the node's existing group, if its credit value falls below that of all other nodes in that group, the node is demoted to the subsequent level within the node grouping. Should the node already reside within the standby node group, it is consequently removed. To avert excessive node deletions that could lead to a diminished node count, a strategy is implemented wherein one node is eliminated every 20 grouping updates.

4 Reward Functions

4.1 Feasibility Analysis of the Reward Function

Practical Byzantine Fault Tolerance (PBFT) serves as a distributed consensus algorithm, ensuring the coherence of nodes within a blockchain network. Within a PBFT network, nodes are required to engage in the consensus process and are rewarded upon successful consensus. The ensuing section presents a succinct design for a PBFT node reward function:

1. First, it is important to recognize that nodes play different roles in the consensus process. Therefore, the rewards of nodes should reflect the proportion

of their contribution to the consensus process. As an example, the workload of a node can be quantified by considering factors such as computational resources, network bandwidth, storage capacity, and related attributes provided by the node in the consensus process.

2. Subsequently, it becomes essential to establish foundational incentives that promote active engagement of nodes in the consensus process while concurrently safeguarding network security. As an illustration, a fundamental approach involves the allocation of a predetermined quantity of base tokens as rewards for each node within each consensus cycle.
3. Moreover, to avert potential system instability arising from a sparse node count, a reward function is employed to incentivize authorized organizations or institutions to allocate nodes to the system. In tandem with the credit value update mechanism crafted within this study, nodes with the lowest credit value within the system are systematically removed every 20 consensus rounds. This collaborative approach not only encourages organizations or institutions to proactively contribute to nodes but also guarantees a consistent presence of an adequate number of nodes participating in the consensus process. This, in turn, serves to enhance both the dependability and stability of the system.
4. Lastly, to uphold the principles of equity and transparency throughout the consensus process, it is imperative to establish a robust monitoring mechanism. This framework would promptly detect any instances of breaches or malicious attacks occurring during the consensus procedure and duly penalize the perpetrators. Consequently, the tokens or rewards allocated to the offending parties shall be subject to reduction or complete revocation.

To sum up, a complete PBFT node reward function should consider node workload, basic rewards, rewards for outstanding performance, penalty mechanism, and supervision mechanism.

4.2 Design of the Reward Function

The reward function takes into account the computational resources, network bandwidth, and storage space of the nodes:

Suppose there are n nodes, of which f are Byzantine nodes. In each consensus cycle, each node can receive a base bonus of B. The length of the consensus cycle is T.

The workload W_i of node i can be calculated by the following Eq. (11):

$$W\_i = w\_c * C\_i + w\_b * B\_i + w\_s * S\_i \tag{11}$$

Among them, w_c, w_b and w_s are weighting coefficients, which indicate the weight of computational resources, network bandwidth, and storage space in the total workload, respectively. These coefficients can be adjusted according to the actual demand. C_i denotes the computational resources (e.g., CPU occupancy, memory usage, etc.) provided by node i in the consensus process, B_i denotes the network bandwidth provided by node i in the consensus process, and S_i denotes the storage space provided by node i in the consensus process.

The total contribution P_i of node i in the consensus process can be calculated by the ratio of its workload W_i to the total workload W of all nodes, Eq. (12) is as follows:

$$P\_i = W\_i \sum_{1}^{n} W\_n \tag{12}$$

The bonus R_i of node i in each consensus cycle can be calculated by the following Eq. (13):

$$R\_i = B * P\_i * (1 - f/n)/T \tag{13}$$

where B is the base bonus, f is the number of Byzantine nodes, and n is the total number of nodes. The more honest nodes in the system out of proportion the more generous the reward.

It is important to note that the various parameters involved in this reward function in terms of the relevant period, base bonus, and credibility should actually be defined according to the specific application scenario and requirements [24].

5 Experiment

In this paper, the performance of PBFT, CPBFT, GPBFT, and CDGPBFT algorithms are simulated using the python language and all the consensus nodes are running on the same host. The hardware configuration of the host is i5-9300H and the graphics card is GTX1660 Ti. The nodes are classified into PBFT nodes, CDGPBFT nodes, and CPBFT and GPBFT nodes by writing different encodings of node behavior. In the following experiments, the performance of CDGPBFT, CPBFT, GPBFT, and PBFT will be compared in the following three aspects.

5.1 Safety Performance Analysis

The security of consensus algorithms is an important criterion. Once attacked, a series of critical problems surface, including irreversible transactions, broken consensus, impaired value, and compromised trust. This section provides a comprehensive security analysis of the CDGPBFT algorithm.

Credibility of Nodes. In the event of a Byzantine primary node, two scenarios may unfold: either it transmits an inconsistent pre-prepared message to all consensus nodes, or it remains unresponsive to client requests. This aberrant occurrence triggers a consensus failure during the preparation stage, necessitating a view change to substitute the primary node. Our approach responds by diminishing the trustworthiness of the malicious primary, subsequently relegating it to the pool of candidate nodes during the consensus node set update. This measure ensures the credibility of nodes within the consensus nodes set. In cases where there are f Byzantine nodes within the consensus node-set, and the total number of consensus node sets surpasses or equals $3f + 1$, any node involved in the prepare and response stages consistently receives no less than $2f$ congruent messages, without impeding the consensus-reaching process. In summation, the CDGPBFT consensus algorithm guarantees verifiable and trustworthy outcomes in every round.

Prevention of Tampering and Forgery. Nodes within PBFT employ a suite of security techniques, including pseudo-identity, hash algorithms, and digital signatures, to establish a robust foundation of security and dependability. Moreover, CDGPBFT integrates asymmetric encryption to thwart any attempts at data manipulation. This means that without the user's private key, no node can falsify its identity or the message signature. By harnessing the power of blockchain technology and digital signatures, the cost of forgery for potential attackers becomes prohibitively high, thus fortifying the overall system's reliability.

Resist DDoS Attack. To thwart DDoS attacks, we implement a scoring mechanism for primary node selection. The true identity of the primary node remains concealed until recommendation. Following each consensus round, nodes in both the consensus node set and the candidate node set undergo reevaluation and reassignment. This ensures the primary node's identity remains covert, preventing adversaries from obtaining this crucial information. Even if adversaries inundate the system with invalid data, the primary node remains insulated as it lacks knowledge of its own designation. Consequently, this safeguard maintains system integrity, effectively fortifying our scheme against DDoS attacks and upholding overall system security.

Resist Selective Attack. In our approach, the total node count surpasses $3f + 1$. Even if all adversarial nodes collaborate, they can only influence f messages at most, which is insufficient to reach a conclusive agreement. A consensus necessitates receiving $2f + 1$ messages. Consequently, all colluding malicious nodes are incapable of reaching a final accord. Thus, our scheme demonstrates resilience against selective attacks.

5.2 Algorithm Complexity Analysis

The communication overhead of the PBFT protocol is proportional to the number of nodes and messages in the network. In PBFT, each node needs to communicate with other nodes by sending a total of $2f + 1$ messages, where f is the maximum number of Byzantine nodes allowed. Since each node has to communicate with other nodes, when the number of nodes increases, the communication overhead increases accordingly. Moreover, in PBFT, three phases are required to reach a consensus among the nodes and each phase requires message exchange, so the number of messages also increases with the increase in the number of phases, which further increases the communication overhead.

To reduce the communication overhead of the PBFT protocol, the commit phase of the consensus process is simplified as well and the reply phase is improved. The stage that originally required every node to confirm was changed to one that required only the primary node to confirm. This can greatly reduce the amount of communication and computation between nodes and improve the efficiency and scalability of the system.

For the merit of PBFT in terms of communication costs, the four algorithms are compared and calculated. The concrete algorithm flow of PBFT is divided into three stages. In the pre-preparation phase, number of connections in the consensus system in this phase is $(N-1)$. In the preparation phase, number of connections in the consensus system in this phase is $(N-1)*(N-1)$ times. In the commit phase, all nodes validate the received preparation messages. When the confirmation result is true, the node sends a confirmation message to all nodes but its own. In this phase, number of connections in the consensus system is $N*(N-1)$. Thus, the count of exchanges in the consensus process of the conventional PBFT consensus algorithm is $2N*(N-1)$.

As can be seen in Fig. 2, the specific calculation process of the CDGPBFT consensus algorithm is divided into three stages. $2N/3$ is the number of consensus nodes. In the pre-preparation phase, the primary node broadcasts pre-preparation messages to all consensus nodes. In this phase, the number of communications in the consensus network is $(2N/3-1)$. In the preparation phase, after verifying that the pre-preparation message sent by the primary node is passed, the node sends the preparation message to the other consensus nodes. In this phase, the number of communications in the consensus network is $(N/3-1)*(N/3-1)*2$. In the response phase, the primary node accepts the verification message from the consensus node and verifies it. In this phase, the number of communications in the consensus network is $(2N/3-1)$. Thus the number of communications for the improved CDGPBFT consensus algorithm to complete a consensus process is $2N^2/9$.

5.3 Communication Overhead

The communication overhead is the communication traffic generated by the consensus of the nodes in the system. The Communication expenses of the PBFT are one of the important performance metrics. Since PBFT uses the solution of the Byzantine General problem, it is an algorithm with very frequent communication between nodes, and message delivery, processing, and verification are required at each stage, so the communication overhead is very high [25].

The communication overhead of PBFT consists of two main components, the overhead of sending and receiving messages, and the overhead of message verification and processing. Specifically, nodes need to send and receive different kinds of messages in the PBFT algorithm, such as pre-prepared messages, prepared messages, source messages, commit messages, etc., and each message contains

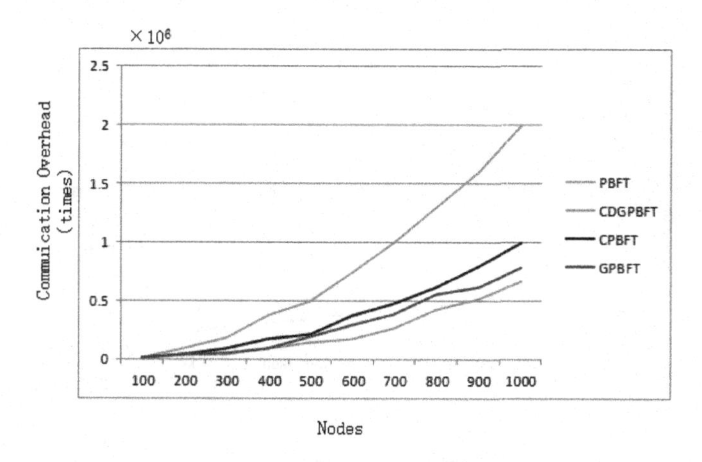

Fig. 4. Communication overhead

several data items, such as sequence number, view number, signature, etc., and the size of these data items affects the communication overhead. At the same time, when verifying and processing messages, nodes need to perform complex operations such as public key encryption, decryption, signature verification, etc. These operations also take up large processing and computing resources, increasing the communication overhead. In this experiment, the transaction traffic of each algorithm is tested separately. As can be seen from Fig. 4, the communication overhead of the algorithms all tend to increase as the number of nodes increases. The more nodes there are, the more obvious the advantage of the scheme becomes.

5.4 Message Throughput

Throughput is defined as the number of transactions executed by the system per unit of time. The level of the system's ability to process transactions relies on the size of the throughput, which is usually indicated in terms of transactions per second (TPS). Information throughput depends on the bandwidth of the network and the processing capacity of the nodes [26].

The Eq. (14) for the calculation of information throughput is shown below:

$$TPS = \frac{transaction\Delta t}{time} \tag{14}$$

where $transaction\Delta t$ is the number of transactions processed by the system during the consensus process and $time$ is the time necessary for the system to deal with the trade.

In this experiment, The client is set to issue 30 claims and record the amount of trades performed per second at various From Fig. 5, it can be seen that the throughput of the CDGPBFT consensus algorithm is higher than the other three consensus algorithms. At the same time, the throughput of the algorithm

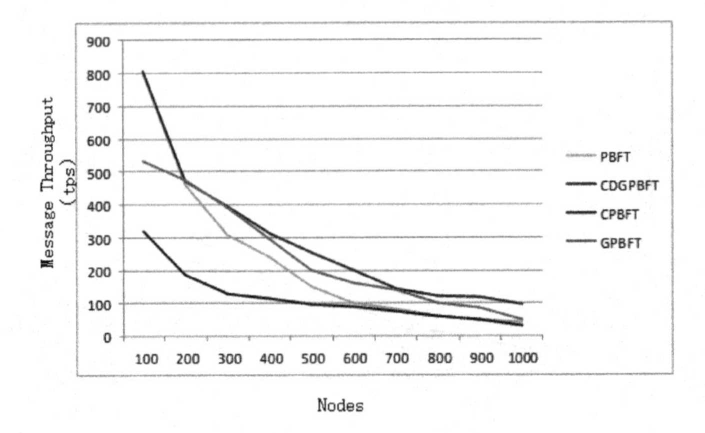

Fig. 5. Message Throughput

decreases as the number of nodes increases. the advantages of CDGPBFT are still evident.

The stability of CDGPBFT also impacts its consensus equality as the system operates due to its divisional groups and credit mechanisms. If a malicious node is employed as a primary node, it will significantly reduce the consensus efficiency. The emulation outcome shows that due to the credit mechanism, the likelihood of malicious nodes participating in consensus is significantly reduced and the incorrect selection probability of primary nodes is also reduced. In terms of long-term operation, CDGPBFT dramatically increases the throughput of the system.

5.5 Transaction Delay

Transaction latency is the time required for a client node to send a transaction request to the primary node until the client receives a transaction receipt confirmation message [27].

The CDGPBFT algorithm is optimized in terms of transaction latency compared to the original PBFT algorithm. The CDGPBFT algorithm simplifies and improves the submission and reply phases, thus reducing the communication overhead and transaction latency. At the same time, the CDGPBFT algorithm uses a reward mechanism based on nodes' computing resources, network bandwidth, and storage space to encourage nodes to participate in consensus correctly, thus increasing the transaction processing speed of the system.

The CDGPBFT algorithm is optimized in the consensus process to simplify the submission and reply phases, reducing the communication overhead and transaction latency between nodes. In the commit phase, the CDGPBFT algorithm adopts a fast commit mechanism, which only needs to wait for the reply from the primary node to upload the transaction to the chain, thus improving the transaction processing speed.

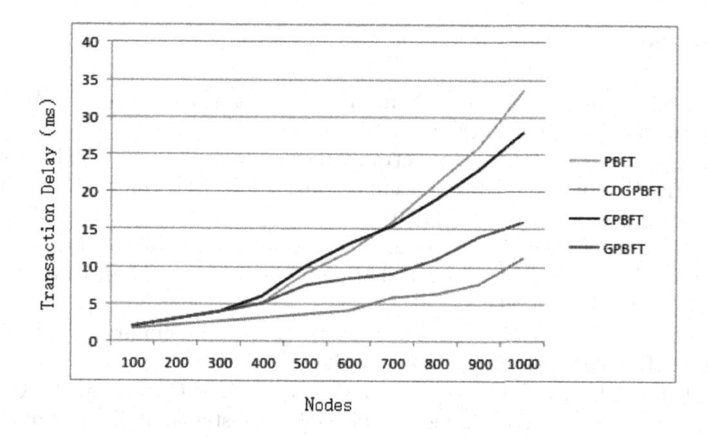

Fig. 6. Transaction Delay

200 tests were conducted, and the average of every 10 transaction latencies was taken as the experimental data. The experimental outcomes show that CDGPBFT significantly outperforms the other three algorithms in terms of transaction latency (Fig. 6).

6 Summary and Future Outlook

This paper presents an improved PBFT consensus algorithm based on credit values and dynamic grouping (CDGPBFT). The algorithm identifies malicious nodes by extending the credit value attribute of nodes and introduces a dynamic node hierarchical grouping design suitable for large-scale federated chain systems [28]. In addition, the algorithm enhances the consensus process to reduce redundant communication between nodes and improves the selection and scope of the master node. Finally, the reward function is carefully designed to ensure the security and stability of the system.

To validate the performance of the CDGPBFT algorithm, comprehensive experimental comparisons are conducted. The results demonstrate that the CDGPBFT algorithm surpasses other consensus algorithms, including PBFT, CPBFT, and GPBFT, in terms of latency, throughput, and communication overhead.

In summary, CDGPBFT stands as an efficient consensus algorithm with potential applications in large-scale coalition chain systems, significantly enhancing system performance. In future work, optimized node selection strategies will be further investigated to further reduce latency and improve consensus efficiency. Meanwhile, it is planned to apply the algorithm to other fields, such as medicine, agronomy, the Internet of Things, and the Internet of Vehicles, to provide more secure and efficient services by utilizing blockchain technology [29].

References

1. Portmann, E.: Rezension blockchain: Blueprint for a new economy. Hmd Praxis Der Wirtschaftsinformatik (2018)
2. Leung, C.H.C.: Semantic image retrieval using collaborative indexing and filtering. In: IEEE/WIC/ACM International Joint Conferences on Web Intelligence & Intelligent Agent Technology (2012)
3. Meng, Y., Cao, Z., Qu, D.: A committee-based byzantine consensus protocol for blockchain. In: 2018 IEEE 9th International Conference on Software Engineering and Service Science (ICSESS) (2018)
4. Zhang, G., et al.: Reaching consensus in the byzantine empire: a comprehensive review of BFT consensus algorithms (2022)
5. Dinh, T.T.A., Liu, R., Zhang, M., Chen, G., Ooi, B.C., Wang, J.: Untangling blockchain: a data processing view of blockchain systems. arXiv e-prints (2017)
6. Luu, L., Chu, D.H., Olickel, H., Saxena, P., Hobor, A.: Making smart contracts smarter. In: The 2016 ACM SIGSAC Conference (2016)
7. Miguel, O.T.D.C.: Practical byzantine fault tolerance (2001)
8. Lao, L., Dai, X., Xiao, B., Guo, S.: G-PBFT: a location-based and scalable consensus protocol for IoT-blockchain applications. In: 2020 IEEE International Parallel and Distributed Processing Symposium (IPDPS) (2020)
9. Hu, Z.: Blockchain improvement scheme based on PBFT consensus algorithm. Comput. Sci. Appl. **11**(3), 643–653 (2021)
10. Hai-Bo, T., et al.: Archival data protection and sharing method based on blockchain. J. Softw. **30**, 2620–2635 (2019)
11. Bentov, I., Lee, C., Mizrahi, A., Rosenfeld, M.: Proof of activity: extending bitcoin's proof of work via proof of stake (2014)
12. Wang, W., et al.: A survey on consensus mechanisms and mining strategy management in blockchain networks. IEEE Access **7**, 22328–22370 (2019)
13. Ongaro, D., Ousterhout, J.K.: In search of an understandable consensus algorithm (2014)
14. Kalajdjieski, J., Raikwar, M., Arsov, N., Velinov, G., Gligoroski, D.: Databases fit for blockchain technology: a complete overview. Blockchain Res. **4**(1), 18 (2023)
15. Tian, H., Tian, C., Li, K., Yuan, C.: Dynamic operation optimization based on improved dynamic multi-objective dragonfly algorithm in continuous annealing process. J. Industr. Manage. Optim. **19**(8), 6159–6181 (2023)
16. Tong, W., Dong, X., Zheng, J.: Trust-PBFT: a peertrust-based practical byzantine consensus algorithm. In: 2019 International Conference on Networking and Network Applications (NaNA) (2019)
17. Ren, X., Tong, X., Zhang, W.: Improved PBFT consensus algorithm based on node role division. Comput. Commun. **11**(2), 20–38 (2023)
18. Denter, N.M., Seeger, F., Moehrle, M.G.: How can blockchain technology support patent management? A systematic literature review. Int. J. Inf. Manage. **68**, 102506 (2023)
19. Kim, J.T., Jin, J., Kim, K.: A study on an energy-effective and secure consensus algorithm for private blockchain systems (PoM: proof of majority). In: International Conference on Information and Communication Technology Convergence (2018)
20. Hu, Q., Wu, X., Dong, S.: A two-stage multi-objective task scheduling framework based on invasive tumor growth optimization algorithm for cloud computing. J. Grid Comput. **21**(2), 31 (2023)

21. Xu, G., et al.: SG-PBFT: a secure and highly efficient distributed blockchain PBFT consensus algorithm for intelligent internet of vehicles. J. Parallel Distrib. Comput. **164**, 1–11 (2022)
22. Jiang, X., Sun, A., Sun, Y., Luo, H., Guizani, M.: A trust-based hierarchical consensus mechanism for consortium blockchain in smart grid. Tsinghua Sci. Technol. **28**(1), 69–81 (2023)
23. Xinjian, M.A., Shiqian, L., Huihui, C.: Civil aircraft fault tolerant attitude tracking based on extended state observers and nonlinear dynamic inversion. Syst. Eng. Electron. (001), 033 (2022)
24. Frauenthaler, P., Sigwart, M., Spanring, C., Sober, M., Schulte, S.: ETH relay: a cost-efficient relay for ethereum-based blockchains. In: 2020 IEEE International Conference on Blockchain (Blockchain) (2020)
25. Qingshui, X., Tianhao, Z., Yue, S.: PBFT algorithm for internet of things. In: 2022 7th International Conference on Computer and Communication Systems (ICCCS), pp. 684–689 (2022)
26. Al-Sumaidaee, G., Alkhudary, R., Zilic, Z., Féniès, P.: Configuring blockchain architectures and consensus mechanisms: the healthcare supply chain as a use case. In: Bouras, A., Khalil, I., Aouni, B. (eds.) Blockchain Driven Supply Chains and Enterprise Information Systems, pp. 135–150. Springer, Cham (2023). https://doi.org/10.1007/978-3-030-96154-1_7
27. Mii, J., Mii, V.B., Chang, X., Qushtom, H.: Multiple entry point PBFT for IoT systems. In: IEEE Vehicular Technology Conference (2021)
28. Zhao, W.: Design and implementation of a byzantine fault tolerance framework for web services. J. Syst. Softw. **82**(6), 1004–1015 (2009)
29. Xu, J., Zhao, Y., Chen, H., Deng, W.: ABC-GSPBFT: PBFT with grouping score mechanism and optimized consensus process for flight operation data-sharing. Inf. Sci. **624**, 110–127 (2023). https://www.sciencedirect.com/science/article/pii/S0020025522015638

Data Exchange and Sharing Framework for Intermodal Transport Based on Blockchain and Edge Computing

Li Wang[1], Xue Zhang[1], Lianzheng Xu[1], Deqian Fu[1,2](✉), and Jianlong Qiu[2](✉)

[1] School of Information Science and Engineering, Linyi University, Linyi, Shandong, People's Republic of China

[2] Key Laboratory of Complex Systems and Intelligent Computing in Universities of Shandong, Linyi University, Linyi, Shandong, People's Republic of China

{fudeqian,qiujianlong}@lyu.edu.cn

Abstract. As an intensive and efficient way of cargo transportation, intermodal transport can take full advantage of various modes of transport through effective combinations. However, there is a problem of lacking effective data connection between collaborators leading to broken business processes, reduced collaborative efficiency, and hindering the development of the intermodal transport industry. In this paper, we propose a data exchange and sharing framework by integrating blockchain technology and edge computing technology to realize efficient, convenient, safe, and credible data exchange, which also conveniently connects different partners of various modes of transport and then promotes cross-sector and cross-system data exchange and sharing in the intermodal transport environment. Furthermore, the ecosystem of intermodal transport data exchange and sharing based on the proposed framework can effectively meet the requirements of scalability, security, believability, and privacy protection.

Keywords: Intermodal Transport · Data Exchange and Sharing · Blockchain · Edge Computing

1 Introduction

With the increasing development of the global logistics business and the improvement of transportation infrastructure, the transportation industry is currently focusing on constructing an integrated transportation system. A crucial objective in the field is to achieve efficient cooperation between different modes of transport so as to improve the overall efficiency of integrated transportation services.

Intermodal transport [1, 2] is an efficient mode of transport that uses multiple modes of transport, such as railways, roads, and water transport, to transport goods jointly. Integrating the advantages of multiple modes can effectively strengthen each mode and solve the issues of a single available mode. Benefitting from cutting social logistics costs, enhancing comprehensive transportation efficiency, and promoting energy conservation

F. Sun and J. Li (Eds.): ICCCS 2023, CCIS 2029, pp. 292–303, 2024.
https://doi.org/10.1007/978-981-97-0885-7_24

and carbon reduction, intermodal transport as a critical support of the global supply chain has become an inevitable trend in the development of the modern logistics industry.

The key to intermodal transport is "connection", which places higher demands on data exchange and sharing between various modes of transport in terms of believability, timeliness, and security. However, the development of informatization in different modes of transport is unbalanced, and a lack of unified data standards. As a result, it is difficult achieving efficient, convenient, safe, and credible data exchange and sharing among these modes of transport [3]. Furthermore, numerous participants were involved in the process, as revealed by analyzing the business scenario of intermodal transport [4]. This gives rise to issues such as trusted identity [5], data ownership, privacy protection, and restricted access to data [6]. These problems ultimately reduce the enthusiasm and possibility of data exchange and sharing.

The intermodal transport data exchange and sharing framework proposed in this paper utilizes technologies such as blockchain [7, 8] and edge computing [9–11] to create a distributed and trusted network for efficient, convenient, secure, and trustworthy data exchange and sharing among intermodal transport participants. The main contributions are abstracted as follows:

(1) Design an architecture of the distributed trusted data exchange and sharing framework integrating blockchain, edge computing, and other technologies. We also explain in detail how the components of each layer in the architecture interact.
(2) Deploy blockchain services into edge computing nodes to form a new distributed computing model binding with security and trustworthiness.
(3) Design a set of blockchain smart contracts for secure, efficient, and trustworthy data exchange and sharing. Examples include a contract for trusted identity application registration and a data exchange sharing contract that utilizes semantic reasoning with ontological knowledge and can connect to external data sources.

This research aims to offer a novel solution for intermodal transport data exchange and sharing, fostering the development and optimization of logistics businesses. Additionally, promote coordination and innovation in intermodal transport, which will ultimately result in a data exchange-sharing system that is more efficient, safe, and reliable [12].

The subsequent sections of this paper are organized in the following manner: Sect. 2 presents a concise summary of the current body of research about intermodal transport, blockchain technology, and edge computing. Section 3 provided a comprehensive exposition of the architectural design of the proposed framework for intermodal transport data exchange and sharing. Ultimately, the closing part encapsulates the entirety and offers perspectives on potential future work.

2 Related Work

Three related primary components are introduced in this section. Initially, the relevant concepts of intermodal transport are described to discuss the requirements of data exchange and sharing. Next, the role of blockchain technology is explored in data exchange with related work in this area. Finally, innovative applications are introduced by integrating blockchain technology and edge computing.

2.1 Intermodal Transport

Intermodal transport refers to the effective connection of two or more modes of transport to complete an integrated and organized cargo transport service. A high resource utilization rate and good comprehensive benefits characterize it. The current forms of intermodal transport include such as road-rail combined transport, rail-water combined transport, road-water combined transport, and air-road combined transport. If a data exchange and sharing framework can be established among all participants, along with an effective mutual trust and cooperation mechanism [13], that will accelerate the process of information sharing and openness and promote data exchange and sharing among participants of intermodal transport.

Currently, data exchange and sharing among information systems of transports can be achieved through traditional Electronic Data Interchange (EDI) [14, 15]. However, this approach can no longer meet the higher standards of intermodal transport for data exchange and sharing. From a management perspective, there is a need to enable real-time data sharing and collaboration across subjects and systems. From a technical standpoint, ensuring the security, reliability, and efficiency of data exchange and sharing is crucial. Integrating blockchain technology and edge computing with the intermodal transport business can solve the current problems in data circulation and inject new vitality into the development of intermodal transport.

2.2 Blockchain

Blockchain is a newly developed distributed framework with the characteristics of decentralization, multi-party consensus, security, and non-tampering. It combines various technologies such as distributed storage, point-to-point transmission, consensus mechanism, and encryption algorithms.

Blockchain is widely studied in many scenarios requiring data exchange and sharing, such as government affairs and medical data. For instance, in [16], the author introduces IoTChain. This model utilizes the Ethereum blockchain as an auditable access control layer to address secure storage and trust data sharing in the IoT environment. Similarly, in [17], the author employs blockchain to maintain traceability throughout the healthcare data exchange process and utilizes smart contracts for controlling data sharing. Furthermore, in [18], the author employs blockchain and a deep learning-based framework to establish a secure platform for digital governance interoperability and data exchange, thereby enhancing believability.

2.3 Blockchain and Edge Computing

Edge computing is a computing model that strategically places computing resources and data at edge nodes in close proximity to the data source to reduce data transmission delays and enhance computing efficiency. In the application of deep integration with edge computing, blockchain technology plays a vital role in enhancing security and believability, providing a reliable foundation for edge computing. Simultaneously, edge computing fosters collaboration among nodes and enables the collaborative sharing of information, resources, and production factors.

A considerable study has been conducted pertaining to the utilization of blockchain technology and edge computing for data exchange and sharing. For instance, in a previous study [19], the authors introduced the MEdge-Chain framework, which combines blockchain with edge computing technology to facilitate fast and secure medical data exchange and storage. Similarly, in [20], the authors introduced EdgeMediChain, a secure and efficient data management framework for sharing health data. In [21], the authors leveraged blockchain technology and smart contracts to establish a logistics resource allocation network while utilizing edge computing to support resource-constrained transportation nodes in performing complex calculations.

3 Architecture Design

In this section, we introduce the layered architecture of the intermodal transport data exchange and sharing framework, as shown in Fig. 1. The architecture consists of five layers: user layer, local edge-mining layer, semantic reasoning layer, blockchain layer, and off-chain distributed storage layer. We will describe in detail how the components of these different layers interact with each other.

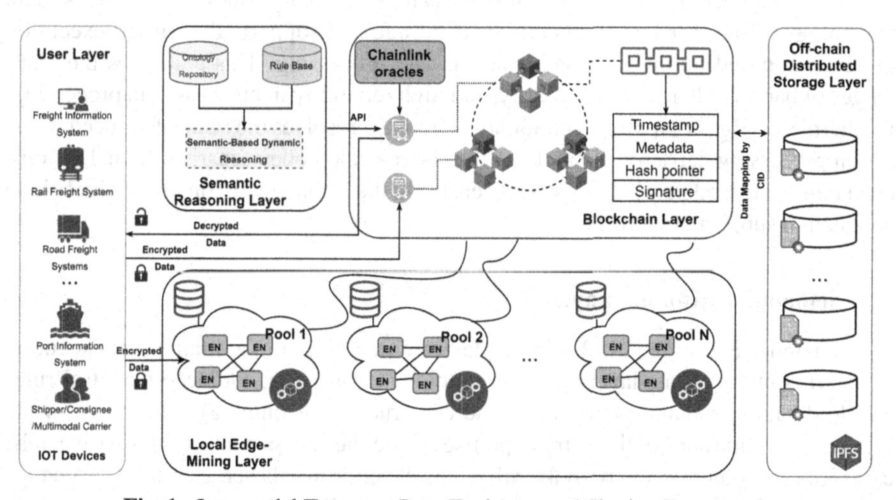

Fig. 1. Intermodal Transport Data Exchange and Sharing Framework

3.1 User Layer

The user layer consists of various information systems involved in the intermodal transport data business process, intermodal transport participants, and IoT Devices. Considering that part of intermodal data needs to be processed and analyzed in real-time. Still, some data remains unchanged from beginning to end. There are two methods for encrypting and uploading the data. The first method encrypts the data and uploads it to the local edge mining layer. For instance, real-time processing of data generated by sensors on logistics vehicles and handheld scanning terminals in multimodal transportation can be efficiently processed in real-time by the edge-mining layer. The processed data can then be securely stored by blockchain service, ensuring trust and reliability. Finally, smart contracts on the blockchain can access the stored data for further analysis and decision-making. The other is to directly upload to the distributed file storage layer under the chain using the smart contract of the blockchain layer after encryption.

3.2 Local Edge-Mining Layer

The core component of the architecture is the edge-mining layer located on the local edge devices of the logistics nodes. This layer consists of distributed data mining pools created by Edge Node Services (ENS), blockchain services, and local databases. Empowering the edge devices with computing power enables horizontal expansion of the network node, addresses data authenticity and believability, enhances terminal heterogeneity capability, and offers real-time computing and verification services.

In addition, combining blockchain and edge computing can create a decentralized edge computing model known as blockchain edge computing. Contrasting with the traditional methods, edge nodes can directly communicate and exchange data with other nodes without needing a centralized server as an intermediary. Additionally, blockchain technology allows for real-time certificate storage and supports the trusted execution of services based on credible data analysis and processing. This addresses the challenges of bandwidth, real-time security, and high reliability in business data processing. Furthermore, edge computing technology effectively resolves the issue of reduced transaction processing capacity of blockchain in large-scale node scenarios. It enables efficient consensus and parallel processing, ensuring the stable operation of the blockchain application support system.

3.3 Semantic Reasoning Layer

The semantic reasoning [22, 23] layer plays a crucial role in facilitating the execution of smart contracts. This layer relies on ontology knowledge, and access control rules enable dynamic semantic reasoning. The construction of ontology knowledge in the logistics domain relies on industry expertise, while the access control rules base can be categorized into public and bespoke rules. This layer primarily serves the data exchange and sharing of smart contracts of the blockchain layer. When executing the contract, the inference service API is called first to determine whether to proceed with the operation.

3.4 Blockchain Layer

The blockchain layer plays an important role in connecting the rest of the layers. The logistics data can exchange and sharing across participants in intermodal transport is allowed, provided that they possess trusted identities. We have developed an effective identity verification mechanism to ensure identity credibility, which will be explained in Sect. 4.1.

The blockchain layer facilitates decentralized data exchange sharing, storage, and verification. The utilization of smart contracts and the Chainlink distributed network achieves automation and conditional execution of data exchange sharing, as discussed in Sect. 4.2. Additionally, by storing the hash of the intermodal data and the URL hash pointer, the detailed intermodal transport data remains privately accessible, thereby safeguarding data security.

3.5 Off-Chain Distributed Storage Layer

The transportation industry generates massive amounts of data, and the blockchain data structure, as shown in Fig. 2, needs improvement work to handle such vast volumes of data. Additionally, storing raw data on the chain can lead to privacy concerns, high storage costs, and scalability issues.

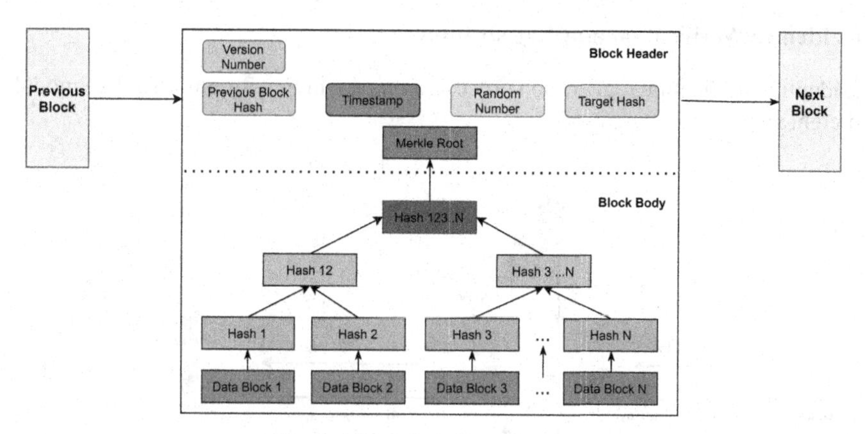

Fig. 2. Blockchain Data Structure

After analyzing the actual requirements, the off-chain storage and on-chain authentication mechanism [24] is adopted here to facilitate off-chain data collaboration as well as on-chain and off-chain separation of different types of data. This approach ensures that the read and write performance of the system will not be affected while achieving efficient storage and distribution. The Interplanetary File System (IPFS), as the database of the off-chain storage layer, adopts the P2P distributed protocol, which can associate the stored data with the blockchain data through a unique CID. The off-chain distributed storage layer provides efficient storage and retrieval mechanisms, and the distribution can effectively avoid single points of failure and ensure the integrity and accessibility of logistics data. In addition, the architecture uses it to store the blockchain user's public key, which is crucial for the data encryption and decryption process.

It is evident that each layer is independent and closely integrated with other layers in the architecture design of the proposed framework, with the significant advantages of scalability, distribution, and efficiency. The architecture can bring more efficient and reliable supporting for data exchange and sharing in the multimodal logistics industry.

4 Smart Contracts

Blockchain smart contracts [25–27] are computer protocols designed to facilitate the communication, verification, and enforcement of agreements efficiently. They enable trusted transactions without needing a trusted third party, ensuring traceability and irreversibility.

In the context of intermodal transport, where scenarios and participants can be complex, the implementation of an efficient smart contract system can enhance the performance and user-friendliness of blockchain-based applications for data exchange and sharing. This section will explore the application of smart contracts in identity verification registration and intermodal transport data exchange.

4.1 Identity Verification and Registration

The identity verification registration process, as shown in Fig. 3, comprises five modules and eight steps.

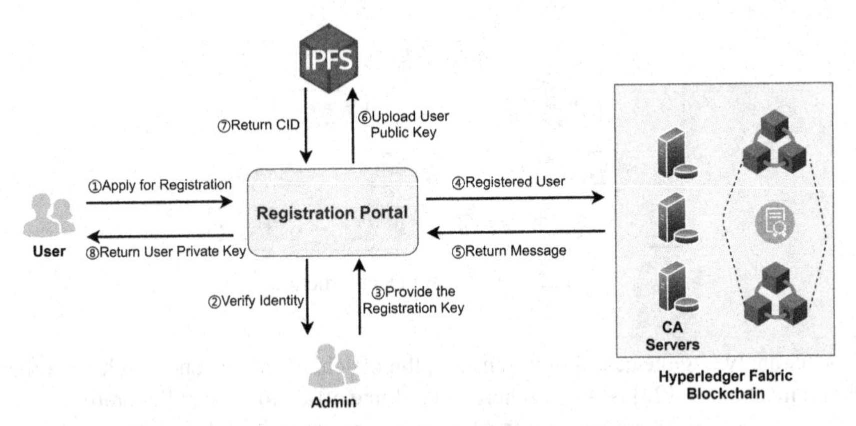

Fig. 3. Identity Verification Registration Process

Initially, the applicant proceeds to submit their application information via the registration portal in order to initiate the registration process (Step 1). The identity registration request is then transmitted to the administrator to verify the user's identification. Upon successful completion of the verification process, the registration key is issued to the user (Steps 2, 3). Upon receipt of the registration key, the registration site initiates a request to the blockchain's certification authority (CA) server. The CA servers execute the registration and activation process of Smart Contract 1, wherein the identity information will document in the blockchain. Additionally, the CA servers return the user with public and private keys, along with other pertinent particulars (Steps 4, 5). The registration portal stores the user's public key within the IPFS to streamline the administration and retrieval of the public key. The IPFS returns a unique CID for the public key (Steps 6, 7). In the final step, the users get their private key and are responsible for its safekeeping (Step 8).

The pseudocode of Smart Contract 1 presents a contract example of implementing identity registration in intermodal transportation. The contract, IntermodalTransportation, includes a structured Participant that stores information such as name, wallet address, and registration status. Additionally, a ParticipantType enumeration is defined to identify the type of participant.

Smart Contract 1 Identity Registration

```
01: contract IntermodalTransportation:
02:     struct Participant:
03:         string name;
04:         address wallet;
05:         bool isRegistered;
06:     enum ParticipantType { Shipper, Carrier, Consignee }
07:     mapping (ParticipantType => mapping (address => Participant)) participants;
08:     function registerParticipant(ParticipantType participantType, string memory n
ame) public {
09:         require (!participants [participantType][msg.sender].isRegistered, "Participa
nt already registered.");
10:         Participant memory newParticipant = Participant ({
11:             name: name,
12:             wallet: msg.sender,
13:             isRegistered: true
14:         });
15:         participants[participantType][msg.sender] = newParticipant;
16:     }
17:     function getParticipantInfo(ParticipantType participantType, address wallet) p
ublic view returns (string memory, bool) {
18:         return (
19:             participants[participantType][wallet].name,
20:             participants[participantType][wallet].isRegistered
21:         );
22:     }
23: }
```

4.2 Data Exchange and Sharing for Autonomous Decision-Making

The smart contract for autonomous decision-making relies on the Chainlink distributed oracle network to obtain external data and use semantic reasoning to determine whether access is authorized. For instance, the smart contract must obtain accurate weather data through the Chainlink oracle network if weather information is required for data exchange and sharing. Once the smart contract's autonomous decision is approved by calling the semantic inference service API, the data exchange and sharing process will be executed, and the relevant information will be recorded on the blockchain to ensure the security and immutability of all transactions.

The pseudocode for this smart contract is described in detail in Smart Contract 2.

Smart Contract 2 Data Exchange and Sharing for Autonomous Decision-making

```
01: import "@chainlink/contracts/src/v0.8/ChainlinkClient.sol";
02: contract WeatherDataSharingContract is ChainlinkClient {
03:     mapping (address => bool) public accessPermissions;
04:     address private oracle;
05:     bytes32 private jobId;
06:     uint256 private fee;
07:     constructor (address _oracle, bytes32 _jobId, uint256 _fee) public {
08:         setPublicChainlinkToken();
09:         oracle = _oracle;
10:         jobId = _jobId;
11:         fee = _fee;}
12:     function authorizeAccess() public {
13:         accessPermissions[msg.sender] = true;}
14:     function revokeAccess() public {
15:         accessPermissions[msg.sender] = false;}
16:     function shareData() public {
17:         require(accessPermissions[msg.sender], "No access permission");
18:         bool hasAccess = callSemanticInferenceAPI(msg.sender);
19:         require (hasAccess, "Access denied");
20:         requestWeatherData();}
21:     function callSemanticInferenceAPI(address requester) private returns (bool) {
22:         // Call the semantic inference service API here, infer based on the provided p
arameters to check access permissions.
23:         // For example, infer based on the requester's identity and request time, etc.
24:         // Return a bool value indicating whether access is granted or not.
25:         // Here, a hypothetical logic is used.
26:         bool hasAccess = true; // Assuming everyone has access permission.
27:         return hasAccess;}
28:     function requestWeatherData() private {
29:         Chainlink.Request memory req = buildChainlinkRequest (jobId, address(thi
s), this.fulfill.selector);
30:         req.add("url", "https://api.weather.cn/data"); // Assume this is the API URL t
o fetch weather data.
31:         req.add("path", "temperature"); // Assume the path to the temperature data in
 the weather API.
32:         sendChainlinkRequestTo(oracle, req, fee);}
33:     function fulfill (bytes32 _requestId, uint256 _temperature) public recordChain
linkFulfillment(_requestId) {
34:         if (_temperature > 39) {
35:             // ...
36:         } else {
37:             // ...
38: }}}
```

The processes of the smart contract are as follows:

1. Import the Chainlink contract.
2. Define the smart contract.
3. Define the mapping of storage access permissions.

4. Define the address and related parameters of the smart contract on the chain.
5. The constructor sets the Chainlink configuration.
6. Authorize access rights.
7. Revoke access rights.
8. Perform data sharing.
9. Call the Semantic Inference Service API.
10. Request weather data.
11. Process the response from the on-chain oracle.

5 Conclusion

The proposed framework of data exchange and sharing for a business with intermodal transport realizes the data exchange and sharing among multiple collaborators safely and controllably by integrating blockchain technology and edge computing technology. It provides a secure, transparent, and reliable data infrastructure, ultimately improving the efficiency and believability of logistics operations and offering better decision support for logistics participants.

In the future, the field of intermodal transport may witness the emergence of multiple blockchains with diverse technical approaches. This could result in the formation of independent value islands, limiting business collaboration and value exchange across different blockchains. Therefore, it is crucial to explore a general, scalable, multi-center, and easily accessible cross-chain solution, which will be the focus of further research based on this paper.

Acknowledgments. This work was supported in part by the Shandong Provincial Natural Science Foundation under Grant ZR2022MF331, the National Natural Science Foundation of China under Grant 61833005, the Development Plan of Youth Innovation Team in Universities of Shandong Province, and the Linyi Institute of Commerce and Logistics Technology Industry under Grant CYY-TD-2022-003.

References

1. Dua, A., Sinha, D.: Quality of multimodal freight transportation: a systematic literature review. World Rev. Intermodal Transp. Res. **8**, 167–194 (2019). https://doi.org/10.1504/WRITR.2019.099136
2. Wenwen, J., Dongli, W., Cui, C.: Research on multimodal transport development in china under the background of internet plus. J. Phys. Conf. Ser. **1486** (2020). https://doi.org/10.1088/1742-6596/1486/2/022027
3. Su, Z., Wang, Y., Xu, Q., Zhang, N.: LVBS: lightweight vehicular blockchain for secure data sharing in disaster rescue. IEEE Trans. Dependable Secur. Comput. **19**, 19–32 (2022). https://doi.org/10.1109/TDSC.2020.2980255
4. Tuchen, S.: Multimodal transportation operational scenario and conceptual data model for integration with UAM. In: Integrated Communications Navigation and Surveillance Conference, ICNS, vol. 2020-Septe (2020). https://doi.org/10.1109/ICNS50378.2020.9223002

5. Zhang, Z., Shao, S., Zhong, C., Sun, S., Lin, P.: Trusted identity authentication mechanism for power maintenance personnel based on blockchain. In: Liu, Q., Liu, X., Shen, T., Qiu, X. (eds.) CENet 2020. AISC, vol. 1274, pp. 883–889. Springer, Singapore (2021). https://doi.org/10.1007/978-981-15-8462-6_102
6. Yu, K., Tan, L., Aloqaily, M., Yang, H., Jararweh, Y.: Blockchain-enhanced data sharing with traceable and direct revocation in IIoT. IEEE Trans. Ind. Inform. **17**, 7669–7678 (2021). https://doi.org/10.1109/TII.2021.3049141
7. Javaid, M., Haleem, A., Pratap Singh, R., Khan, S., Suman, R.: Blockchain technology applications for industry 4.0: a literature-based review. Blockchain Res. Appl. **2** (2021). https://doi.org/10.1016/j.bcra.2021.100027
8. Monrat, A.A., Schelén, O., Andersson, K.: A survey of blockchain from the perspectives of applications, challenges, and opportunities. IEEE Access **7**, 117134–117151 (2019). https://doi.org/10.1109/ACCESS.2019.2936094
9. Cao, K., Liu, Y., Meng, G., Sun, Q.: An overview on edge computing research. IEEE Access **8**, 85714–85728 (2020). https://doi.org/10.1109/ACCESS.2020.2991734
10. Hassan, N., Yau, K.L.A., Wu, C.: Edge computing in 5G: a review. IEEE Access **7**, 127276–127289 (2019). https://doi.org/10.1109/ACCESS.2019.2938534
11. Luo, Q., Hu, S., Li, C., Li, G., Shi, W.: Resource scheduling in edge computing: a survey. IEEE Commun. Surv. Tutor. **23**, 2131–2165 (2021). https://doi.org/10.1109/COMST.2021.3106401
12. Humayun, M., Jhanjhi, N., Hamid, B., Ahmed, G.: Emerging smart logistics and transportation using IoT and blockchain. IEEE Internet Things Mag. **3**, 58–62 (2020). https://doi.org/10.1109/iotm.0001.1900097
13. Malik, V., et al.: Building a secure platform for digital governance interoperability and data exchange using blockchain and deep learning-based frameworks. IEEE Access (2023). https://doi.org/10.1109/ACCESS.2023.3293529
14. Klapita, V.: Implementation of electronic data interchange as a method of communication between customers and transport company. Transp. Res. Procedia **53**, 174–179 (2021). https://doi.org/10.1016/j.trpro.2021.02.023
15. Sarabia-Jacome, D., Palau, C.E., Esteve, M., Boronat, F.: Seaport data space for improving logistic maritime operations. IEEE Access **8**, 4372–4382 (2020). https://doi.org/10.1109/ACCESS.2019.2963283
16. Ullah, Z., Raza, B., Shah, H., Khan, S., Waheed, A.: Towards blockchain-based secure storage and trusted data sharing scheme for IoT environment. IEEE Access **10**, 36978–36994 (2022). https://doi.org/10.1109/ACCESS.2022.3164081
17. Koscina, M., Manset, D., Negri, C., Kempner, O.P.: Enabling trust in healthcare data exchange with a federated blockchain-based architecture. In: Proceedings - 2019 IEEE/WIC/ACM International Conference on Web Intelligence Workshop, WI 2019 Companion, pp. 231–237 (2019). https://doi.org/10.1145/3358695.3360897
18. Al-Rakhami, M.S., Al-Mashari, M.: A blockchain-based trust model for the internet of things supply chain management. Sensors. **21**, 1–15 (2021). https://doi.org/10.3390/s21051759
19. Abdellatif, A.A., et al.: MEdge-chain: leveraging edge computing and blockchain for efficient medical data exchange. IEEE Internet Things J. **8**, 15762–15775 (2021). https://doi.org/10.1109/JIOT.2021.3052910
20. Akkaoui, R., Hei, X., Cheng, W.: EdgeMediChain: a hybrid edge blockchain-based framework for health data exchange. IEEE Access **8**, 113467–113486 (2020). https://doi.org/10.1109/ACCESS.2020.3003575
21. Chen, J., Zhang, J., Pu, C., Wang, P., Wei, M., Hong, S.: Distributed logistics resources allocation with blockchain, smart contract, and edge computing. J. Circ. Syst. Comput. **32** (2023). https://doi.org/10.1142/S0218126623501219

22. Muhammad, S., Admodisastro, N., Osman, H., Ali, N.M.: The dynamic web services adaptation framework in context-aware mobile cloud learning using semantic-based method. Int. J. Eng. Adv. Technol. **9**, 2353–2357 (2019). https://doi.org/10.35940/ijeat.A2652.109119
23. Chen, S., Xiao, L., Cheng, M.: A semantic-based multi-agent dynamic interaction model. In: ACM International Conference on Proceeding Series, pp. 101–108 (2020). https://doi.org/10.1145/3425329.3425332
24. Hong, Z., Guo, S., Zhou, E., Chen, W., Huang, H., Zomaya, A.: GriDB: scaling blockchain database via sharding and off-chain cross-shard mechanism. Proc. VLDB Endow. **16**, 1685–1698 (2023). https://doi.org/10.14778/3587136.3587143
25. Weiqin, Z., et al.: Smart contract development: challenges and opportunities. IEEE Trans. Softw. Eng. (2019)
26. Zheng, Z., et al.: An overview on smart contracts: challenges, advances and platforms. Futur. Gener. Comput. Syst. 475–491 (2020)
27. Khan, S.N., Loukil, F., Ghedira-Guegan, C., Benkhelifa, E., Bani-Hani, A.: Blockchain smart contracts: applications, challenges, and future trends. Peer-to-Peer Netw. Appl. **14**, 2901–2925 (2021). https://doi.org/10.1007/s12083-021-01127-0

Internet Of Rights(IOR) in Role Based Blockchain

Yunling Shi[1](✉), Jie Guan[1], Junfeng Xiao[1], Huai Zhang[2], and Yu Yuan[3]

[1] Beijing Tsingzhi Shuyuan Technology Co., Ltd., Beijing, China
yunlingdshi@gmail.com
[2] Tsinghua University, Beijing, China
[3] IEEE, Beijing, China

Abstract. A large amount of data has been accumulated. with the development of the Internet industry. Many problems have been exposed with data explosion: 1. The contradiction between data privacy and data collaborations; 2. The contradiction between data ownership and the right of data usage; 3. The legality of data collection and data usage; 4. The relationship between the governance of data and the governance of rules; 5. Traceability of evidence chain. To face such a complicated situation, many algorithms were proposed and developed. This article tries to build a model from the perspective of blockchain to make some breakthroughs. The Internet Of Rights(IOR) model uses multi-chain technology to logically break down the consensus mechanism into layers, including storage consensus, permission consensus, role consensus, transaction consensus, etc., thus building new infrastructure, that enables data sources with complex organizational structures and interactions to collaborate smoothly on the premise of protecting data privacy. With blockchain's nature of decentralization, openness, autonomy, immutability, and controllable anonymity, the Internet Of Rights(IOR) model registers the ownership of data and enables applications to build an ecosystem based on responsibilities and rights. It also provides cross-domain processing with privacy protection, as well as the separation of data governance and rule governance. With the processing capabilities of artificial intelligence and big data technology, as well as the ubiquitous data collection capabilities of the Internet of Things, the Internet Of Rights(IOR) model may provide a new infrastructure concept for realizing swarm intelligence and building a new paradigm of the Internet, i.e. intelligent governance.

Keywords: Internet Of Rights · IOR · multi-level consensus · multi-chain storage consensus · permission consensus · role consensus · privacy protection · data confirmation · legality of data usage · cross-domain information processing

F. Sun and J. Li (Eds.): ICCCS 2023, CCIS 2029, pp. 304–321, 2024.
https://doi.org/10.1007/978-981-97-0885-7_25

1 Introduction

In the data era, information technology can help people better perceive their surroundings, build smooth communication channels, and assist decision-making. Industries such as the Internet of Things, cloud computing, big data, and the Internet are generating data all the time, and the accumulation of data accelerates. To provide better data services for applications such as artificial intelligence and big data, the ownership, legality, and privacy of data resources, as well as the relationship between rule makers and regulators, rules and data, need to be sorted out. Therefore, the Internet of Rights (IOR) came into being.

Considering the loose correlations and contradictory data, the Internet of Rights(IOR) model attempts to set up a new data-sharing mechanism based on blockchain and build a swarm intelligence ecosystem that incorporates the Internet of Things and cloud computing. It has multiple features such as high efficiency, fairness, transparency, privacy protection, right protection, and supervision, therefore providing a solid data foundation for artificial intelligence and big data.

2 Background

2.1 Data Privacy and Collaboration Contradiction

High-quality data is the foundation of artificial intelligence and big data. Data collaboration requires high quality and as much data as possible to provide support for artificial intelligence and big data. However, the owners of the data do not want data to be misused, especially the data that needs privacy protection. Therefore, they only want to provide as little data as possible. This is the contradiction between data privacy and data collaboration.

2.2 Data Legality

Under vague terms of user agreements, user data is often over-collected or used without notice, so the collectors will have legal risks. Collectors often do not have the capabilities of artificial intelligence and big data processing, and they need to entrust a third party to process data and do the calculations. This requires third parties and even more participants to have the right to legally use data, which introduces more complexity and security requirements to the system. On the other hand, it is usually difficult for artificial intelligence and big data to obtain legal data, which leads to the slow development of such technologies.

2.3 Conflict Between Ownership and the Rights of Usage

The "free of charge" mode on the Internet is based on a centralized system, where the user data in the system is used for free. More and more users gradually realize the value of their data, and require a declaration of data ownership or even monetization. The use of data will gradually change from free mode to payment mode.

Currently, data is stored in a centralized way on the Internet. A large amount of data is averaged before it reaches date consumers, and the public has fewer and fewer opportunities to obtain raw data. Therefore, the value of personalized data is obscured. Data is driven by data centers instead of the requirements of data producers and data consumers. With the development of data, more users are willing to discover the value of data on a paid basis, so they need a new infrastructure to make this possible.

2.4 Separation of Data Governance and Rule Governance

Since data privacy is often violated and data confirmation cannot be guaranteed, the common method is to centralize the data on a trusted management platform. This may cause the management system to be too large to manage. The responsibility of such a system may also become a large burden. On the other hand, it caused the data to be used in a daunting manner and the value of data could not be efficiently discovered. Therefore, the data responsibilities, rights, and benefits need to be consistent. The governance of data, which refers to the behaviors of data owners and data users; and the governance of rules, which refers to the behaviors of rule makers and data flow regulators; are two different dimensions. The governance of data and the governance of rules, need to be separated logically.

The trusted management platform is usually a rule maker or regulator. The aggregation of data to a trusted party breaks the logical relationship between data and rules. How to ensure that data and rules operate separately and that data flows effectively within the infrastructure is extremely important.

2.5 The Traceability of the Evidence Chain

The rule itself and the processes of rule execution need to be auditable for a while and can be used as evidence to verify that these have indeed happened. Therefore, the evidence should be traceable and immutable. A mechanism should be provided to demonstrate how the data and rules are created and updated. The original data and rules cannot be modified on logs, and can only be updated with explanations and signatures if necessary.

2.6 The Cost of Preventing Infringement

The common solution to prevent infringement is to detect infringement through comparison by human beings or by artificial intelligence. Infringement would be punished using legal means. In many scenarios, the cost of infringement detection and reduction is higher than the compensation itself from the lawsuit. Therefore, infringement is dealt with case by case, and cannot be banned totally.

Ideally, in addition to legal protection, mechanisms should also be introduced to make legal usage cheaper than infringement, and the benefits of legal usage higher than infringement.

3 Overview

IOR, as a decentralized infrastructure, provides a feasible methodology for the five issues mentioned above. Next, we will describe how the Internet of Rights(IOR) builds a model of intelligent governance.

3.1 Internet of Rights(IOR) Model

In a centralized system, each center communicates via interfaces. This manner of data usage is not conducive to user communication and privacy protection. The Internet of Rights(IOR) model, as a decentralized infrastructure, supports controllable anonymity, data sharing, collaboration, supervision, and traceability. It is a trusted bridge to connect data, and to connect data consumers with data producers, thus optimizing the allocation of resources (Fig. 1).

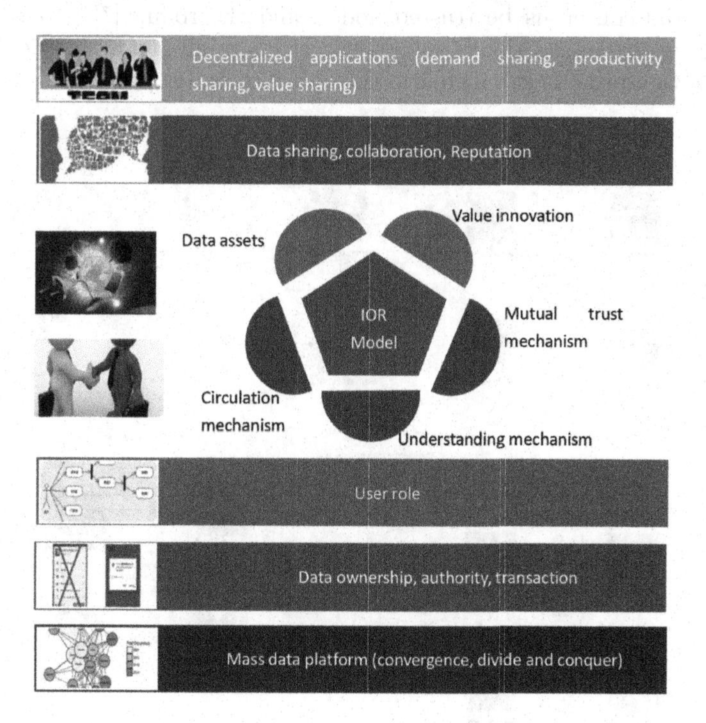

Fig. 1. Internet of Rights(IOR) model

Using converging data assets, the Internet of Rights(IOR) model establishes mutual trust and understanding mechanisms. The Internet of Rights(IOR) model introduces roles and permissions into the traditional blockchain, and standardizes data governance and rule governance, to make legal usage of data throughout its life cycle. With identity management within the blockchain, self-sovereign identity would be made possible, and a safe and efficient identity authentication and authority management mechanism would be created [13].

3.2 Topology

Internet of Rights (IOR) model supports the characteristics of multi-chain and chain separation. It evolves from a flat structure to a tree structure. The main chain and sub-chains are logically separated chains that can be expanded and upgraded independently. Each sub-chain is connected to the upper chain through a group of leader nodes, which can speed up transaction submissions and ensure credibility [6].

Multi-chain is divided into three types: role chain, data access control chain or permission chain, and service chain. The service chain can be separated recursively, and the separated sub-chain contains at least a role chain and a permission chain. The fewer chain nodes submit transactions, the faster the consensus can be achieved. Therefore, the number of consensus nodes in the chain does not need to be large. To increase the reliability, we suggest that the nodes participating in the consensus be trusted nodes, and the rollup [7] process can be selected after the transaction is submitted, with a minimum amount of data for the transaction proof is submitted to the upper chain [8] (Fig. 2).

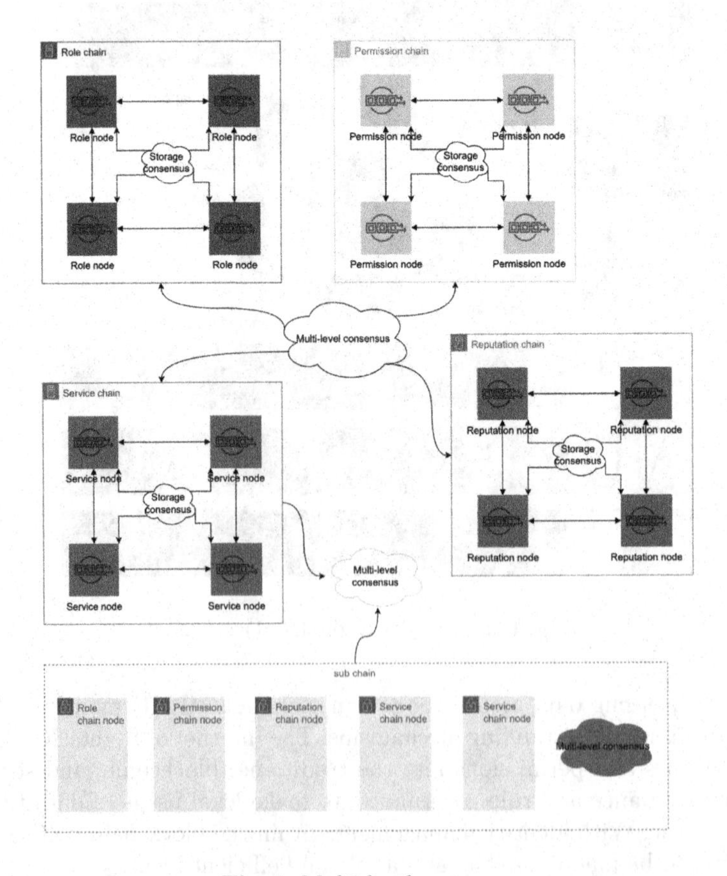

Fig. 2. Multi-level consensus

Big data technology focuses on massive data processing, storage, and calculation, while blockchain technology focuses on decentralization, immutability, controllable anonymity, and cross-domain processing of information. They are perceived as two complementary strategies. Mass data should be processed by the big data system, but the fingerprints and metadata of permissions of the data are on the chain. The role chain and permission chain are the standardization of data governance rules, allowing an effective integration of blockchain technology and big data technology. The integration of blockchain technology with distributed file systems, big data analysis, cloud computing, artificial intelligence, and other technologies is crucial for future development [14].

3.3 Role

The role refers to the nature of the user, which classifies users with similar permissions. Management of all users and data by a centralized system will inevitably lead to data concentration and unnecessary responsibility. The role clearly defines the responsibility boundaries of various participants, so that everyone can obtain the necessary data in compliance with rules and regulations. Participant can only access and operate data within the scope of their roles. The role definition is defined by the user manager and the metadata is submitted to the role center. The user manager assigns role identifiers on the role chain to ensure uniqueness in the role chain. The role chain does not provide role application services, but only role consensus services, that is, role data on the chain.

Role data mainly includes the relationship between users and roles. Not all role relationships of a user are stored in one role data structure, and one role data structure only contains one user. The modification of role data must be done through blockchain transactions.

3.4 Permission

Permission refers to the collection of data access and use capabilities, which has subordinate permissions and is a tree structure. Permissions refer to dimensions of data categorizing and the ways of using data. Permission is an attribute of data and does not change with the changes of users.

Permissions and roles are separated. The authority center is only responsible for the definition of data usage attributes, not for those who might have the permissions. The permission definition is defined by the regulators and metadata is submitted to the permission center. The regulators assign the authority identifiers on the authority chain to ensure that they are unique within the chain. The permission chain does not provide permission application services, but only permission consensus services.

Permission data includes two types of data: role and permission mapping, and data permission. The mapping of roles and permissions defines which permissions a role has. The permissions contained in a role are not necessarily contained in only one permission data structure, and one permission data structure only contains the permissions owned by one role. A data permission only includes all permission definitions for data. The modification of data permissions must be done through blockchain transactions.

3.5 Traceability

One of the basic capabilities of the blockchain-based on UTXO technology is traceability, which increases the cost of fraud and maintains a trusted collaboration environment. The traceability of UTXO is realized by three features:

1. The vins field contains the hash value of the pre-UTXO (hash) and the owner's signature.
2. The vouts field contains the allocation of subsequent UTXOs, and the sum of their values is equal to the sum of the values of all preceding UTXOs. The hash value (hash) of the subsequent UTXO is represented by the hash value (hash) of the entire tx.
3. UTXO cannot be double spent. It can only be spent once.

Therefore, the UTXO technology makes the source paths of all UTXO traceable through the vins and vouts fields and has the corresponding UTXO owner's signature to ensure the legality of UTXO usage. UTXO also guarantees that there can only be one source path in the life cycle of a UTXO by prohibiting double-spending.

4 Multi-level Consensus

The introduction of smart contracts has brought many opportunities to the ecology of the blockchain, and the blockchain's support for data has made a big step forward. However, simply assigning all data logic to smart contracts will undoubtedly increase the complexity, and decrease the robustness of smart contracts. We use multi-level consensus to extract non-business logic in smart contracts, thus managing data and rules more effectively and securely.

Multi-level consensus is bottom-up and has good scalability, including but not limited to storage consensus, role consensus, and permission consensus. The scope of multi-level consensus will be broadened in the future.

4.1 Storage Consensus

The basis of multi-level consensus is storage consensus. Participants in different scenarios are different, so each scenario has an independent chain to provide consensus services. The sharing of information between multiple chains requires mutual authentications between the chains, i.e. the storage consensus.

Storage consensus requires the following capabilities:

1. Multi-chain addressing capability: Since the storage between two chains is not shared, a communication and addressing mechanism must be provided so that the chain can understand each other's location.

2. Distributed storage capacity: The storage consensus itself relies on distributed storage to allow the data stored by each participant to reach an agreement.
3. Distributed verification capability: When receiving a verification request that is not signed by the original chain, the authenticated data needs to be found and sent to the source chain for verification, based on the identification information of the source chain contained in the data itself, using addressing capability.
4. Distributed transaction capability: The transaction in a multi-chain system is distributed, and the participants of the transaction are not limited to one chain. In the UTXO transaction scenario, all input UTXOs are recovered by their respective source chains, and the recycled UTXOs will generate a corresponding new UTXO in the transaction chain. All output UTXOs are generated from these new UTXOs. Both recycled and newly generated UTXO is encrypted and signed by the private key designated by the system.
5. Distributed security capability: All communication channels need to be encrypted, and the identities of the participating parties are mutually verified. The security of the verification process is solved by cryptography, the communication channel uses SSL, and the identity verification uses an asymmetric encryption algorithm for abstract and signature.

4.2 Role Consensus

In a distributed decentralized network, data ownership belongs to users. The user saves the fingerprint of the role information through the storage consensus in the role chain.

The role data is accessed through the blockchain network as the signaling network communication. After the signaling network handshake, the private information is exchanged through the temporary trusted channel, and the access log is stored through hash collision. When needed in the future, it can be compared and verified on the blockchain network to ensure the authenticity of information exchange.

Role consensus is based on storage consensus. Role data can be stored on any role chain, be authenticated on any role chain, and participate in distributed transactions of any role chain.

4.3 Permission Consensus

The nature of the multiple roles of producers and consumers determines that their permissions are also multi-dimensional. Among the data in each domain, there is a need for confidential data to be shared with other domains and subject to supervision. Therefore, the data provided by each domain must be honest and reliable, and there must be no data conflicts on different occasions. However, it does not want to be publicly visible, nor does it want the administrator to have excessive authority.

The fingerprint of the permission data is uploaded to the chain to form a permission consensus. According to the multidimensional authority standard.

The definition and verification of distributed authority are realized under the framework of distributed roles, and data assembly and packaging capabilities are provided. Permission data is transmitted through a temporary encrypted channel and is not stored on the blockchain network, preventing immutable information from being cracked by more advanced cryptographic schemes in the future, avoiding data leakage, and protecting user privacy.

Permission consensus is based on storage consensus. Permission data can be stored in any permission chain, authenticated on any permission chain, and participate in distributed transactions in any permission chain.

4.4 Reputation Consensus

Reputation refers to the influence of an entity calculated based on user data, user behavior, intellectual property, and digital assets. Reputation includes but is not limited to reward points and credits. Reputation reflects the entity's influence, which can be extended to its derivatives and interaction with other entities.

Reputation consensus refers to the consensus of influence of a subject reached within a scope and is recorded in blockchain. Reputation generally agreed that blockchain would be a better replacement for page ranking for search engines. A good reputation will gain high visibility and will gain more potential benefits, while a poor reputation will reduce potential benefits. Through reputation consensus, the subject will be encouraged to act in the manner promoted by the rules of consensus within its scope.

5 Protocol

5.1 Protocol Data Structure

In the Internet of Rights(IOR) model, blockchain is based on the UTXO extended model [2], uses the dyeing model as the domain division and also adds the ability to verify historical scenes. The transaction data structure is divided into several parts: vins, vouts, color, reference, cooperation, tx data, tx signature, and block data. These designs ensure that all data is verifiable, and responsibilities, rights, and benefits are bounded.

The "color" field represents the color model, and different colors represent different scenes. The "reference" field indicates the entity association and can be used for entity indexing. The "cooperation" field is a multi-party collaboration signature. The "tx" data field is transaction data. The "tx" signature field is the transaction data signature field, and the tx data field is protected from modification. Block data contains the block number and transaction number, which are used for historical scene verifications.

Due to the multi-chain technology, the number of consensus nodes in each chain is reduced, and 51% of attacks would become easier. The GHOST [5] algorithm puts forward the principle of maximizing the number of blocks after the fork, not only the principle of the longest block of the fork, which increases the difficulty of 51% of attacks. The Conflux [9] algorithm also adopts the principle of maximizing the number of bifurcation blocks, combined with directed acyclic graph (DAG) calculations to improve performance and security. The intelligent governance model multi-chain technology may use the advantages of these two technologies, combined with trusted node verification and the Rollup mechanism [7], to further enhance performance and security. The attacker has to forge the longest chain and the number of nodes [5], which would only break through the protection of the first layer of the chain. It also needs to break through the second layer of the Rollup chain to make its forgery take effect, so the degree of difficulty is doubled.

5.2 Authorization Process

The owner A of entity E manages the authority of entity E in the blockchain system and uses UTXO U1 as the proof of ownership, where the value of U1 is the number of authorizations. User B of entity E needs to apply for permission from A before using it. A converts his U1 into two new UTXOs through the UTXO model. One is to provide B with UTXO U2 for the usage of E, and the other is A's ownership of UTXO U3. The sum of the value of U2 and U3 is equal to the value of U1.

Similarly, when user B uses multiple entities to generate a new entity F, he will obtain multiple UTXOs in turn. When user B publishes a new entity, he converts these multiple UTXOs into the ownership UTXO of a new entity. The conversion of the right to use UTXO to the ownership of UTXO needs to be signed by multiple parties, including the process party, the right to use UTXO party, and the owner of the UTXO party.

When user C uses entity F, he needs to apply for authorization from owner B, but user C does not need to apply for authorization from owner A. Since user B has applied for authorization from owner A, the traceability of the authority is guaranteed.

In this process, each entity's permission UTXO (including usage rights and ownership) is equivalent to a point, and each authorization is equivalent to an edge, which forms a directed graph, and all authorization models can be calculated through graph calculations. The value of UTXO represents the number of permissions and is separable. It is modified every time it is converted, and the life cycle management of the permission can be carried out. The directed graph records all compliance and permission transfer information, and its relationship can be calculated based on the graph to get a reasonable result, ensuring the legitimate rights and interests of owners and users (Fig. 3).

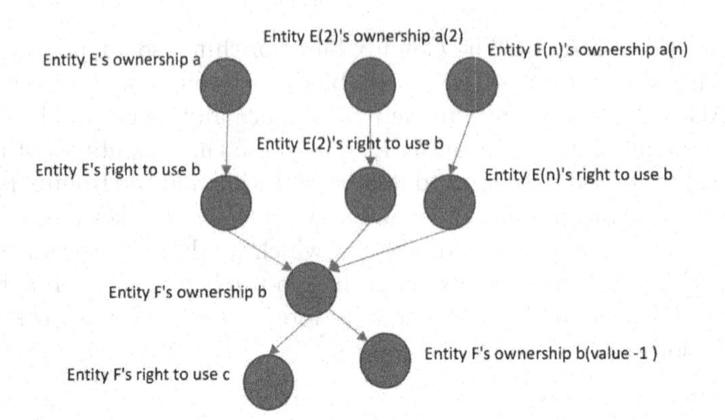

Fig. 3. Authorization process

5.3 Multi-party Collaborative Signature

The multi-party cooperative signature adopts the proxy signature [3] mechanism, and requires the orderliness of the cooperative signature. First, the reference field stores the hash value (Hash) of the multi-party collaboration process UTXO, and the UTXO can verify the authorization of the participants and the collaboration sequence of the multi-party collaboration. Second, each participant provides its UTXO and signature. Finally, the transaction committer signs the entire transaction data, saves it in the TX signature field, and submits it. Multi-party collaborative signatures support each step of signing data on the chain, but the intermediate process cannot be authenticated and is only stored as a log.

The signature algorithm adopts verifiable random function $VRF(x, sk)$ algorithm [1], input string (x) and private key (sk), VRF algorithm returns a hash value $(hash)$ and proof the value is $(proof)$. The proof value $(proof$, denoted by $\pi)$ can be calculated from the public key (pk) and hash value $(hash)$ for verification. Since the private key (sk) is only owned by the user, it can prove the validity of the signature (Fig. 4).

```
procedure MultiSignatureTx(tx,sk_num,sk_array,sk):
  while i < sk_num do
    sk_i ← sk_array[i]
    Tx.cooperation[i].sign ← VRF(tx.cooperation[i].
        data,sk_i)
    i++
  (hash,π) ← VRF(tx,sk)
  Cache(hash,π)
  If tx.reference is entity identity then
    Cache(tx.reference, hash)
  Return (hash, tx)
```

Fig. 4. Signature algorithm function

Since the cooperation will add a hash of the participant each time the multi-party collaborative signature is transferred, the cooperation part is only used to prove that the participant owns the previous data. When data is on the chain, the blockchain consensus node fills in the current block number and its signature for historical validity verification.

5.4 Data Authentication

Transaction data (TX) can be authenticated through the inverse operation process of multi-party collaborative signatures, and is divided into two types of authentications:

1. Current validity authentication.
2. Historical validity verification.

Log data is all historical data in the blockchain system, and historical validity can be used to verify whether it is legal. The valid history does not indicate the validity as of now, and the validity as of now indicates a valid history.

Validity check of current data: First, find the transaction data itself on the chain according to the transaction hash value from the blockchain, check whether it is unspent, and then verify whether all the multi-party cooperative signatures of the transaction data are legal and valid.

Validity check of historical data: Find the corresponding block through the block number in the transaction data to be verified, then find the corresponding transaction data from the block, check whether the transaction to be verified is included, and finally verify all the multi-party cooperation of the transaction data, in turn, to see if the signature is legal and valid.

5.5 Rules of Protocols

The data is effectively segmented through the role and permission system, and different participants use different data according to the role definition so that the necessary data is available when the data is needed. Data beyond the scope of authority will not be accessible, which fully protects data privacy. Users no longer worry about excess data being used at will, and the promised data usage range will not be exceeded.

All the fingerprints of role and permission data need to be chained, and the collected behavior data needs to be chained after data collection. Collecting behavioral data refers to who, when, how, and the signature of the data. This mechanism is adopted to ensure that privacy is not violated. Evidence is provided to check if the data is used in compliance and lawfulness.

When data, roles, and permissions are on the chain, they must be signed by multiple parties. The application process and results of roles and permissions are all signed. These signatures are the authentication of the ownership and use rights of the data. If a new right to use is needed, it must have the owner's signature.

The chain separation technology separates data governance and rule governance. The manager is only responsible for rule governance, and all participants are required to upload activity logs according to the rules. Therefore, the intelligent governance model provides a way to clarify the boundaries of each participant's responsibilities and legitimize the activities of the participants within their responsibilities.

5.6 Evaluation Method

For the entire system to have an inherent self-driving force for automatic optimization, each chain needs to be adjusted and optimized based on the results after evaluation. Rewards are provided to the chains with high results penalties to the chains with poor results. Economic principles are adopted to allow market forces to drive the entire system toward the optimal allocation of resources. There are currently two assessment methods:

1. The performance of the chain, and the speed of consensus propagation, describe the time from the submission of consensus to the consensus of the whole chain of a transaction.
2. The trust degree of the chain describes the trust relationship between the chain and other chains, which can be expressed by a weighted adjacency matrix.

6 Evaluation Calculation

6.1 Propagation Speed Assessment

Due to the complexity of the application scenario and the characteristics of the blockchain system, throughput and performance are inversely proportional to the number of participating consensus nodes. The chain separation technology reduces the number of consensus nodes in a single chain, improves the consensus speed, and reduces the time to reach the consistency of the entire chain. Since the consensus is only within the chain, and other chains are used as read-only nodes to verify data, the data propagation of other chains becomes read-only propagation, the network topology has changed from a mesh structure to a snowflake structure.

Due to the large size of the transaction data structure designed by the protocol framework, the block data size is also large. The size of the transmission block data message is related to the transmission delay cost. The smaller the message, the more the additional load. To avoid excessive additional load and subsequent additional calculations, a very small inv message is added before each message is transmitted to detect whether a specified block is to be transmitted, and the node that has received it. It no longer needs to obtain the block [4].

Suppose S is a node that has not obtained a block, I is a node that has obtained a block, and the total number of nodes in the chain is N.

Definition: t is the number of transmissions, which is a positive integer.

Then the ratios of the two types of nodes to the number of sub-chain nodes at the time of t are recorded as $s(t)$, $i(t)$, and the number of two types of nodes are: $S(t)$, $I(t)$.

At the initial moment $t = 0$, the initial ratio of the number of acquired nodes to the number of unacquired nodes is s_0 and i_0.

The average number of nodes (the average number of node addresses held by the network) of each node in a propagation cycle is λ.

Each propagation marks the nodes that have not obtained blocks in the ratio of $\lambda * s(t)$ as "obtained" ones, so the number of nodes that have obtained blocks is $N * i(t)$, so there is every day $\lambda * s(t) * N * i(t)$ nodes to obtain blocks, that is, the number of nodes that have been obtained blocks newly added every day, the differential equation can be obtained:

$$N * di(t)/dt = \lambda * s(t) * N * i(t) \tag{1}$$

$$s(t) + i(t) = 1 \tag{2}$$

$$\begin{aligned} di(t)/dt = \lambda{*}(1 - i(t)) * i(t) \\ \textbf{, and } i(0) = i_0 \end{aligned} \tag{3}$$

The total number of obtained block nodes:

$$I(t) = N * i(t) \tag{4}$$

Solved according to the natural logarithm:

$$\begin{aligned} i\left(t\right) &= \frac{i_0 e^{\lambda t}}{1 + i_0(e^{\lambda}t + 1)} \\ &= \frac{1}{1 + (\frac{1}{i_0} - 1)e^{-\lambda t}} \end{aligned} \tag{5}$$

From the nature of the natural logarithm, it can be obtained that when $i(t) = 1$, t is infinite. Since I, S and t are all positive integers, we find that t makes $i(t) = N - 1/N$, the above formula is transformed into: $t = 1/\lambda ln((N - 1)(i_0 - 1)/i_0)$

Because one block transmission is two API calls (one inv), we may set the average delay of each API call to p,

Then the delay of a propagation: $delay = \lambda * 2 * p$

From $i(t) = N - 1/N$, $i(t)$ is infinitely close to 1, but not equal to 1.

Premise: N, S, I, t are all positive integers

Corollary: $i(t) >= 1$ when $t + 1$. Since it is impossible for $i(t)$ to be > 1, it can be deduced that $i(t) = 1$.

Set the final duration: $T = (t + 1) * delay$

Calculation result: $T = 2p\ ln((N - 1)(i_0 - 1)/i_0) + 2p\lambda$

6.2 Trust Evaluation

Trust. The degree of trust refers to the degree to which an entity trusts another entity, where the trust degree of the entity to itself is fixed at 1. The degree of trust can be calculated from the three indicators of credibility, reliability, and intimacy.

Suppose the trust degree of entity i to entity j is $Cred_{ij}$ (the degree to which entity j trusts entity i)

$$[Cred_{ij}] = [NC_i \times DPR(G_i) \times I^*_{ij}] \tag{6}$$

Credibility. Suppose the credibility is NC, and NC^{init} is the initial credibility [10]. Reliability is a constantly changing value that is weighted and calculated based on feedback.

Note: t_{cur} and t_{pre} represent the time gap between the current credibility and the last credibility.

$T(i)$ and $RP(i)$ indicate the number of weighting calculations and adjustment values.

Then, $NC(i)$ is calculated as follows:

$$\begin{aligned} NC_i &= NC_i^{init} \\ &+ (t_{cur} - t_{pre} + 1) \sum_i^j T(i) \\ &* RP(i) \end{aligned} \tag{7}$$

Network Reliability. Suppose entity i is a distributed network, then the reliability of entity i is recorded as $DPR(G_i)$ [11].

$$\begin{aligned} DPR(G_i) = p_s[&p_e p_v DPR1(G * e) \\ &+ (1 - p_e p_v) DPR1(G - e)] \end{aligned} \tag{8}$$

Among them, p_s is the probability of normal operation of the starting node s, e is the network transmission from s to v, and p_e is the probability of normal operation of the network transmission from s to v. p_v is the probability of normal operation of node v. $DPR1(G * e)$ represents the original network G fusion e means: the distributed reliability of the merged s and v, $DPR1(Ge)$ represents the original network G after cropping e distributed reliability.

Intimacy Between Entities. The intimacy I_{ij} between entity i and entity j is defined by the frequency of messages sent by the two entities [12], and the total number of entity i sent to entity j is recorded as q_{ij}, the total number of messages received by the entity is recorded as q^{in}, and the total number of messages sent by the entity is recorded as q^{out}. Then, the formula for calculating intimacy between entities:

$$I_{ij}^{*} = \frac{q_{ij}}{\sqrt{q_i^{out} q_j^{in}}} \tag{9}$$

$$q_i^{out} = \sum_{j=1}^{n} q_{ij} \tag{10}$$

$$q_j^{in} = \sum_{i=1}^{n} q_{ij} \tag{11}$$

6.3 Credit Result

According to the above formula, the network composed of five sub-chains is calculated and evaluated. The result of the trust degree between them is (See Table 1).

Table 1. Result of the trust degree

Chain	1	2	3	4	5
1	1	0.053785	0.164643	0.212927	0.086383
2	0.275296	1	0.068036	0.289826	0.072169
3	0.018092	0.003233	1	0.014873	0.041154
4	0.498808	0.157861	0.136248	1	0.158461
5	0.084898	0.114817	0.027762	0.033741	1

7 Conclusion

The Internet of Rights(IOR) model is a blockchain application model that tries to solve certain problems faced by the data era with considerations of distributed governance. It formulates governance rules through consensus on roles and permissions, allowing each participant to conduct activities by the rules, and put the fingerprints of the activity log on the chain.

The idea is to reach a consensus on data without knowing what the data is. Therefore data privacy and collaboration no longer conflict. Participants of the chain obtain only the data they need according to their roles and permissions, which reduces data security issues and reduces liability risks so that the entire link from data collection to use is compliant and legal.

The storage consensus of the Internet of Rights(IOR) model makes mutual verification between multiple chains possible and provides a way to resolve the conflict between owners and users. Data owners can specify the scope of users based on roles and permissions, and reasonably divide the rights of owners and users through data authorization and authentication services.

The chain separation feature provided by the Internet of Rights(IOR) model separates the role chain, the permission chain, and the service chain, and provides a method of separating data governance and rule governance, allowing managers to formulate rules and users to use data according to the rules. It reduces the impact of a single point of failure, improves the availability of the system, reduces the risk of managers who used to manage too much data before, and creates an intelligent work model.

The Internet of Rights(IOR) model increases the cost of fraud through immutable and traceable accounting capabilities, retains log records, can be traced when needed, promotes the relationship of mutual trust between various chains, and enhances the participation in the creation of data value. Therefore, evaluating the participants and making continuous adjustments based on historical records is necessary. Participants with higher scores are rewarded, which helps everyone to form a fair and just environment.

Under the multi-chain consensus mechanism of the Internet of Rights(IOR) system, the UTXO model and the account model will be used together as the two modes of bookkeeping in the future, and the transactions between them can be converted to each other [15]. Multichain consensus integrates dual-ledger inclusive design, which not only adopts the maturity and stability of traditional technology but also leaves room for new distributed ledger technology, making the two distributed technologies compatible with each other, parallel, and complementary [14].

References

1. Gilad, Y., Hemo, R., Micali, S., Vlachos, G., Zeldovich, N.: Algorand: scaling byzantine agreements for cryptocurrencies. In: SOSP '17: Proceedings of the 26th Symposium on Operating Systems Principles, pp. 51–68, October 2017
2. Chakravarty, M.M.T., et al.: Native custom tokens in the extended UTXO model. In: Margaria, T., Steffen, B. (eds.) Leveraging Applications of Formal Methods, Verification and Validation: Applications. ISoLA 2020. LNCS, vol. 12478, pp. 89–111. Springer, Cham (2020). https://doi.org/10.1007/978-3-030-61467-6_7
3. Bamasak, O.: A collusion-resistant distributed agent-based signature delegation (CDASD) protocol for e-commerce applications. Intell. Inf. Manag. **2**(4), 262–277 (2010)
4. Decker, C., Wattenhofer, R.: Information propagation in the Bitcoin network. In: IEEE P2P 2013 Proceedings. Trento, Italy (2013)
5. Sompolinsky, Y., Zohar, A.: Secure high-rate transaction processing in bitcoin. In: Bohme, R., Okamoto, T. (eds.) Financial Cryptography and Data Security. FC 2015. LNCS, vol. 8975, pp. 507–527. Springer, Berlin, Heidelberg (2015). https://doi.org/10.1007/978-3-662-47854-7_32
6. Pass, R., Shi, E.: Thunderella: blockchains with optimistic instant confirmation. In: Nielsen, J., Rijmen, V. (eds.) Advances in Cryptology – EUROCRYPT 2018. EUROCRYPT 2018. LNCS, vol. 10821, pp. 3–33. Springer, Cham (2018). https://doi.org/10.1007/978-3-319-78375-8_1

7. Konstantopoulos, G.: How does Optimism's Rollup really work? (2021). https://www.paradigm.xyz/2021/01/how-does-optimisms-rollup-really-work
8. ethhub-io. Optimistic Rollups. https://github.com/ethhub-io/ethhub/blob/master/docs/ethereum-roadmap/layer-2-scaling/optimistic_rollups.md
9. Li, C., Li, P., Zhou, D., Xu, W., Long, F., Yao, A.: Scaling Nakamoto consensus to thousands of transactions per second, distributed. Parallel. Clust. Comput. https://doi.org/10.48550/arXiv.1805.03870
10. Hu, Z., Zhong, L., Yang, Z., Liu, W.: Blockchain improvement scheme based on PBFT consensus algorithm. Comput. Sci. Appl. https://doi.org/10.12677/CSA.2021.113066
11. Sun, Y., Zhang, X.: Efficient algorithm for computing reliability of acyclic directed networks. Northeastern University, CN (2004)
12. Yao, L., Xiaohui, K., Hong, G., Qiao, L., Zufeng, W., Zhiguang, Q.: A community detecting method based on the node intimacy and degree in social network. J. Comput. Res. Dev. (2015). https://doi.org/10.7544/issn1000-1239.2015.20150407
13. 姚前, 区块链技术十周年——回眸与前瞻, CSDC, CN, 2018
14. 姚前, 数字货币的缘起、发展与未来, CSDC, CN, 2018
15. Zahnentferner, J., Ledgers, C.: Translating and Unifying UTXO-based and Account-based Cryptocurrencies, Input Output HK, Hong Kong (2018)

Tibetan Jiu Chess Intelligent Game Platform

Yandong Chen[1,2], Licheng Wu[1,2], Jie Yan[3], Yanyin Zhang[1,2], Xiali Li[1,2], and Xiaohua Yang[4](✉)

[1] Minzu University of China, Beijing 100081, China
[2] Key Laboratory of Ethnic Language Intelligent Analysis and Security Governance of MOE, Minzu University of China, Beijing, China
[3] Hangzhou Hikvision Digital Technology Co., Ltd., Hangzhou 310051, China
[4] China Industrial Control Systems Cyber Emergency Response Team, Beijing, China
yangxiaohua@infoip.org

Abstract. The scarcity of gaming platforms for Tibetan Jiu Chess is not conducive to the dissemination of Tibetan Jiu Chess culture, and even more so, it is impossible to efficiently collect Jiu Chess game data to serve scientific research. In order to solve the above problems, and considering that most of the Jiu Chess masters are herdsmen distributed in Tibetan areas, a WeChat mini-program applicable to most of the Jiu Chess players has been developed. The intelligent platform has the functions of online game and game analysis and is realized by using cloud development and real-time data push technology. In the National Tibetan Chess Competition in 2023, the online test of the platform was successfully completed. Entrusted by the organizing committee of the competition, the platform collected more than 10 games data of winning masters and completed online replay. At present, the platform has more than 200 players playing online on a daily basis, and its operation is stable.

Keywords: Intelligent Platform · Tibetan Jiu Chess · Jiu Chess Game Data · National Tibetan Chess Competition · Cloud Development · Real-Time Data Push

1 Introduction

Tibetan Jiu Chess is a kind of folk chess sport with complex and unique rules, which is based on the "Square Chess" in the northwest, and is formed by the innovative development of multi-ethnic cultures such as Chinese, Tibetan, Mongolian, and Hui. It has dozens of variants and is both competitive and interesting [1]. Since 2010, national Tibetan Chess tournaments have been held annually in Sichuan and Tibet. In 2019, Tibetan Chess was included in the list of items for the annual China University Computer Gaming Competition and China Computer Gaming Championship.

The chessboard of Tibetan Jiu Chess has different sizes such as 8 and 14-way [2] and has a variable chessboard. The game process is divided into two stages: layout, battle (move and fly), each of which has unique rules and affects each other, and the determination of victory and defeat is complicated [3]. Tibetan Jiu Chess game belongs

F. Sun and J. Li (Eds.): ICCCS 2023, CCIS 2029, pp. 322–333, 2024.
https://doi.org/10.1007/978-981-97-0885-7_26

to the complete information game in machine games, and the research of its game algorithms and game platforms is still in the initial stage. At present, the lack of game data and expert knowledge is one of the common problems faced by Tibetan Jiu Chess games and the preservation and inheritance of Tibetan Jiu Chess culture. The lack of Jiu Chess game data will not be able to provide strong data support for Jiu Chess game intelligence, nor will it be able to provide basic support for AI research as well as applications. By developing a intelligent digital game platform for Jiu Chess, we can collect game data and extract expert knowledge, which is of great significance for the protection and development of traditional Tibetan culture in China and research of high-level Tibetan Jiu Chess artificial intelligence.

In the era of mobile Internet, cell phone, as the most versatile type of terminal tool, has become an essential tool for people's life [4]. Considering the fact that the game platform of Tibetan Jiu Chess has to satisfy the factors of convenient collection of game data and ease to be used by most of the Tibetan herdsmen at the same time, we have chosen WeChat cloud development technology to build a mini-program of Tibetan Jiu Chess. WeChat mini-program [5, 6] allows us to develop an application that runs fast, is easy to operate, and can be accessed without downloading. Users can open the mini-program by simply searching or sweeping, which is convenient for chess players and conducive to the collection of chess game data. In addition, the mini-program can also serve the National Tibetan Chess Tournament by providing the game platform needed for the tournament, collecting game data between masters, and using it for display. Therefore, it is very feasible to design and develop a Tibetan Jiu Chess mini-program.

Section 2 analyzes the current status on the Tibetan Jiu chess and game platform. Section 3 describes the development environment and technical solutions for the mini-program. Section 4 describes the design scheme of the mini-program, including the interface design of the mini-program, the design of the game flow, the design of the data table, the design of the game data, the replay of chess game data, the representation of the Jiu Chess board and chess shape, the design specification of the Jiu Chess game data, and the win-lose judgment. Section 5 introduces test results of the platform. The work and its prospects are summarized in the last.

2 Research Status

2.1 Related Works on Game Platform

Chess, as a kind of educational, fun, and gaming game, is loved and enthusiastic by players [7]. With the development of new Internet technology, chess games have also turned to the web and mobile. At present, the functions of popular chess platforms have been perfected, and the development of the platform has reached saturation. For example, Yike Weiqi has attracted a large number of users by virtue of its functionally perfect platform and has reached more than 24 million games of chess against each other in 2019 alone. Tencent's QQ game platform covers a wide range of board games such as Go, Chinese Chess, Chess, Gomoku, Checkers, Mahjong, and so on, with a large number of users and a wide range of dissemination, which has greatly promoted the development of board games. The general-purpose gaming platform Ludii can be used to experience and design all kinds of games. The variety of games covers board games, dice games, mathematical games, etc. [8] The Ludii system provides a new intuitive approach to game design and gameplay, and the language used for describing games opens up many new avenues of research, especially in the area of general game generation and analysis [9].

2.2 Researches on Tibetan Jiu Chess

Currently, the research on Tibetan Jiu chess focuses on the development of search algorithms, deep reinforcement learning models, and intelligent game programs. The attack and defense strategy based on chess shapes [10] is the first Tibetan Jiu chess program that needs to extract chess patterns from 300 games of human expert game data in advance and then designs different stages of attack and defense strategies based on expert knowledge. The game program based on the segmental evaluation method [11] constructs a method for evaluating the state of the game at different phases. The small-sample machine learning approach [12] uses Bayesian networks with high inference power from small samples of data to extract chess shapes. The phased game strategy [13, 14] applies the Alpha-Beta pruning search algorithm in the battle phase of Tibetan Jiu chess, which obtains the optimal action solution by constructing a huge game tree, but requires a good evaluation function and pruning conditions. The hybrid deep reinforcement learning model [15] used a hybrid time-difference algorithm combined with a deep neural network to train the self-play of Tibetan Jiu chess without relying on human knowledge. The hybrid update deep reinforcement learning self-play model [16] combines the Sarsa(λ) algorithm and the Q-Learning algorithm with deep neural networks to improve the efficiency of the model self-play training.

In contrast, less research work has been done on the Tibetan Jiu chess gaming platform. In 2016, a research team from the Central University for Nationalities developed a stand-alone version of the Tibetan Jiu chess program [17]. It can be regarded as the initial stage of Tibetan chess software development. In 2022, the research team of Central University for Nationalities innovated again and used the game development engine Cocos Creator to create a web-based version of the Tibetan Jiu chess online gaming platform. The platform was used as the designated gaming platform for the Tibetan Jiu chess program of the 2022 National Student Computer Gaming Competition and successfully completed the task of the competition. Due to the complex rules of Tibetan Jiu chess and the late attention of researchers, the types and functions of the game platform of Tibetan Jiu chess are not perfect. With the high-speed development of new technology of mobile Internet, it has become a common trend to make board games shift to mobile. Therefore, it is necessary to design and develop the mobile network gaming platform of Tibetan Jiu chess.

3 Development Environment and Technical Solutions

The development and deployment of the Tibetan Jiu Chess mini-program is done by using WeChat developer tools. As shown in Fig. 1, the front end of the mini-program uses programming languages such as Wxml, Wxss, and JavaScript to complete the design of the user interface. In addition, the development of the Jiu chess mini-program uses cloud development technology [18], which only requires writing and calling cloud functions to achieve real-time dynamic interaction with the cloud database, providing great convenience for the development work. Some static image resources of the mini-program are saved in the cloud storage environment to reduce the size of the compiled resource package of the mini-program and improve the initial loading speed when the user opens the mini-program.

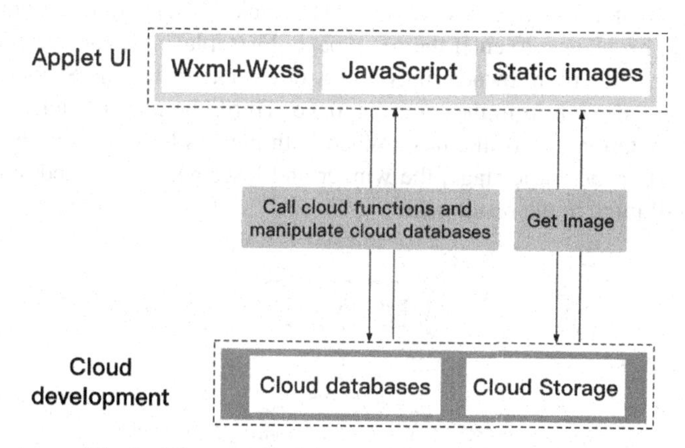

Fig. 1. Diagram of mini-program cloud development.

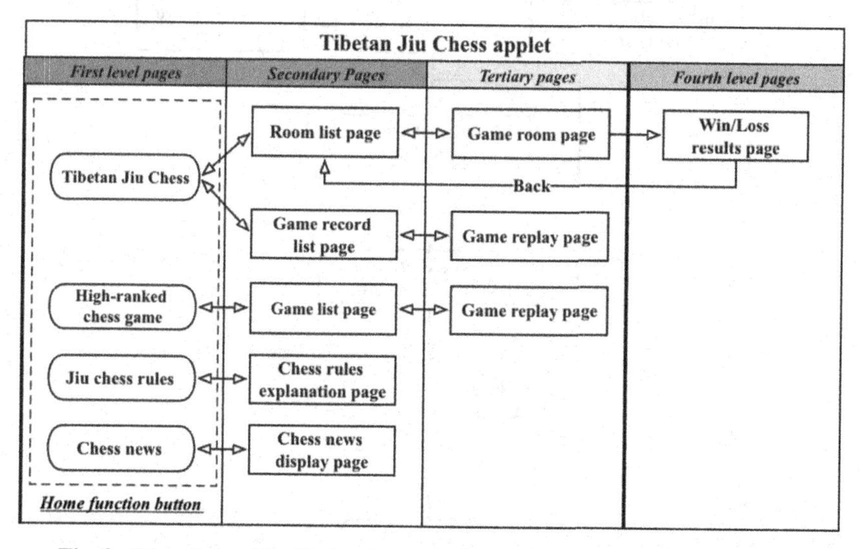

Fig. 2. The relationship diagram between the pages of Jiu Chess mini-program.

4 Design Solutions for the Mini-program

4.1 Interface Design

The mini-program contains pages such as the home page, Jiu Chess room list, game room, game record, high-ranked chess game, chess rules explanation, chess news display, etc. The relationship between each page is shown in Fig. 2. When on the home page, players can enter the secondary page they want to view by clicking on the function button. There are three ways for players to enter the room page: First, players create a room and then enter the room as the room owner and wait for other players to join. Secondly, players select and join the room from the room list page. Third, the player enters the room by clicking the invitation link of the host. When both players have entered the room, the game starts. After the battle stage, the winner and loser will be decided, and then the room page will jump to the result page.

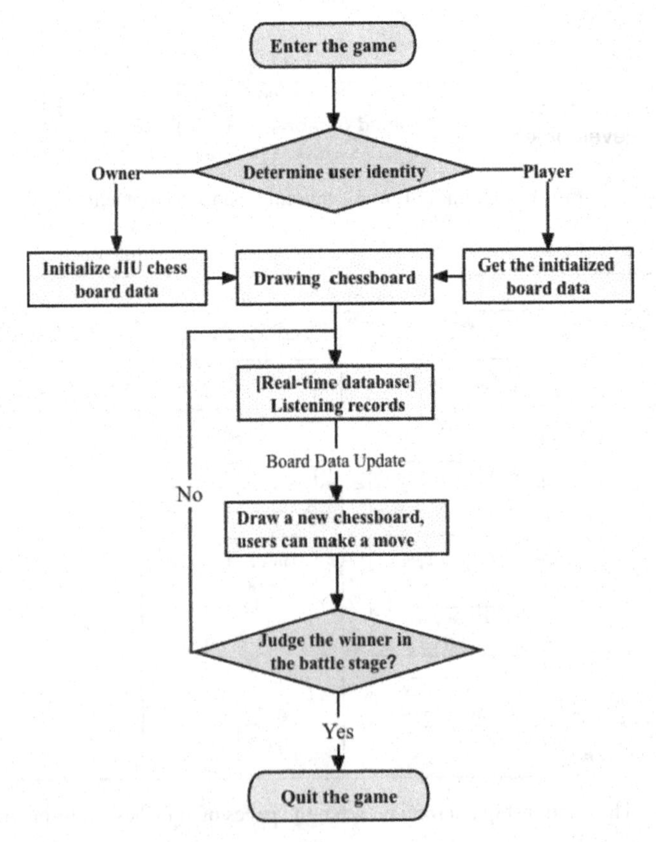

Fig. 3. Flow chart of Jiu Chess game based on real-time data push.

4.2 Flow Design of Jiu Chess Game

The Jiu Chess mini-program uses the real-time data push technology developed by WeChat Cloud to interact with data from both sides of the Jiu Chess match by dynamically listening to the changes in the corresponding fields of the cloud database. The game process can be summarized as the following steps:

(a) **Game entry and identity judgment**

As shown in Fig. 3, the identity of a player is divided into owner (the creator of the room) and player (the player who joins the room afterward). If a player actively creates a new free room, then the player's identity is the owner of the newly created room.

If there is a free room, then the player who joins the free room later is the player.

(b) **Listen to real-time database updates**

When a player with the identity of owner enters a newly created game room, it is necessary to listen for the player with the identity of the player to enter the game immediately. In other words, it is necessary to listen to the changes in the "people" field of the corresponding record in the "room" table through the function of the real-time database.

(c) **Update database fields**

When a player enters a game room, the "people" field in the game room data table is updated. When the "people" field is updated from 1 to 2, it means that the number of players in the room is full and the game will start.

(d) **Update of the chessboard**

Players of both Player and Owner status are required to listen for updates to the remote chessboard. When the remote chessboard is updated, the local chessboard is redrawn according to the latest chessboard state. This is followed by a win/loss determination for the game, and if a win/loss can be determined, the game is exited. Otherwise, the local lock is opened and the player is allowed to play.

(e) **End of game and exit**

If the battle phase of Jiu Chess has been able to determine the winner or loser, then it will automatically enter the exit logic of the game. The exit logic is divided into two parts, the first is to jump to the game's win/loss result page. Then the cloud function is called to store the game data.

4.3 Design of Data Tables

As shown in Table 1, the cloud database stores user information, game room information, and other data. Among them, the "userinfo" table stores "openid" as an identification field to distinguish users, and user nicknames, avatars, and other information are used for user login, leaderboards, and other functions. The "state" field in the "room" table indicates the state of the game. The number "−1" indicates that the game is terminated in the middle of the game, the number "1" indicates that the player is in the layout phase, the number "2" indicates that the player has won in the move phase, and the number "3" indicates that the player has won in the fly phase.

Table 1. Data tables stored in the cloud database.

Data table	Stored data fields	Description Information
userinfo	"_openid":The unique identifier of the user "avatarUrl":The user's avatar "nickname":The user's nickname	Used to store player information
rooms	"roomid":The room number "owner":The identifier of the room creator "player":The identifier of the player who joins the room afterward "chessmen":Chessboard array "state":Phases of the game "manual": Chess manual "nowcolor":The color of the current chess player "people":Number of people in the room	Stores information about the rooms created by players during a game

4.4 Design of Game Data

In order to facilitate more efficient processing of the Jiu Chess manual data and extraction of expert knowledge at a later stage, the design specifications for the Jiu Chess manual were developed with reference to the Go game format (Table 2). In the chess manual data, the letters "W" and "B" are used to denote white and black pieces respectively, and the coordinates (x, y) denote the position of the pieces on the chessboard. For example, W(3, 3) represents the white piece in the third row and third column.

Table 2. Design specifications for Jiu Chess manual data.

Characters	Description Information
"[ChessJiu]"	Indicates Jiu chess
"[2023.05.01 10:00]"	Indicates the time of the start of the game
"Stage[1]"	Indicates entry into the layout phase
"Stage[2]"	Indicates entry into the battle phase
"W(x,x)"	Indicates the coordinates of the white piece
"B(x,x)"	Indicates the coordinates of the black piece
"O(x,x)"	Indicates that there are no pieces at this coordinate
"TC:"	Indicates jump capture

(continued)

Table 2. (*continued*)

Characters	Description Information
"FC:"	Indicates square capture
"W(x,x)-O(y,y)"	White piece moves from (x, x) to (y, y)
";"	Separator for the order of game play

4.5 Representation of the Chessboard and Chess Shape

The chessboard of Tibetan Jiu Chess can be represented by a 2-dimensional array of 8 × 8 [14]. Where black, white, and empty spaces are stored with –1, 1, and 0 respectively. "Dalian" chess shapes (Fig. 4) are distributed in different arrangements within a rectangular area, so a two-dimensional array can be used to represent the chess shapes. By traversing the entire chessboard in a two-dimensional array, the number of all shapes can be queried and used to determine the winner of the Jiu Chess game.

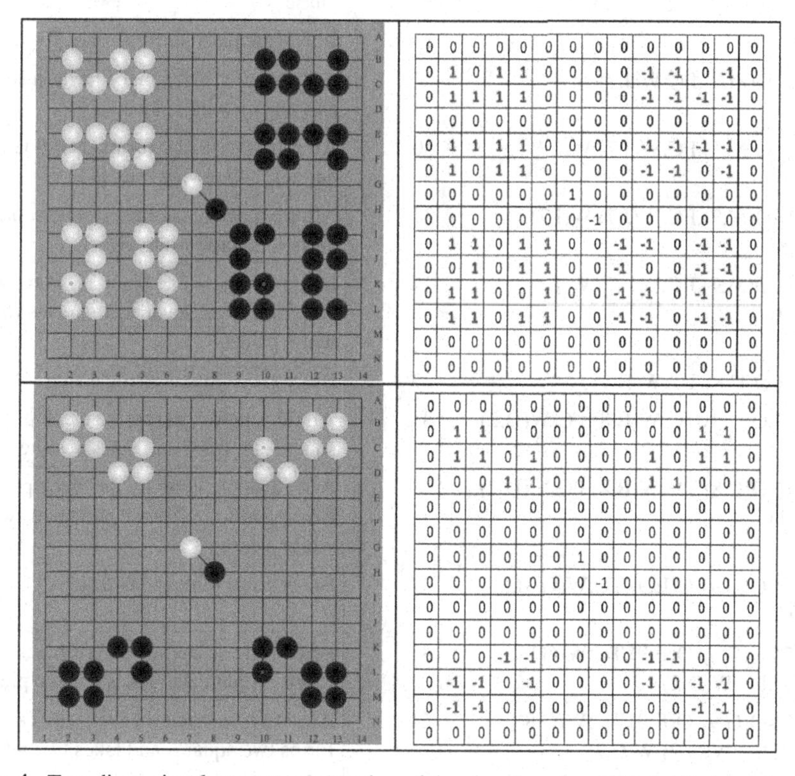

Fig. 4. Two-dimensional array representation of the chessboard and "Dalian" chess shapes.

4.6 Chess Game Replay

The game data generated by the game is dynamically stored in the cloud database, so the first step to replay the game is to get the game data from the cloud database. The game data is actually stored as a string. The game data is used to represent the order of black and white pieces, their position on the chessboard, and their movement. The game replay is implemented by parsing the game data in a fixed format (as shown in Table 3) step by step according to the number of rounds played, and converting the strings into coordinates according to the fixed position labels, so as to obtain the chessboard array after each move. By reading the chessboard array for each move and redrawing the pieces, the chessboard is recovered.

Table 3. Possible chess move options in the game. (Take black piece as an example)

Possible chess move options represented by a string	Length	Description Information
B(J,8)	6	Placing the black piece in the layout phase
B(J,8)-O(H,8)	13	The black piece moves a step
B(L,7)-O(M,7),FC:W(J,7)	23	After a move black piece forms a square and takes a white piece
B(L,7)-O(M,7),FC:W(J,7),W(I,7)	30	After a move black forms two squares and takes two white pieces
B(L,7)-O(M,7),FC:W(J,7),W(I,7), W(K,7)	37	After a move black forms three squares and takes three white pieces
B(L,7)-O(M,7),FC:W(J,7),W(I,7), W(K,7),W(A,1)	44	In the fly phase, the black piece moves and forms four squares, and takes four white pieces
B(H,J)-O(H,8),TC:W(H,9)	23	Black piece jumps a single step
B(A,8)-O(C,8),TC:W(B,8),FC:W(D,2)	33	After a single jump by the black piece, a square is formed and takes a white piece
B(A,8)-O(C,8),TC:W(B,8),FC:W(D,2), W(D,1)	40	After a single jump by the black piece, two squares are formed and takes two white pieces
B(G,7)-O(G,9)-O(E,9),TC:W(G,8), W(F,9)	37	Black piece jumps in two steps
B(G,7)-O(G,9)-O(E,9),TC:W(G,8), W(F,9),FC:W(D,2)	47	After two consecutive jumps black forms a square and takes a white piece
B(G,7)-O(G,9)-O(E,9),TC:W(G,8), W(F,9),FC:W(D,2),W(C,2)	54	After two consecutive jumps black forms two squares and takes two white pieces
…	…	…

4.7 Ways to Achieve a Win-Lose Judgement

In the move phase, the first step is to count the number of "Square" shapes formed by the opponent by traversing the chessboard array. When the opponent has no "Square" shape, then count the number of "Dalian" shapes formed by your side. When the condition that your side has formed two or more "Dalian" shapes and your opponent has no "Square" shape is met at the same time, then you can decide that your side has won the game in the move phase. In the fly phase, the player with fewer than 4 pieces loses.

5 Test Results of the Platform

5.1 Tests for Online Gaming

Twenty national Tibetan chess masters were invited to participate in the actual testing process of the platform in the National Tibetan Chess Tournament in 2023. The stability of the developed platform was verified after testing the platform's online game and battle-watching functions. The platform can run smoothly on the Android and Apple systems, and all functions can be displayed normally.

5.2 Tests for Game Data Replay

Entrusted by the organizing committee of the competition, the platform collected more than 10 games data of winning masters and completed online replay. Figure 5(a) illustrates the list of game data generated by chess masters. Figure 5(b) shows the replay page of the game. At the top of the page, the room ID and the number of white and black pieces on the current chessboard are recorded. The nicknames and avatars of both players are displayed at the top of the chessboard. The bottom of the chessboard shows the current progress, the total progress, and the track record of the current moves. By clicking on the "forward" or "back" buttons at the bottom of the page, you can dynamically display the game history.

The author of this article has collected the games of senior players, which will be used to prepare a book on Tibetan Jiu Chess together with the organizing committee to further promote the culture of Tibetan chess.

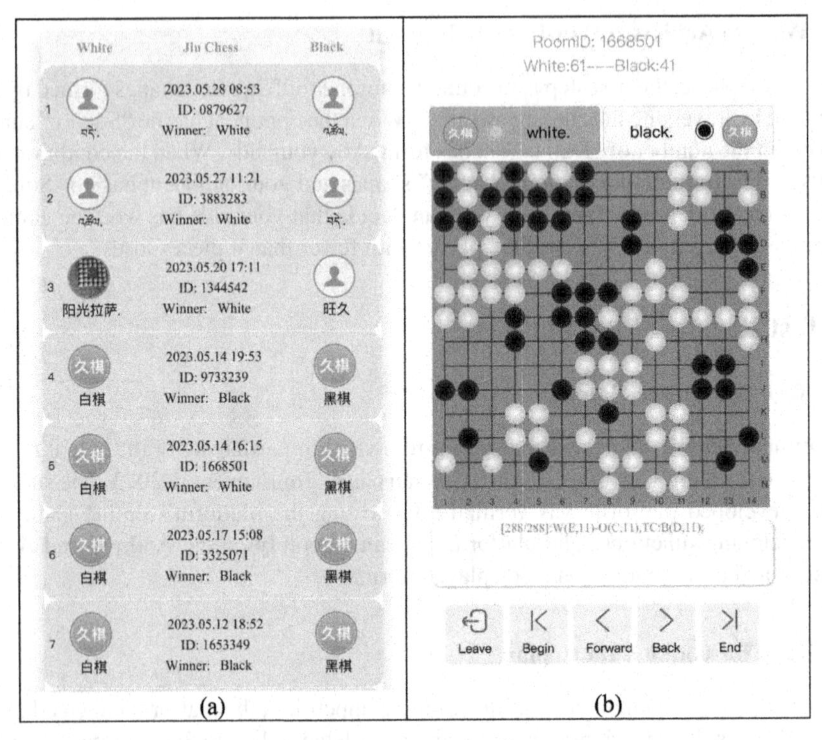

Fig. 5. Chess game replay.

6 Summary and Outlook

The mini-program developed in this paper has the functions of collecting game data and analyzing games, which can serve the national Tibetan chess competitions. This platform provides a practical solution for the collection of game data and the dissemination of the gaming culture of Jiu Chess. In future work, we will further improve the functionality of the Jiu Chess mini-program by adding features such as chessboard image recognition. By developing a digital platform for Tibetan Jiu Chess, we can collect Jiu Chess game data for building an open dataset of Tibetan Jiu Chess and develop Jiu Chess AI programs with a high level of chess ability, which is of great significance to the preservation and inheritance of Tibetan Chess culture.

Acknowledgment. This work was supported in part by the National Natural Science Foundation of China under Grant 62276285, Grant 62236011 and in part by the Major Projects of Social Science Fundation of China under grant 20&ZD279.

References

1. Liu, Q.: Research on the origin of Jiu Qi of Tibetan chess. Tibetan Studies, **6**, 105–109 (2017)

2. Liu, Q.: The attempting improvement and innovation of Jiu Qi. Tibetan Stud. **4**, 126–130 (2020)
3. Li, X.L., Wang, S., Lyu, Z.Y., Li, Y.J., Wu, L.C.: Strategy research based on chess shapes for Tibetan JIU computer game. J. ICGA 318–328 (2018)
4. Liu, W.Q.: Culture + creativity + technology: the characteristics of traditional cultural WeChat mini-games. New Media Res. **6**(09), 80–82 (2020)
5. Li, X., Zhang, Y.J., Yu, J.: Design and development of information WeChat mini program. Sci. Technol. Innov. **31**, 106–108 (2021)
6. Duan, J.W.: New features presented by WeChat mini-games. New Media Res. **5**(07), 66–68 (2019)
7. Ding, W.: Research and realization of key technology of Android-based chess game system. Master's thesis, Shanghai Jiao Tong University (2017)
8. Piette, E., Soemers, D.J., Stephenson, M., Sironi, C.F., Winands, M.H., Browne, C.: Ludii–the ludemic general game system. arXiv preprint arXiv:1905.05013 (2019)
9. Stephenson, M., Piette, E., Soemers, D.J., Browne, C.: An overview of the Ludii general game system. In: 2019 IEEE Conference on Games (CoG), pp. 1–2 (2019)
10. Li, X., Wang, S., Lyu, Z., Li, Y., Wu, L.: Strategy research based on chess shapes for Tibetan Jiu computer game. ICGA J. **40**(3), 318–328 (2018)
11. Zhang, X.C., Liu, L., Chen, L., Tu, L.: An evaluation method for the computer game agent of the intangible heritage Tibetan Jiu chess item. J. Chongqing Univ. Technol. (Nat. Sci.) **35**(12), 119–126 (2021)
12. Deng, S.T.: Design and implementation of Jiu game prototype system and multimedia courseware. Master's thesis, Minzu University of China (2017)
13. Shen, Q.W., Ding, M., Du, W.T., Zhao, W.L.: Research on the phased algorithm of Tibetan chess Jiu. Intell. Comput. Appl. **11**(2), 88–92 (2021)
14. Li, X.L., Chen, Y.D., Yang, Z.Y., Zhang, Y.Y., Wu, L.C.: A two-staged computer game algorithm for Tibetan Jiu Chess. J. Chongqing Univ. Technol. **36**(12), 110–120 (2022)
15. Lyu, Z.Y.: A study of chess games based on deep reinforcement learning. Master's thesis, Minzu University of China (2020)
16. Li, X.L., Lyu, Z.Y., Wu, L.C., Zhao, Y., Xu, X.: Hybrid online and offline reinforcement learning for Tibetan Jiu chess. Complexity (2020)
17. Yang, W.X.: Research on Tibetan Go Game Software and Its Educational Application Technology. Master's thesis, Minzu University of China (2016)
18. Lin, Y.Q., Yan, X.R., Ding, J.X.: Fun game planning and design based on wechat cloud development——big words eat small words recite words light games. Software **44**(01), 35–38 (2023)

Author Index

A
Arai, Tatsuo 28

B
Bao, Xin 87
Bu, Yin 212

C
Cao, Jian 3
Cao, Wujing 96
Chen, Jingda 258
Chen, Shuming 147
Chen, XiangYong 87
Chen, Yandong 322
Chen, Zhuo 28
Cheng, Ke 174
Chu, Zhongyi 48

D
Dai, Lican 224
Deng, Mingfang 192
Ding, Xiaokun 212
Duan, Zhiqiang 123

F
Fang, Jiali 61
Fang, Kairen 174
Feng, Ce 74
Feng, Saisai 15
Fu, Deqian 292
Fu, Jian 34, 159

G
Gao, Hongbo 74
Gao, Rui 247
Guan, Jie 304
Guo, Zhifeng 15

H
Han, Weixin 233
Hou, Bochao 212
Hu, Cuixin 147
Hu, Shichao 174
Huang, Feifan 74
Huang, Qiang 28

J
Jiang, Ruhai 74
Jiang, Tao 3

K
Kang, Bingbing 3
Kang, Zengxin 48

L
Li, Jinke 96
Li, Shanshan 132
Li, Xiali 322
Li, Xiaolong 34, 159
Liang, Jie 3
Liao, Longlong 123
Liao, Yanzhen 74
Liu, Chang 212
Liu, Chunfang 61
Liu, Dan 28
Liu, Jiaqi 107
Liu, Lekang 123
Liu, Wangwang 132, 174
Liu, Xiaoming 28
Lu, Lingyun 192

M
Ma, Yue 96

F. Sun and J. Li (Eds.): ICCCS 2023, CCIS 2029, pp. 335–336, 2024.
https://doi.org/10.1007/978-981-97-0885-7

Q
Qian, Xuewu 87
Qin, Pengjie 96
Qiu, Jianlong 292
Qiu, JianLong 87

S
Shi, Yunling 304
Shou, Yingxin 212
Su, Yanqin 107
Sun, JianQiang 87

T
Tang, Xiaoqing 28
Tong, Xiangrong 272
Tseng, H. Eric 201

W
Wang, Donglin 87
Wang, Jingjing 74
Wang, Jintao 233
Wang, Kai 247
Wang, Lei 132, 174
Wang, Li 292
Wang, Lin 15
Wang, Xiangyang 96
Wang, Xuming 3
Wang, Yan 123
Wang, Ying 192
Wang, Yukun 233
Wen, Lu 201
Wu, Licheng 322
Wu, Ruilin 123
Wu, Xinyu 96

X
Xiang, Dai 224
Xiao, Junfeng 304
Xu, Bin 212
Xu, Lianzheng 292

Y
Yan, Jie 322
Yang, Hanqing 74
Yang, Ming 258
Yang, Xiaohua 322
Yang, Zhu 34, 159
Yi, Zhengkun 96
Yin, Meng 96
Ying, Cui 224
Yu, Pan 61
Yuan, Xiaohu 123, 192, 247
Yuan, Yu 304

Z
Zhai, Haonan 272
Zhang, Changchun 132
Zhang, Huai 304
Zhang, Longlong 247
Zhang, Mingchuan 15
Zhang, Songan 201
Zhang, Xue 292
Zhang, Yanyin 322
Zhao, Huailin 192
Zhao, Qingjie 132, 174
Zheng, Shaoqiu 247
Zhu, Jun 123
Zhu, Zhiyuan 247
Zhuang, Hanyang 201, 258
Zou, Xiaohui 147
Zou, Yuanbing 132

Printed in the United States
by Baker & Taylor Publisher Services